Chemistry

for NCEA Level Three

May Croucher

NELSON
CENGAGE Learning

Australia • Brazil • Japan • Korea • Mexico • Singapore • Spain • United Kingdom • United States

Chemistry for NCEA Level Three
1st Edition
May Croucher

Cover design: Cheryl Rowe, Macarn Design
Text designer: Cheryl Rowe, Macarn Design
Production controller: Siew Han Ong

Any URLs contained in this publication were checked for currency during the production process. Note, however, that the publisher cannot vouch for the ongoing currency of URLs.

Acknowledgements

This book is based on *Pathfinder Chemistry Year 13* by May Croucher and Paul Croucher. It has been completely updated and revised for the NCEA Level 3 Curriculum.

The author wishes to thank those whose help, inspiration and enthusiasm in chemistry has influenced much of the content of this book. These are teachers and students at Rotorua Girls High School and Taylors College Auckland, colleagues at DSIR Wellington, Forest Research Institute Rotorua and PSM Healthcare Auckland, and co-authors, in particular Dr John Packer and Dr Paul Croucher. The tolerance of family is also much appreciated.

Images: Shutterstock for images on pages 6, 7, 10, 16, 19, 28, 30, 31, 38, 40, 41, 42, 45, 50, 51, 54, 59, 62, 63, 64, 65, 66, 73, 77, 78, 79, 80, 81, 92, 93, 97, 105, 114, 118, 125, 130, 149, 150, 151, 152, 153, 155, 158, 160, 167, 172, 173, 174, 176, 180, 190, 199, 210, 211, 213; Leçons de Physique ; Éditions Vuibert et Nony (public domain) for image on page 67; United States Department of Agriculture, Agricultural Research Service for image on page 138.

© 2015 Cengage Learning Australia Pty Limited

Copyright Notice

Copyright: This book is not a photocopiable master. No part of the publication may be copied, stored or communicated in any form by any means (paper or digital), including recording or storing in an electronic retrieval system, without the written permission of the publisher. Education institutions that hold a current licence with Copyright Licensing New Zealand, may copy from this book in strict accordance with the terms of the CLNZ Licence.

For product information and technology assistance,
in Australia call **1300 790 853**;
in New Zealand call **0800 449 725**

For permission to use material from this text or product, please email
aust.permissions@cengage.com

National Library of New Zealand Cataloguing-in-Publication Data
A catalogue record for this book is available from the National Library of New Zealand.

ISBN: 978 0 17035554 4

Cengage Learning Australia
Level 7, 80 Dorcas Street
South Melbourne, Victoria Australia 3205

Cengage Learning New Zealand
Unit 4B Rosedale Office Park
331 Rosedale Road, Albany, North Shore 0632, NZ

For learning solutions, visit **cengage.co.nz**

Printed in Australia by Ligare Pty Limited.
2 3 4 5 6 7 8 22 21 20 19 18

Contents

IV   ACHIEVEMENT STANDARD 91387   Internal assessment · 4 credits

Chemistry 3.1
Carry out an investigation in chemistry involving quantitative analysis

V   ACHIEVEMENT STANDARD 91389   Internal assessment · 3 credits

Chemistry 3.3
Demonstrate an understanding of chemical processes in the world around us

VI   ACHIEVEMENT STANDARD 91391   External assessment · 5 credits

Chemistry 3.5
Demonstrate an understanding of the properties of organic compounds

VII   ACHIEVEMENT STANDARD 91388   Internal assessment · 3 credits

Chemistry 3.2
Demonstrate an understanding of spectroscopic data in chemistry

3.4

Demonstrate an understanding of thermochemical principles and the properties of particles and substances

External assessment   **5 credits**

Unit	Content	Aim
1	**Atomic properties and electron configuration**	To write the **electron configuration** of the first 36 elements using s, p, d notation
2	**Periodic properties**	To discuss and explain **periodic trends** in atomic and ionic radii, ionisation enthalpy and electronegativity
3	**Bonding and molecular shape**	To understand the **attractive forces between atoms** in molecules (**intramolecular bonds**) **and ions** in ionic compounds
4	**More molecular structures**	To write and use **Lewis structures** of molecules to explain their shapes and polarities
5	**Bond enthalpy**	To relate bond enthalpy data to **bond strength**
6	**Intermolecular attractions**	To **relate** the **physical properties** of a range of groups of substances to **structure** and the **attractive forces** between particles (**intermolecular bonds**)
7	**Thermochemical principles**	To understand the concepts of **enthalpy of formation** and **enthalpy of combustion** To apply **Hess's Law** to calculate **enthalpy changes in reactions** To relate **enthalpy** and **entropy changes** to **spontaneity of chemical reactions**

Unit 1 | Atomic properties and electron configuration

Learning Outcomes — on completing this unit you should be able to:

- describe atomic particles using standard notation
- write the electron configuration of the first 36 elements using *s, p, d* notation.

Everything and everyone is made up of atoms.

Atomic structure

- **Atoms** consist of a central nucleus with negatively charged electrons moving in the space around it. In a chemical reaction, only the electrons in the outermost shell (the valence electrons) are involved.
- The **nucleus** of the atom is made up of positively charged protons *and* uncharged neutrons. Most nuclei are stable but some disintegrate and emit **radiation.**
- An atom can be described using symbols which give the **atomic number** (number of protons) and the **mass number** (number of protons + neutrons). A proton or neutron has almost 2000 times the mass of an electron.
- For example, a sodium atom has 11 protons and 12 neutrons. It is represented by $^{23}_{11}\text{Na}$. The number **23** is the mass number, the sum of protons and neutrons (p + n). The number **11** is the atomic number, the number of protons (p). This number is equal to the total number of positive charges in the atom.
- In an electrically neutral atom, the number of electrons (e) equals the number of protons (p). So $^{23}_{11}\text{Na}$ has 11p, 12n and 11e.
- An atom becomes an **ion** (a charged atom) by gaining or losing electrons. For example, the sodium atom **Na** becomes an ion by losing one electron. The sodium ion therefore has an overall positive charge. It is written **Na⁺** and has 11p, 12n and 10e.
- **Isotopes** of an element have different mass numbers. They differ in their number of neutrons. For example, naturally occurring hydrogen has three isotopes:

$^{1}_{1}\text{H}$ (**hydrogen**) $^{2}_{1}\text{H}$ (**deuterium**) $^{3}_{1}\text{H}$ (**tritium**)

- These isotopes, with mass numbers 1, 2 and 3, can also be written as hydrogen-1, hydrogen-2 and hydrogen-3, respectively.

FACTS

The hydrogen atom has one electron orbiting around one proton. The diameter of the hydrogen nucleus is 1.75×10^{-15} m while the diameter of the atom is 1.06×10^{-10} m. The nucleus weighs 1840 times the mass of the electron.

Suppose we expanded the atom so that the diameter of the nucleus is 1 mm. Place yourself 290 m away from the nucleus and there will be a high probability that an electron cloud will pass you by (although you won't notice it!). Most of the atom is unoccupied space.

    ISBN: 9780170355544   PHOTOCOPYING OF THIS PAGE IS RESTRICTED UNDER LAW.

- Electrons move around the dense nucleus at high speeds.
- *The electrons have differing amounts of energy. The energy of an electron is affected by its distance from the nucleus and by the surrounding electrons. Its energy depends on the attractive force from protons in the nucleus and the repulsive forces of the electrons around it.*

Electron energy

- Electrons moving around the nucleus of an atom exist in certain **energy levels**. Up to 2 electrons can be in the first level, 8 in the second level, 18 in the third, 32 in the fourth. *In general, the nth level can accommodate $2n^2$ electrons.*
- Electrons in the same energy level can have slightly different energies, so each main level is divided into energy **sublevels**. The first level has one sublevel, the second has two, the third has three. *In general, the nth level has n sublevels.*
- Electrons in the same sublevel may be found in different regions in the space around the nucleus. They are said to occupy different **orbitals**.
- At any level, the first sublevel consists of one orbital called an *s* orbital; the second sublevel consists of three orbitals called *p* orbitals; the third sublevel consists of five orbitals called *d* orbitals. So the first level, which has only one sublevel, has only one *s* orbital. The second level has two sublevels, and therefore has one *s* and three *p* orbitals.
- Each orbital can be occupied by one or two electrons.
- *Electrons in an atom are found in discrete energy levels. Each level has sublevels. Each sublevel consists of a number of orbitals. Each orbital can contain one or two electrons. Electrons not in an atom can have any amount of energy.*

Orbital shapes

- The simplest orbital shapes are those of the *s* and *p* orbitals. The *s* orbitals are spherical in shape, and the *p* orbitals are bi-lobed or dumbbell-shaped.

Energy of electron levels and sublevels

- The label of a sublevel gives the energy level and the orbital shape. For example, 2*s* indicates that the energy level is 2 and that the shape is that of an *s* orbital (i.e. spherical).
- In the diagram on the right, an orbital is indicated by a box. The higher the orbital is on the diagram, the more energy the electrons in that orbital have.
- In order of increasing energy, the sublevels are:

  1*s* < 2*s* < 2*p* < 3*s* < 3*p* < 4*s* < 3*d*

  The empty 4*s* sublevel is lower in energy than the 3*d* sublevel.

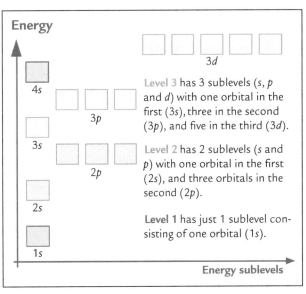

Level 3 has 3 sublevels (*s*, *p* and *d*) with one orbital in the first (3*s*), three in the second (3*p*), and five in the third (3*d*).

Level 2 has 2 sublevels (*s* and *p*) with one orbital in the first (2*s*), and three orbitals in the second (2*p*).

Level 1 has just 1 sublevel consisting of one orbital (1*s*).

ISBN: 9780170355544   PHOTOCOPYING OF THIS PAGE IS RESTRICTED UNDER LAW.

- An *s* sublevel is made up of one *s* orbital and it can have up to 2 electrons (e.g. $1s^2$). The superscript gives the number of electrons.
- A *p* sublevel is made up of three *p* orbitals and it can have up to 6 electrons (e.g. $2p^6$).
- A *d* sublevel is made up of five *d* orbitals and it can have up to 10 electrons (e.g. $3d^{10}$).
- The diagram on the right shows the relationship between levels. The darker the shaded area, the higher the probability of finding an electron with the corresponding energy.

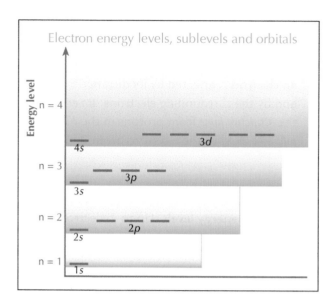

Electron energy levels, sublevels and orbitals

Electron configurations

- The **electron configuration** of an atom or ion can be written in a way that shows:
  — the energy levels of the electrons
  — the orbitals occupied by the electrons
  — the total number of electrons in that atom or ion.
- For example, Na has 11 electrons. Its electron configuration is: $\mathbf{1s^2 2s^2 2p^6 3s^1}$.
- The superscripts ^{2}, ^{2}, ^{6}, ^{1} give the number of electrons, so the total number of electrons is 11 (2 + 2 + 6 + 1 = 11). The letters ***s*** and ***p*** describe the orbitals involved, and the large numbers **1**, **2**, **3** describe the energy levels.
- *Electrons fill orbitals of lowest energy first.* This is because systems are stable in their lowest energy states. The order in which electrons fill sublevels is:

$$1s \rightarrow 2s \rightarrow 2p \rightarrow 3s \rightarrow 3p \rightarrow 4s \rightarrow 3d \rightarrow 4p$$

- If there are orbitals of the same energy in a sublevel, the electrons will *fill them singly* until there is one electron in each orbital. In this way, electron-electron repulsion is minimised. This is described in the following examples.

Electron configuration of nitrogen

- Nitrogen, N, has 7 electrons and its electron configuration is $\mathbf{1s^2 2s^2 2p^3}$. Its electron energy diagram is as shown on the right.
- The three $2p$ electrons occupy three separate orbitals; they are said to occupy the orbitals *singly*.

Electron configuration of oxygen

- Oxygen, O, has 8 electrons and its electron configuration is $\mathbf{1s^2 2s^2 2p^4}$. Its electron energy diagram is shown on page 9.
- Of the four $2p$ electrons, two occupy orbitals singly, and the other two are paired.

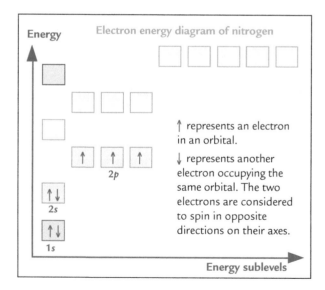

Electron energy diagram of nitrogen

↑ represents an electron in an orbital.

↓ represents another electron occupying the same orbital. The two electrons are considered to spin in opposite directions on their axes.

ISBN: 9780170355544   PHOTOCOPYING OF THIS PAGE IS RESTRICTED UNDER LAW.

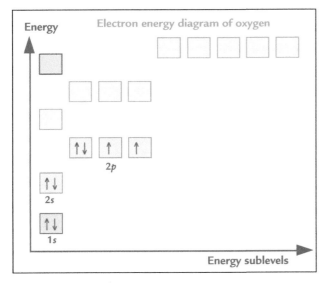

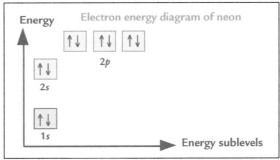

Electron configuration of neon

- Neon, Ne, has 10 electrons and its electron configuration is $1s^22s^22p^6$.
- All six $2p$ electrons of neon are paired. So there are no partially filled orbitals.

SPECIAL SKILL: WRITING ELECTRON CONFIGURATIONS

Steps:

1 First find the total number of electrons in the atom or ion.
   - For an atom, the number of electrons equals the atomic number given in the Periodic Table.
   - For a positive ion (cation), the number of electrons equals the atomic number *minus* the charge on the ion. For example, Mg^{2+} has 12 – 2 = 10 electrons.
   - For a negative ion (anion), the number of electrons equals the atomic number *plus* the number of negative charges on the ion. For example, S^{2-} has 16 + 2 = 18 electrons.
2 Allocate the electrons to sublevels.
   Remembering that an s sublevel can hold up to 2e, a p sublevel up to 6 and a d sublevel up to 10, write the first 2 electrons in the $1s$ sublevel as $1s^2$, the next 2 in the $2s$ sublevel as $2s^2$, the next 6 in the $2p$ sublevel as $2p^6$, until all the electrons have been allocated.
   The order in which the sublevels fill is:

$$1s \rightarrow 2s \rightarrow 2p \rightarrow 3s \rightarrow 3p \rightarrow 4s \rightarrow 3d$$

Exceptions to the rule are:

1 Cr, which has 24 electrons, has the configuration
$$1s^22s^22p^63s^23p^63d^54s^1 \text{ and not } 1s^22s^22p^63s^23p^63d^44s^2.$$
The first configuration has only one electron in each of the $3d$ and $4s$ orbitals. The second has one electron in each of four $3d$ orbitals but a pair of electrons in the $4s$. Electron-electron repulsion is less in the first configuration so it is more stable.

$$\underset{3d^5}{\cdot\ \cdot\ \cdot\ \cdot\ \cdot} \quad \underset{4s^1}{\cdot} \text{ is more stable than } \underset{3d^4}{\cdot\ \cdot\ \cdot\ \cdot\ \_} \quad \underset{4s^2}{\cdot\cdot}$$

2 Cu, which has 29 electrons, has the configuration

$$1s^22s^22p^63s^23p^63d^{10}4s^1 \text{ and not } 1s^22s^22p^63s^23p^63d^94s^2.$$

Both configurations have one unpaired electron but the first has a slightly lower energy than the second.

All other elements in the row from Sc to Zn in the Periodic Table have two $4s$ electrons. For an atom which has more than 18 electrons, fill the $4s$ sublevel before the $3d$. When forming a cation, the atom loses the $4s$ electrons before the $3d$ (see Fe and Fe^{2+} in the table at the top of page 10).

Atoms or ions with electrons in the lowest energy levels are in their **ground state**. Some ground state electron configurations are given below.

Atom/Ion	Number of electrons	Electron configuration
H	1	$1s^1$
B	5	$1s^22s^22p^1$
S	16	$1s^22s^22p^63s^23p^4$
S^{2-}	16 + 2 = 18	$1s^22s^22p^63s^23p^6$
Fe	26	$1s^22s^22p^63s^23p^63d^64s^2$
Fe^{2+}	26 − 2 = 24	$1s^22s^22p^63s^23p^63d^6$
Cu	29	$1s^22s^22p^63s^23p^63d^{10}4s^1$

Electrons are in discrete energy levels.
An electron in an atom can have only certain values of energy. To move to the next level, an electron must gain a definite amount of energy.

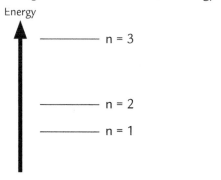

Each rung of the ladder corresponds to a certain gravitational potential energy.
To move to the next rung (**level**), a person must have enough energy to do so.

Electron configurations and the Periodic Table

- Electrons in the same energy level are said to be in the same **electron shell**. Electrons in the highest energy level are said to be in the **valence shell**.
- In the Periodic Table, atoms are placed in order of increasing atomic number. Atoms with similar outer electron configuration are in the same column. For example, Li, Na and K, which are all in the same column, have the following electron configurations (valence electrons are shown in red):
  Li: **$1s^22s^1$**   Na: **$1s^22s^22p^63s^1$**   K: **$1s^22s^22p^63s^23p^64s^1$**
- These can also be written as [He]**$2s^1$**, [Ne]**$3s^1$** and [Ar]**$4s^1$** respectively, where [He] stands for the electron configuration of helium (**$1s^2$**), etc.

KEY POINTS SUMMARY

- Electrons in an atom are in discrete energy **levels**.
- Energy levels are made up of **sublevels**. The first level has one sublevel, the second has two, the third has three, etc.
- Each sublevel consists of a number of **orbitals**. The first sublevel, *s*, has one, the second sublevel, *p*, has three, the third sublevel, *d*, has five, etc.
- Orbitals can contain one or two **electrons**.
- **Electron configurations**, such as **$1s^22s^22p^5$**, indicate the energy levels involved, the types of orbitals occupied, and the number of electrons in them.

ISBN: 9780170355544   PHOTOCOPYING OF THIS PAGE IS RESTRICTED UNDER LAW.

Chemistry and Technology: Energy Absorption

- An atom with its electrons in the lowest possible energy levels is said to be in its 'ground state'. Electrons can absorb energy such as heat and light, and move to higher energy levels. The atom is now said to be in an 'excited state'.

- Many coloured compounds contain elements between Sc and Zn in the Periodic Table. These elements have 3*d* electrons, which can absorb wavelengths from visible light to move to higher energy orbitals. Copper sulphate solution is blue. Electrons in copper ions absorb the red-orange-yellow part of the spectrum. We see the remainder of the spectrum. The intensity of colour depends on the concentration of copper ions. By comparing the colour of a solution of unknown concentration with colours of solutions of known concentrations, the unknown concentration can be found. This can be done accurately with a colorimeter. Electrons of most elements need more energy than that available in the visible light spectrum to move to higher levels. Because their solids reflect all the light, they appear white. Their solutions transmit all the visible light and appear colourless.

- When atoms are heated in a flame, the amount of energy absorbed is directly related to the number of atoms. By allowing a small amount of solution to vaporise in a flame, the amount of energy absorbed by the atoms can be measured. Thus the concentration of the solution can be determined. This is the basis of *Flame Atomic Absorption Spectroscopy* (Flame AAS). Ions discharge in the flame and become atoms.

ASSESSMENT ACTIVITIES

1 Matching terms with descriptions

a lowest energy state	electrons in atoms
b space in an atom filled by up to two electrons	orbital
c shows energy levels, sublevels and orbitals occupied by electrons	electron configuration
d electrons in highest energy level	ground state
e another term for an energy level	electron shell
f electrons having only certain discrete energy values	valence electrons

2 Writing electron configurations for atoms

Use a Periodic Table to get the atomic number and therefore the number of electrons.

a He	**b** B	**c** C	**d** O
e K	**f** Al	**g** P	**h** Cl
i Cr	**j** Fe	**k** Cu	**l** Zn
m Br	**n** Ar		

3 Writing electron configurations for ions

Use a Periodic Table and the charge on the ion to get the number of electrons.

a H^+ b Mg^{2+} c Na^+

d O^{2-} e K^+ f Al^{3+}

g P^{3-} h Cl^- i Cr^{3+}

j Fe^{2+} k Cu^{2+} l S^{2-}

4 Writing and interpreting electron configurations

Copy and complete the following table for atoms and ions. Use a Periodic Table to find any atomic numbers you require.

Atomic number	Charge	Configuration	Symbol
			B
			C
			N
			O
			O^{2-}
			H
			H^+
			As
			Ne
7	–3		
9	–1		
10	0		
12	0		
		$1s^2\ 2s^2\ 2p^6$	Na+
	–2	$1s^2\ 2s^2\ 2p^6$	
	+3	$1s^2\ 2s^2\ 2p^6$	
24	+3		
30			Zn
16	–2		
26	+3		
28	+2		
33	0		
35	–1		

ISBN: 9780170355544
PHOTOCOPYING OF THIS PAGE IS RESTRICTED UNDER LAW.

Unit 2 | Periodic properties

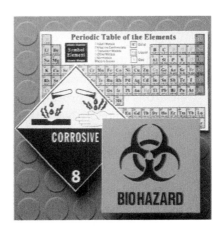

Learning Outcomes — on completing this unit you should be able to:

- describe periodic trends in atomic and ionic radii, ionisation enthalpy and electronegativity
- compare atomic and ionic radii.

Atom arrangements on the Periodic Table

- In 1869, the Russian chemist Mendeleev arranged the atoms of all known elements in order of increasing atomic number **Z** (number of protons in nucleus). Elements with similar chemical properties were placed in the same column.
- On the modern Periodic Table the elements with atomic numbers 58 to 71 and 90 to 103 have been removed and placed in a separate block as shown below. There are now 118 known elements.
- The Periodic Table below shows the atoms arranged into four blocks in total. In each block the same electron sublevel (*s*, *p*, *d* or *f*) is being filled. In lanthanum, La (atomic number 57), and hafnium, Hf (atomic number 72), a *d* sublevel (5*d*) is being filled. So La and Hf are placed together in the *d* block elements. Elements 58 to 71 and elements 90 to 103, in which the *f* sublevel is being filled, are placed in a separate *f* block.

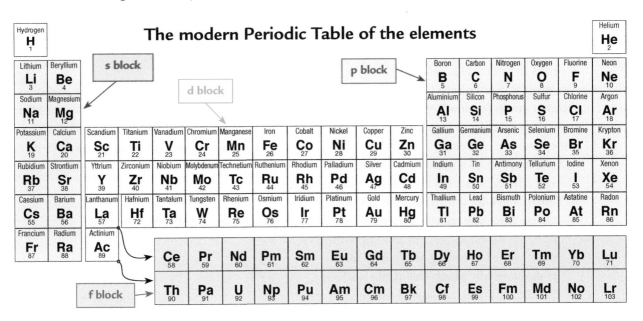

Sizes of atoms and ions

- *Down a group*, the number of electron shells increases. Thus **atomic radii increase**.
- *Across a period*, the nuclear charge increases. The atoms have the same number of inner electron shells so the shielding of the nuclear attraction for electrons is approximately the same. The increasing nuclear charge results in stronger attraction for electrons and the **atom size decreases**.

- Consider the ions formed by elements in Period 3, Na^+, Mg^{2+}, Al^{3+}, P^{3-}, S^{2-}, Cl^-.

  Na^+, Mg^{2+} and Al^{3+} have the same electron configuration $1s^2 2s^2 2p^6$. Na^+ attracts the electrons with 11 protons, Mg^{2+} with 12 and Al^{3+} with 13. Therefore Na^+ is the largest and Al^{3+} is the smallest.

  P^{3-}, S^{2-} and Cl^- have the same electron configuration $1s^2 2s^2 2p^6 3s^2 3p^6$.

  P^{3-} attracts the electrons with 15 protons, S^{2-} with 16 and Cl^- with 17. Therefore P^{3-} is the largest of the three and Cl^- is the smallest.

  Because P^{3-}, S^{2-}, Cl^- have one more electron shell than Na^+, Mg^{2+} and Al^{3+}, they are larger.

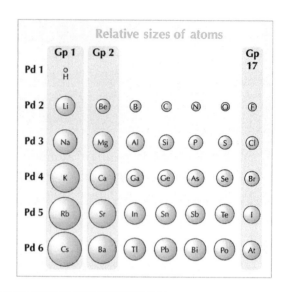

Relative sizes of atoms

Group	Group 1	Group 2	Group 13	Group 14	Group 15	Group 16	Group 17	Group 18
Period 3 ions	Na^+	Mg^{2+}	Al^{3+}	Si	P^{3-}	S^{2-}	Cl^-	Ar
Relative ionic size	1+	2+	3+		3–	2–	1–	

Effect of stable electron shells and attraction of the nucleus

- Other properties of atoms and ions that can be explained by their position on the Periodic Table are *ionisation enthalpy* and *electronegativity*. As we go down a group, the atoms have more electron shells. As we go across a period, the number of electron shells is the same, but the number of protons in the nucleus increases.

- The most important effect is the increase in **shielding of the nucleus by stable electron shells**. Down a **group (column), the number of electron shells increases so the electrons are more shielded from the attraction of the nucleus.**

- *If the number of electron shells is the same*, then the effect of **increasing nuclear charge** needs to be considered. Across a **period** (row), the nuclear charge is increasing.

- Overall effects for atoms are summarised as follows:

  1 **Down a group**, atoms have an *increasing number of electron energy shells* as well as *increasing nuclear charge*, but the important effect is the **increasing number of filled or stable shells which shield the nucleus**. *The nucleus has decreasing attraction for electrons.*

  2 **Across a period**, the *nuclear charge increases*, but as all outer electrons are in the same shell (energy level), the shielding effect of electrons does not increase. So across a period, there is **increasing attraction between the nucleus and electrons.**

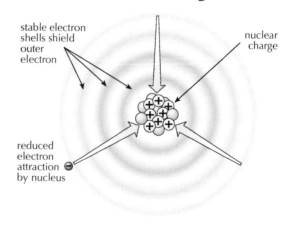

Nuclear shielding

stable electron shells shield outer electron

nuclear charge

reduced electron attraction by nucleus

ISBN: 9780170355544
PHOTOCOPYING OF THIS PAGE IS RESTRICTED UNDER LAW.

Ionisation enthalpy

- The **first ionisation enthalpy** (potential) of an atom is the energy required to remove an electron from the isolated atom to give a +1 ion. Isolated atoms are in the gas phase. For example, the first ionisation enthalpy of a sodium atom is the energy required for the reaction:

$$Na(g) \rightarrow Na^+(g) + e$$

- Energy is required because the attraction between the positively charged nucleus and the negatively charged electron is disrupted. *The stronger the attraction an atom has for its electrons, the higher the ionisation enthalpy.* An atom which loses an electron easily has a low ionisation enthalpy.
- The **second ionisation enthalpy** for any atom is the energy required to remove an electron from the isolated +1 ion, so it is the energy needed to remove the second electron. For example, the second ionisation enthalpy for sodium, Na, is the energy required for the reaction:

$$Na^+(g) \rightarrow Na^{2+}(g) + e$$

- For Na, with the electron configuration $[Ne]3s^1$, the first ionisation enthalpy is relatively small. Only a small amount of energy is needed to remove Na's first electron from the third electron shell. The second electron to be removed is in the second electron shell and is therefore closer to the nucleus, so this electron is more strongly attracted to the nucleus. Also, this electron is removed from a +1 ion, not a neutral atom. Thus Na's second ionisation enthalpy is much larger than the first.
- Down a group, for example Group 1 (Li, Na, K, Rb and Cs), the outer electrons become more shielded from the attraction of the nucleus. Less energy is needed to remove an electron. Less energy is needed for the reaction:

$$K(g) \rightarrow K^+(g) + e$$

than for

$$Na(g) \rightarrow Na^+(g) + e$$

So down a group, the first ionisation enthalpies decrease (look at Li, Na, K, Rb and Cs in the graph at the bottom of this page).

- Across a period, increasing nuclear charge makes it more difficult to remove electrons. *In general, first ionisation enthalpies increase across a period.*

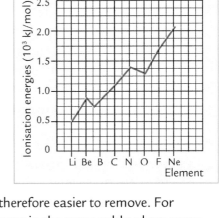

Period 2 first ionisation energies

Note, however, that electrons in the *p* sublevel are slightly further from the nucleus than those in the *s* sublevel and are therefore easier to remove. For example, B has a lower first ionisation energy than Be. Electrons in the same sublevel are more stable if they are not paired in orbitals. For N ($1s^22s^22p^3$), each of the three *p* orbitals has one electron. For O ($1s^22s^22p^4$), one of the *p* orbitals must have two electrons. Small repulsions between these electrons occupying the same region in space make it easier to remove an electron from O. So O has a lower first ionisation energy than N (see graph above right for Period 2 first ionisation energies).

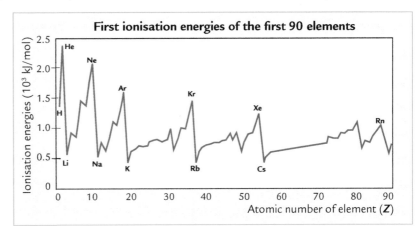

First ionisation energies of the first 90 elements

Electronegativity

- The **electronegativity** of an atom is a measure of the attraction that a bonded atom has for the bonding electrons. A table of electronegativity values is given below.

Table of electronegativity values

Group 1	Group 2	Group 13	Group 14	Group 15	Group 16	Group 17
(H) 2.1						
Li 1.0	Be 1.5	B 2.0	C 2.5	N 3.0	O 3.5	F 4.0
Na 0.9	Mg 1.2	Al 1.5	Si 1.8	P 2.1	S 2.5	Cl 3.2
K 0.8	Ca 1.0	Ga 1.6	Ge 1.8	As 2.0	Se 2.4	Br 2.8
Rb 0.8	Sr 1.0	In 1.7	Sn 1.8	Sb 1.9	Te 2.1	I 2.5
Cs 0.7	Ba 0.9	Tl 1.8	Pb 1.9	Bi 1.9	Po 2.0	At 2.2

Note that the atom with the greatest electronegativity is fluorine (F), while the atom with the lowest is caesium (Cs). Notice their positions on the Periodic Table.

Fluorine, the most electronegative atom, has been used as a superoxidiser in rocket propellant.

- Atoms with higher electronegativity values have a greater attraction for bonding electrons.
- Due to the same factors that affect ionisation energy trends, *electronegativities generally decrease down groups and increase across periods.*
- The most electronegative element, fluorine, gains electrons very easily. It is the most reactive non-metal and causes severe chemical burns on contact with the skin.
- The least electronegative element, caesium, loses electrons very easily. Visible light supplies enough energy to remove electrons from it and it is used in photoelectric cells. It is the most reactive of the non-radioactive metals and reacts violently with water.

Oxidation number

- Atoms tend to lose, gain or share the number of electrons that will result in a stable outer energy level or shell.
- The **oxidation number** of an atom gives the number of electrons that the atom gains, loses or shares in its compounds.
- When electrons are shared, the bonded atom with the higher electronegativity value is assigned a negative oxidation number and the other atom is given a positive oxidation number.
- Oxidation numbers for atoms in groups 1, 2, 13 and 14 include +1, +2, +3 and +4, respectively.
- Oxidation numbers for atoms in groups 15, 16 and 17 include –3, –2 and –1, respectively.

ISBN: 9780170355544 PHOTOCOPYING OF THIS PAGE IS RESTRICTED UNDER LAW.

KEY POINTS SUMMARY

- Atomic radii increase down a group because of the increasing number of electron shells. They decrease across a period (where the number of electron shells is unchanged) because of the increasing number of protons.
- Ionic radii increase down a group. Across a period, anions are larger than cations as they have one more electron shell.
- Ionisation enthalpy and electronegativity decrease down a group because of increasing shielding of the nucleus by stable electron energy shells (despite large increases in nuclear charge).
- Ionisation enthalpy and electronegativity generally increase across a period because of increasing nuclear charge with no increase in nuclear shielding by filled shells (except group 18 have low electronegativity).

Increasing electronegativity and ionisation enthalpy →

Decreasing electronegativity and ionisation enthalpy ↓

H 2.1																	He -
Li 1.0	Be 1.5											B 2.0	C 2.5	N 3.0	O 3.5	F 4.0	Ne -
Na 0.9	Mg 1.2											Al 1.5	Si 1.8	P 2.1	S 2.5	Cl 3.0	Ar -
K 0.8	Ca 1.0	Sc 1.36	Ti 1.5	V 1.6	Cr 1.6	Mn 1.5	Fe 1.8	Co 1.9	Ni 1.9	Cu 1.9	Zn 1.6	Ga 1.6	Ge 1.8	As 2.0	Se 2.4	Br 2.8	Kr -
Rb 0.8	Sr 1.0	Y 1.2	Zr 1.4	Nb 1.6	Mo 1.8	Tc 1.9	Ru 2.2	Rh 2.2	Pd 2.2	Ag 1.9	Cd 1.7	In 1.7	Sn 1.8	Sb 1.9	Te 2.1	I 2.5	Xe -
Cs 0.7	Ba 0.9	La 1.0	Hf 1.3	Ta 1.5	W 1.7	Re 1.9	Os 2.2	Ir 2.2	Pt 2.2	Au 2.4	Hg 1.9	Tl 1.8	Pb 1.9	Bi 1.9	Po 2.0	At 2.0	Rn -
Fr 0.7	Ra 0.9																

ASSESSMENT ACTIVITIES

1 Matching terms with descriptions

a	measure of the attraction a bonded atom has for the bonding electrons	period of elements
b	number of electrons that an atom gains, loses or shares in its compounds	group of elements
c	elements found in the same column down the Periodic Table	first ionisation enthalpy
d	elements found in the same row across the Periodic Table	second ionisation enthalpy
e	energy required to remove an electron from an isolated atom	electronegativity
f	energy required to remove an electron from an isolated +1 ion	oxidation number

For the following questions, have a Periodic Table beside you to refer to.

2 **Reviewing sizes of atoms and ions**
    a Write the following atoms in order of increasing size (the smallest first) and justify your answer: Al, Br, P.
    b Explain which is larger, Ca or Ca^{2+}.
    c Write the following ions in order of increasing size (the smallest first) and justify your answer: Cl^-, K^+, S^{2-}.

3 **Reviewing ionisation enthalpies**
Ionisation enthalpies in kilojoules per mole ($kJ\ mol^{-1}$) for sodium, magnesium and aluminium are listed in the following table.

Atom	First ionisation enthalpy	Second ionisation enthalpy	Third ionisation enthalpy
Na	502	4569	6919
Mg	744	1457	7739
Al	584	1823	2751

Study the data carefully then answer the questions below.
    a Explain the difference between the first, second and third ionisation energies of an atom.
    b Explain why the first ionisation enthalpy of sodium is lower than that of magnesium.
    c Explain why the first ionisation enthalpy of aluminium is lower than that of magnesium (look at their electron configurations).
    d Explain why for any atom, the first ionisation enthalpy is always the lowest.
    e Explain why for sodium, the second ionisation enthalpy is much greater than the first.
    f Explain why for magnesium, the third ionisation enthalpy is much greater than the second, and the second is larger than the first.
    g Explain why the fourth ionisation enthalpy for aluminium can be predicted to be a very large value.

4 **Graphing ionisation enthalpies**
The first ionisation enthalpies (in $kJ\ mol^{-1}$) for the elements Ne to Ar are given below.

Atom	Ne	Na	Mg	Al	Si	P	S	Cl	Ar
First IE	2090	502	744	584	793	1020	1010	1260	1530

    a Draw a graph of ionisation enthalpy (vertical axis) against atomic number Z.
    b Explain the general trend in the values Na to Ar.
    c Explain the unexpectedly low values for Al and S.
    d Why is the ionisation enthalpy of Ar lower than that of Ne?

5 **Explaining electronegativity trends**
Explain in terms of the attraction of a nucleus for electrons why fluorine is the most electronegative element while caesium is the least electronegative.

6 **Relating trends in ionisation enthalpy and electronegativity**
    a Explain why ionisation enthalpy and electronegativity generally increase across a period but decrease down a group.
    b Why do elements at the sides of the Periodic Table (Groups 1, 2, 16 and 17) tend to form ions, while those in other groups tend to form covalent bonds?

 ISBN: 9780170355544 PHOTOCOPYING OF THIS PAGE IS RESTRICTED UNDER LAW.

Unit 3 | Bonding and molecular shape

Learning Outcomes — on completing this unit you should be able to:

- use electronegativity values to predict whether atoms will form ionic or covalent bonds
- draw Lewis (electron dot) structures of molecules and polyatomic ions
- apply VSEPR theory to Lewis structures to predict shapes of molecules
- relate the polarity of molecules to molecular shape and valence electrons.

Shapes and structures of the molecules that make up DNA enable them to form a double helix.

Ionic and covalent bonds

- Atoms bond together to form compounds in which the atoms are usually in a more stable energy state.
- In these compounds, the atoms often have eight electrons in their outermost (valence) energy level (e.g. **...$3s^2 3p^6$**). However, Period 1 elements (H and He) are stable with just two electrons (**$1s^2$**).
- The bonding between atoms in compounds is either ionic or covalent.
- To form an **ionic bond**, one atom completely gains electrons from another atom. The atom that has gained electrons will be a negatively charged ion called an **anion**. The atom that has lost electrons will be a positively charged ion called a **cation**. Cations are attracted to anions by **electrostatic attraction**. This attraction between oppositely charged ions is called ionic bonding.
- In forming a **covalent bond**, one atom attracts electrons from another atom but does not gain them completely. So these electrons are attracted to two nuclei. The attractive forces hold the atoms together to form a molecule. This **sharing of bonding electrons** by two nuclei is called covalent bonding.
- The type of bonds that atoms form depends on their ability to attract electrons. **Electronegativity** values give a measure of the relative attraction that bonded atoms have for bonding electrons in a compound. Values for atoms in Groups 1 and 2 as well as Groups 13 to 17 are given in the table on page 16. Group 18 atoms seldom form compounds and are omitted from the table.
- Atoms of metallic elements (such as Mg) tend to form ionic bonds with non-metallic elements (such as S). Non-metal elements bond together with covalent bonds.
- In general, metals have low electronegativity values and non-metals have high electronegativity values.

Ionic compounds are found in minerals or as dissolved salts.

Bonds in molecules

- When atoms form covalent bonds, no electrons are completely gained or lost, and **molecules** are formed. Unlike ions, molecules have no net charge.
- The bonds within molecules are said to be intramolecular.
- *If the two bonded atoms are identical, the bonding electrons are equally shared.* For example, in the oxygen molecule O_2, each O atom has six valence electrons and needs another two to be stable. Each O atom will give a share of two electrons to the other O atom and in return it gets a share of two.
- The equal sharing of electrons results in a **non-polar covalent bond**.
- The sharing of two pairs of electrons is a double covalent bond. This is more stable than a single covalent bond (one pair of electrons).
- Each O atom can be drawn with its valence electrons, and the O_2 molecule drawn with its outer electrons as shown below. These diagrams are called **Lewis structures** (or electron dot structures).

$$:\ddot{O}:\ +\ :\ddot{O}: \longrightarrow :\ddot{O}::\ddot{O}: \longrightarrow :\ddot{O}=\ddot{O}:$$ Draw a line for each bond pair of electrons.

The electrons are represented by dots.

- If the two bonded atoms are different, the bonding electrons are not shared equally. For example, in the molecule hydrogen bromide, HBr, H with one outer electron and Br with seven outer electrons, both need one electron to achieve stability. The H atom gives the Br atom a share of its electron. In return, Br gives H a share of one of its electrons.
- But Br has a greater attraction for the shared electrons (bonding electrons); its electronegativity value is greater than that of H's.
- The atom with the greater share of electrons (Br) will have a slight negative charge ($\delta-$) and the other atom (H) will have a slight positive charge ($\delta+$). The bond formed is called a **polar covalent bond**.

$$H\cdot\ +\ \cdot\ddot{\underset{\cdot\cdot}{Br}}: \longrightarrow H:\ddot{\underset{\cdot\cdot}{Br}}: \longrightarrow H-\ddot{\underset{\cdot\cdot}{Br}}:$$
$$ \delta+\ \delta- \delta+\ \delta-$$

- As the HBr molecule has a slightly positive ($\delta+$) end and a slightly negative ($\delta-$) end, HBr is a **polar molecule**.

Lewis structures and molecules

- Lewis structures show the order in which atoms are bonded in a molecule and the arrangement of shared and unshared (lone) **valence** electrons.
- The arrangement of valence electrons and the positions of the atoms allow us to predict the shape and stability of the molecule. For example, triple bonds are very stable. Knowing the shape and stability of a molecule also provides information on how it will react.
- Bonding and shapes of molecules will be studied in the remainder of this unit and the next. In the following skills, ammonia is used to show how Lewis structures can be used to predict molecular shape.

Valence shell electron pair repulsion theory

- Repulsions between regions of high electron density (or electron sets) determine the way in which they orientate themselves in the space around an atom's nucleus.
- **VSEPR (valence shell electron pair repulsion) theory** describes how the bonding and non-bonding pairs of valence electrons will be orientated in space.
- This theory for pairs of electrons can be extended to regions (or sets) of electrons.

 ISBN: 9780170355544 PHOTOCOPYING OF THIS PAGE IS RESTRICTED UNDER LAW.

- **Regions of high electron density** (or sets of electrons) repel each other and will be arranged with the maximum possible distance between them.
- The two electrons in a single bond, the four electrons of a double bond, the six electrons of a triple bond, the two electrons in a non-bonding pair, or even an unpaired electron can each be regarded as a region of high electron density.

SPECIAL SKILL 1: WRITING LEWIS STRUCTURES

A Lewis structure can be used to show the arrangements of valence electrons of covalently bonded atoms in a molecule such as ammonia, NH_3, or in a polyatomic ion such as ammonium, NH_4^+. In both cases, N is covalently bonded to H.

Steps:
1 Draw the atoms in the order in which they are bonded. If there is only one of a particular atom, that atom is usually the central atom. (Further rules for the order of atoms are found in the next unit.)
2 Use dots • or crosses × to represent the valence electrons of the atoms.
3 Try to give each atom a stable number of electrons and, where possible, draw the electrons in pairs in the molecule (a pair of electrons can occupy the same orbital, i.e. the same region in space).

Example: Draw the Lewis structure of an ammonia molecule, NH_3.
1 Drawing the atoms in the order they are bonded: as there is only one N, it must be the central atom.

$$\begin{array}{ccc} H & N & H \\ & H & \end{array}$$

2 Using dots to show valence electrons: nitrogen has five valence electrons; each H has one electron.

$$H\times \quad \cdot \overset{\cdot\cdot}{\underset{\cdot}{N}}\cdot \quad \times H$$
$$\overset{\times}{H}$$

3 Giving each atom a stable number of electrons and drawing electrons in pairs: to obtain a stable set of eight electrons, nitrogen must obtain a share of one electron from each H atom; in return, each H atom must get a share of one of N's electrons to have a stable set of two.

$$H\overset{\times}{}\overset{\cdot\cdot}{N}\overset{\times}{}H$$
$$\overset{\cdot\times}{H}$$

In the above diagram, × represents an electron from H and • an electron from N. The Lewis diagram can also be drawn using the same symbol • for all electrons. For example:

$$H:\overset{\cdot\cdot}{\underset{\cdot\cdot}{N}}:H$$
$$H$$

Draw a line for each bonding pair of electrons.

$$H-\overset{\cdot\cdot}{N}-H$$
$$\underset{H}{|}$$

ISBN: 9780170355544   PHOTOCOPYING OF THIS PAGE IS RESTRICTED UNDER LAW.

Rules for placing regions of electrons

The placement of regions of valence electrons around an atom can be determined by applying the following rules.

- **Two regions** of electrons will be found at 180° to one another.

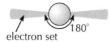

electron set

The bond angle will be 180° and the arrangement is called **linear**.

- **Three regions** of electrons will be directed to the vertices of an equilateral triangle.

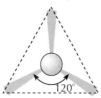

The bond angle will be 120° and the arrangement is called **planar**.

- **Four regions** of electrons will be directed towards the corners of a tetrahedron.

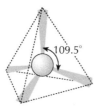

The bond angle will be 109.5° and the arrangement is called **tetrahedral**.

- **Five regions** of electrons will be directed towards the vertices of a triangular bi-pyramid.

The bond angles will be 120° and 90° and the arrangement is called **trigonal bi-pyramidal**.

- **Six regions** of electrons will be directed towards the vertices of a square bi-pyramid.

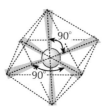

The bond angle will be 90° and the arrangement is called **octahedral**.

SPECIAL SKILL 2: DETERMINING MOLECULAR SHAPE

It is the position of the atoms in a molecule that determines the shape and polarity of the molecule. Electrons are not 'seen'.

Steps:

1 Starting with the Lewis structure, identify the number of regions of valence electrons around the central atom.
2 Decide on the location of those electron regions around the central atom using the rules above.
3 Draw the other atoms in around the central atom.

 ISBN: 9780170355544   PHOTOCOPYING OF THIS PAGE IS RESTRICTED UNDER LAW.

Example: Determine the shape of the ammonia molecule, NH_3.
The Lewis structure is shown here.

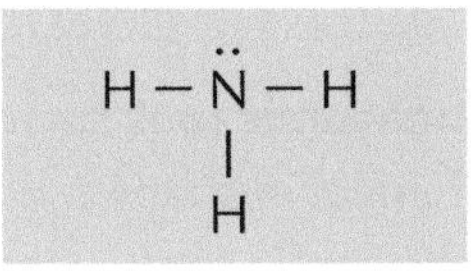

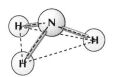

1 Identifying the number of regions of electrons: there are four electrons regions around the central N atom.

2 Deciding on the placement of electron regions: according to the rules, the four electron regions will be directed towards the corners of a tetrahedron.

3 Drawing in the other atoms: put the three H atoms at three of the vertices of the tetrahedron.
This gives a **trigonal pyramid** shape. The non-bonding electrons are not 'seen' and do not contribute directly to the shape of the molecule.

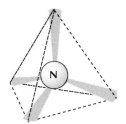

SPECIAL SKILL 3: DETERMINING POLARITY

Electronegativity values can be used to predict whether bonded atoms in molecules will be slightly positive or slightly negative.

Steps:
1 Compare the electronegativity values of the atoms.
2 Find out whether the centres of positive and negative charge coincide, then decide what the polarity of the molecule will be.

Example: Determine the polarity of the ammonia molecule, NH_3.

1 Comparing electronegativity values: N at 3.0 is more electronegative than H at 2.1. So N has a greater attraction for the bonding electrons than H. N will be slightly negative ($\delta-$) and the three H's slightly positive ($\delta+$).

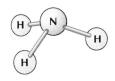

2 Locating centres of charge: the centre of positive charge is in the middle of the triangle formed by the three H's. The centre of negative charge is on the N atom. Because the centre of negative charge does not coincide with the centre of positive charge, NH_3 is a **polar molecule**.

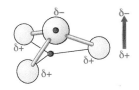

Monatomic and polyatomic ions

- **Monatomic ions** such as Cl^- and Na^+ consist of a single atom. These ions are considered to be spherical in shape.
- **Polyatomic ions** such as NH_4^+, NO_3^- and SO_4^{2-} contain more than one atom. These ions have covalent bonds between the atoms. For example, the bonds between S and O in SO_4^{2-} are covalent. Drawing Lewis structures gives information about the shape of polyatomic ions.

SPECIAL SKILL 4: POLYATOMIC ION SHAPES

Steps:

1 Obtain the total number of electrons by adding up the valence electrons of all atoms, then add 1 for each –ve charge but subtract 1 for each +ve charge.

2 Decide which is the central atom. If there is only one of a particular atom in the ion, then it is usually the central atom. The central atom is considered to be the atom that has lost or gained electrons giving the overall charge on the ion.

3 Draw the Lewis structure, then determine the shape of the polyatomic ion.

Example: Determine the shape of the polyatomic ion ammonium, NH_4^+.

1 Totalling the number of electrons: N has 5 valence electrons, the 4 H's have 4 electrons. Subtract 1 electron for the positive charge. Total number is therefore 8 electrons arranged in 4 pairs.

2 Identifying the central atom: as there is only one N atom, it must be the central atom.

3 Drawing the Lewis structure:

$$\left[\begin{array}{c} H \\ | \\ H-N-H \\ | \\ H \end{array}\right]^+$$

Determining the shape of the ion: because of electron repulsions, the four pairs of electrons around N will be directed towards the corners of a tetrahedron. As the four H atoms are on the corners of the tetrahedron, the ion is **tetrahedral**.

The ion does have an overall positive charge through the loss of one electron from the N atom.

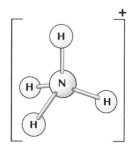

Summary for determining the polarity of a molecule

1 Draw the Lewis (electron dot) structure to show the placement of atoms and valence electrons.

2 Apply the VSEPR theory to find the approximate bond angles.

3 Look at the positions of the atoms to determine shape.

4 Compare the electronegativity values of the bonded atoms to determine the bond polarities.

5 Look at the centre of distribution of +ve charge. Look at the centre of distribution of –ve charge. Are they at the same point? Or is the distribution of one type of charge symmetric about the other charge?
A 'yes' in either case means the molecule is non polar.

Polar water molecules are attracted to a charged rod.

The table on the opposite page gives a list of molecules, their shapes and polarities.

 ISBN: 9780170355544 PHOTOCOPYING OF THIS PAGE IS RESTRICTED UNDER LAW.

Electron placement and molecular geometry

3.4

Electron regions	Bonding regions	Lone pairs	Electron region geometry	Molecular geometry	Examples		Polarity of the structure shown
2	2	0	linear	linear	BeF_2, CO_2 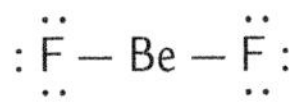	$:\!\ddot{F}-Be-\ddot{F}\!:$	Non polar
	1	1		linear	CO, N_2	$:C\equiv O:$	Polar
3	3	0	trigonal planar	trigonal planar	BF_3, CO_3^{2-} 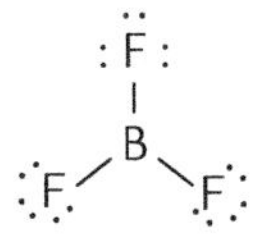		Non polar
	2	1		bent	O_3, SO_2 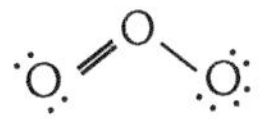		Polar
	1	2		linear	O_2 	$\ddot{O}=\ddot{O}$	Non polar
4	4	0	tetrahedral	tetrahedral	CH_4, SO_4^{2-} 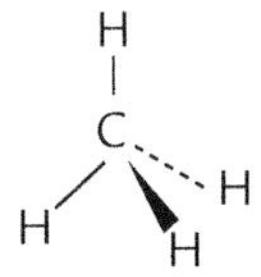		Non polar
	3	1		trigonal pyramidal	NH_3, H_3O^- 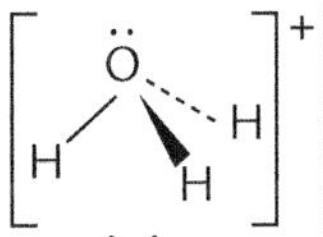		(+ve ion)
	2	2		bent	H_2O, ICl_2^+ 		Polar
	1	3		linear	HF, OH^- 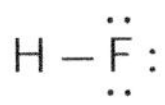	$H-\ddot{F}:$	Polar
5	5	0	trigonal bipyramidal	trigonal bipyramidal	PF_5 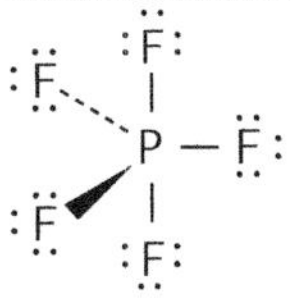		Non polar
	4	1		seesaw or distorted tetrahedral	SF_4, $TeCl_4$, IF_4^+ 		Polar
	3	2		T-shaped	ClF_3, BrF_3, IF_3 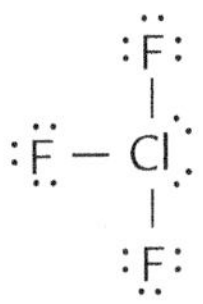		Polar
	2	3		linear	I_3^-, XeF_2 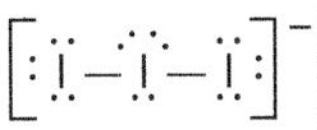		(−ve ion)

(table continued over page)

Electron regions	Bonding regions	Lone pairs	Electron region geometry	Molecular geometry	Examples		Polarity of the structure shown
6	6	0	octahedral	octahedral	SF_6, PF_6^-, SiF_6^{2-} 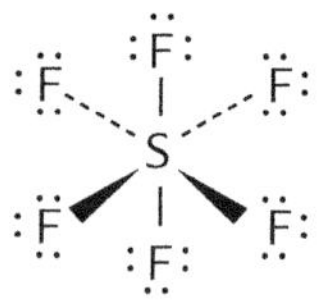		Non polar
	5	1		square pyramidal	BrF_5, $SbCl_5^{2-}$ 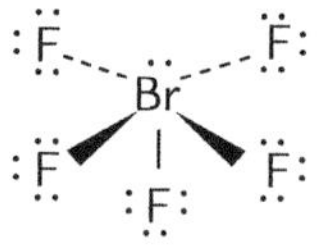		Polar
	4	2		square planar	XeF_4, ICl_4^- 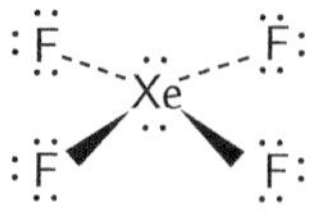		Non polar

KEY POINTS SUMMARY

- Bonding between atoms is either **ionic** or **covalent**. If one bonded atom is metallic and the other non-metallic, the bond in general will be ionic. If the two bonded atoms are non-metallic, the bond will be covalent.
- **Intramolecular bonds** are always covalent. Molecules may be **polar**, but they have no net charge.
- **Polyatomic ions** have covalent bonds between atoms.

ASSESSMENT ACTIVITIES

1 Matching terms with descriptions

a	shows arrangement of atoms and valence electrons in covalently bonded atoms		ionic bond
b	negatively charged ion		cation
c	electrostatic attraction between two oppositely charged ions		anion
d	molecule whose centres of positive and negative charge do not coincide		covalent bond
e	bond in which atoms share electrons but not equally		electronegativity
f	bond formed between atoms when both nuclei attract bonding electrons		non-polar covalent bond
g	positively charged ion		Lewis structure
h	valence shell electron pair repulsion theory		polar covalent bond
i	ion made up of covalently bonded atoms		polar molecule
j	bond in which atoms share electrons equally		VSEPR theory
k	measure of a bonded atom's ability to attract electrons		polyatomic ion

ISBN: 9780170355544
PHOTOCOPYING OF THIS PAGE IS RESTRICTED UNDER LAW.

2 Identifying bond types

Use the table of electronegativities on page 16 to name the type of bond (ionic, polar covalent or non-polar covalent) that will form between the following pairs of atoms:

a Li and Br
b O and Cl
c two S atoms
d Mg and S
e Ca and F
f two Br atoms
g O and S
h Br and Cl
i K and S
j Si and C

3 Drawing Lewis structures

Complete Lewis structures for the following:

a NF_3
b OF_2
c BrF_3
d O_3
e SCl_2
f SO_2
g CO_2
h SiH_4
i SF_4
j IF_3
k XeF_4
l NH_4^+
m NO_3^-
n SO_4^{2-}
o PO_4^{3-}
p CO_3^{2-}
q PF_6^-

4 Using VSEPR theory

Use the valence shell electron pair repulsion theory to predict how the regions of electrons will be placed around the central atom for each of the molecules and ions listed in question **3**.

5 Describing molecular and polyatomic ion shapes

By looking at the positions of the atoms involved, describe the shapes of each of the ions and molecules listed in question **3**.

6 Determining polarity

For the molecules in question **3** (**3a** to **3k**), state whether each is polar or non-polar.

Unit 4 | More molecular structures

Learning Outcomes — on completing this unit you should be able to:

- write Lewis structures for oxyacid molecules and describe their shapes
- describe the effect of lone pair electrons on the tetrahedral angle
- describe molecules whose atoms have incomplete or expanded octets
- use VSEPR theory to predict shapes of molecules disobeying the octet rule
- describe some of the limitations of Lewis structures.

Sulfuric acid is important in New Zealand agriculture. It is used to make fertilisers.

Oxyacid molecules

- It is not possible to use the rules described in the last unit to predict the Lewis structure of all molecules. Some additional rules are required.
- Acids such as nitric acid, HNO_3, and sulfuric acid, H_2SO_4, are known as **oxyacids** because they contain oxygen atoms. They are considered to be made up of an oxide plus water. For example:

nitric acid $H_2O + N_2O_5 = 2HNO_3$
sulfuric acid $H_2O + SO_3 = H_2SO_4$

- In oxyacids, *the H atoms are always bonded to O atoms*. The single atom that is neither H nor O is the central atom. So HNO_3 can be written $(HO)NO_2$, where **N** is the central atom, and H_2SO_4 as $(HO)_2SO_2$, where **S** is the central atom.

SPECIAL SKILL: WRITING LEWIS STRUCTURES FOR OXYACIDS

Steps:
1. Write the atoms in the correct order. The H atoms are bonded to O atoms, and the O atoms are bonded to the central atom.
2. Put one electron in for all H atoms and six valence electrons in for all O atoms bonded to the H atoms.
3. Put the electrons in for the central atom, placing them in pairs between atoms wherever possible.
4. Put electrons in for terminal O atoms to satisfy the **octet rule** (an atom tends to form bonds until it is surrounded by eight valence electrons).

Example: Write the Lewis stucture for nitric acid, HNO_3.
1. Writing atoms in the correct order:

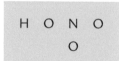

ISBN: 9780170355544
PHOTOCOPYING OF THIS PAGE IS RESTRICTED UNDER LAW.

2 Putting in electrons for H's and the O's bonded to them: H and O share an electron with each other.

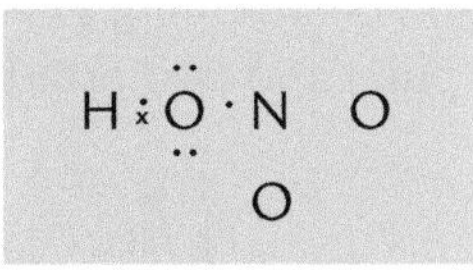

$$H \overset{..}{\underset{..}{\overset{x}{O}}} \cdot N \quad O$$
$$O$$

3 Putting in the electrons for the central atom and pairing up: N supplies five electrons and now has six.

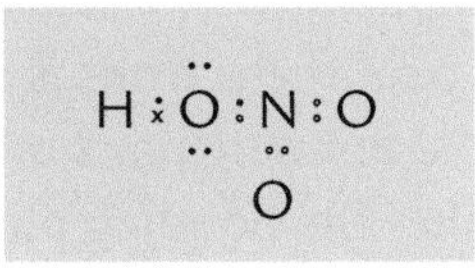

$$H \overset{x}{\underset{..}{O}} : \overset{..}{N} : O$$
$$O$$

4 Putting electrons in for the terminal O's: N needs two electrons. It can gain a share of two electrons from an O and give that O a share of two in return.

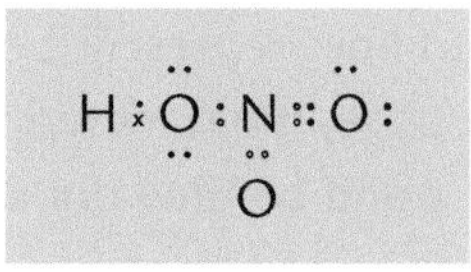

$$H \overset{x}{\underset{..}{O}} : \overset{..}{\underset{..}{N}} :: \overset{..}{\underset{..}{O}} :$$
$$O$$

The final O needs two from N. N does not need any more. N gives a share of two electrons to O without getting any in return.

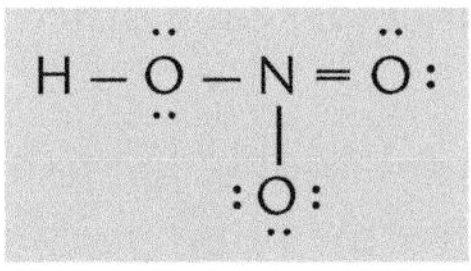

$$H - \overset{..}{\underset{..}{O}} - N = \overset{..}{O} :$$
$$|$$
$$: \overset{..}{\underset{..}{O}} :$$

The bonding in nitric acid is:

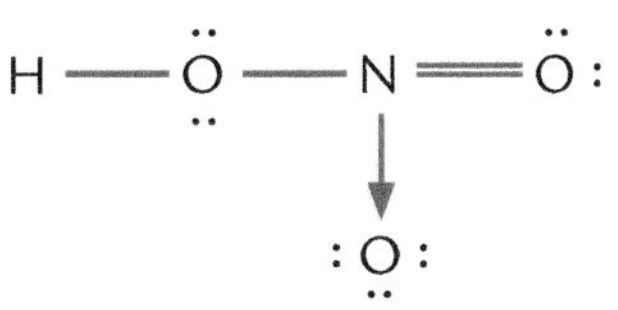

$$H \longrightarrow \overset{..}{\underset{..}{O}} \longrightarrow N = \overset{..}{O} :$$
$$\downarrow$$
$$: \overset{..}{\underset{..}{O}} :$$

—— represents a single bond involving one pair of electrons
══ represents a double bond involving two pairs of electrons
⟶ represents a **dative covalent bond** in which one atom donates both bonding electrons. The arrow goes from the atom supplying both electrons to the atom receiving them.

- By considering electron repulsions between electron regions, the shapes of oxyacids can be deduced. Consider the Lewis structure of nitric acid:

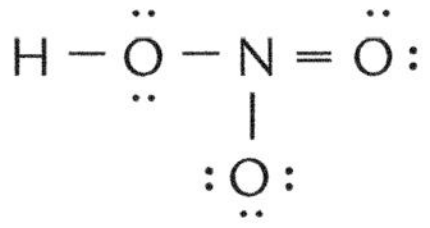

  $$H - \overset{..}{\underset{..}{O}} - N = \overset{..}{O} :$$
  $$|$$
  $$: \overset{..}{\underset{..}{O}} :$$

  a The three regions of electrons around the N atom will be at approximately 120° to one another. Therefore N and the three O atoms will all lie in the same plane.
  b The four regions of electrons around the O bonded to the H will be directed to the corners of a tetrahedron with a bond angle close to 109°. The N and the H atoms will be found at two vertices of the tetrahedron.

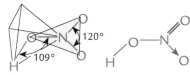

- In nitric acid, the two terminal bonds between N and O are found to be identical in bond length and bond strength. The Lewis structure above shows that one bond is a double bond and the other is a single (dative) bond.
- This is a limitation of Lewis structures. The actual molecular structure is regarded as having equal contributions from the following two structures:

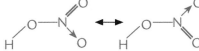

Deviations from the tetrahedral angle

- The three molecules methane, CH_4, ammonia, NH_3 and water, H_2O all have four pairs of electrons around the central atom.
- In CH_4, all four electron pairs are bond pairs. In NH_3, there are three bonding pairs and one non-bonding electron pair (lone pair). In H_2O, there are two bond pairs and two non-bonding pairs.

Note: The red lines represent bonds in the plane of the page, the dark red triangle represents a bond coming out of the page and the dashed line represents a bond going behind the page.

- Electrons in a bond pair are attracted to two nuclei. For example in NH_3, electrons in the N–H bond are attracted to both N and H. The non-bonding pair of electrons is attracted to N only and so are closer to N. The non-bonding pair repels the bonding pairs more than the bonding pairs repel each other. So the non-bonding pair pushes the bonding pairs slightly closer together.
- The more non-bonding pairs there are, the closer together the bond pairs are found and the smaller the bond angle.

Fewer than eight electrons

- An **octet** is the name given to a stable arrangement of eight electrons in an electron energy level. In some molecules it is not possible for every atom to have an octet of electrons.

Insufficient pairs of electrons around central atom

- For example, in boron trifluoride, BF_3, B has three valence electrons. Each F will supply one to B in return for a share of one from B. B now has a total of three pairs of electrons around it. The arrangement is planar.

Total number of valence electrons
= 3 [from B] + (3 x 7) [from 3 F] = 24

Molecules with unpaired electrons

NO_2 is a brown gas.

- If the sum of all the valence electrons of the atoms in a molecule is an odd number, it is not possible to have all the electrons paired up.
- For example, consider nitrogen dioxide, NO_2. N has five valence electrons and O six. The sum of the valence electrons, $5 + (2 \times 6) = 17$, is an odd number. The atom that has the odd number of valence electrons (N) is the one with the unpaired electron in the molecule. The Lewis structure of NO_2 is shown. N ends up with only seven electrons.

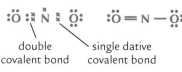

double covalent bond   single dative covalent bond

- One bond in NO_2 is a *single dative covalent bond* and the other is a *double covalent bond*.

ISBN: 9780170355544 PHOTOCOPYING OF THIS PAGE IS RESTRICTED UNDER LAW.

- In experiments, the bonds in nitrogen dioxide are found to be identical. They are stronger than N–O single bonds but not as strong as N=O double bonds. The Lewis structure does not predict this. The actual structure is regarded to have equal contributions from the two structures shown at right.

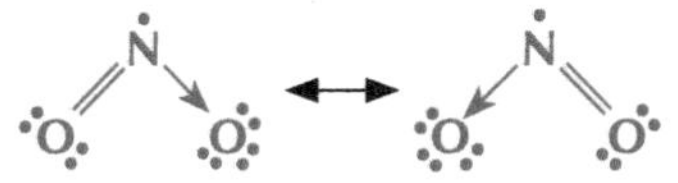

Because there are three regions of electrons around the N atom, the molecule is bent (V-shaped). The single electron on N makes it reactive. In the reaction $2NO_2(g) \longrightarrow N_2O_4(g)$, the two single electrons become paired.

More than eight electrons

- Some atoms acquire more than eight electrons in their outer shells when bonding with other atoms. They have **expanded octets**.
- These are the larger atoms that belong to the third row or lower on the Periodic Table.
- Oxygen belongs to the second row of the Periodic Table. In ozone (O_3), each O atom has six valence electrons and needs to gain two more. The Lewis structure for ozone is:

$$:\ddot{O} = \ddot{O} - \ddot{O}:$$

- Oxygen's valence shell can only accommodate eight electrons. This is the only possible Lewis structure for O_3.
- There are three regions of electrons around the central O atom. The shape of O_3 is bent (or V-shaped) with a bond angle close to 120°.
- Sulfur belongs to the third row of the Periodic Table. In sulfur dioxide (SO_2), each atom also has six valence electrons and needs to gain two more. Although S is stable with eight electrons in the valence shell, the valence shell of sulfur can accommodate up to 18 electrons. Also, the sulfur atom is large enough to fit many atoms around it. So SF_4 and SF_6 can form, but the corresponding compounds for O cannot form.
- There are two possible Lewis structures for SO_2:

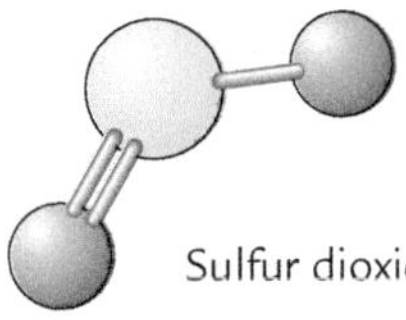

$$:\ddot{O} = \ddot{S} - \ddot{O}: \quad \text{and} \quad :\ddot{O} = \ddot{S} = \ddot{O}:$$

In the second structure, S has an expanded octet.

- Like O_3, the shape of SO_2 is bent (V-shaped), with the bond angle indicating that there are three regions of electrons around the central atom. Both of the above structures comply.

Sulfur dioxide

Limitations of Lewis structures

- As discussed above, sometimes there can be more than one Lewis structure for a molecule or polyatomic ion and the actual structure may not be accurately represented by any of them. An approximation that to some extent overcomes this problem is made by saying that the actual structure is contributed to by all the possible structures.

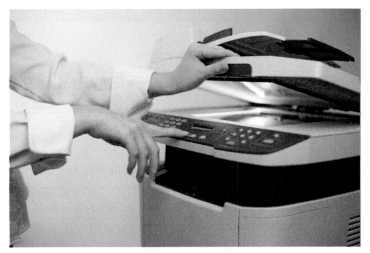

The sharp smell of ozone is noticed when the bright light of the photocopier causes O_2 to form O_3.

- The Lewis structure for ozone shows that one bond is a double bond, while the other is a single bond. However, the bonds in ozone are found to be identical in bond length, strength, etc. The actual structure is said to be a combination of the two shown here:

$$:\ddot{O}=\ddot{O}-\ddot{O}: \longleftrightarrow :\ddot{O}-\ddot{O}=\ddot{O}:$$

- In SO_2, the two bonds are also found to be identical with properties between those of a single and a double bond. The bond angle in both O_3 and SO_2 is approximately 120°.
- Oxygen, O_2, has the Lewis structure $:\ddot{O}=\ddot{O}:$ in which all the electrons are paired. However, the behaviour of oxygen molecules in magnetic fields suggests that O_2 has two unpaired electrons. The Lewis structure $:\dot{\ddot{O}} : \dot{\ddot{O}}:$ is also not satisfactory as the bond has the properties of an O=O double bond and not of an O–O single bond.

Placement of atoms

- If there is only one of a particular atom (other than H) in a molecule, that atom will be the central atom. For example, N is the central atom in NH_3, S is the central atom in SCl_2, B is the central atom in BF_3, and P is the central atom in PCl_5.
- An exception is N_2O, in which the order of atoms is N–N–O.

 KEY POINTS SUMMARY

- In **oxyacids**, the hydrogens are bonded to oxygen.
- **Repulsions** between **regions of electrons** determine how they are placed in space and therefore give the approximate sizes of **bond angles**. Repulsions by non-bonding pairs can push bond pairs together.
- **Lewis structures** have limitations in their ability to predict the observed properties of molecules.
- Exceptions to the **octet rule** occur. When an atom has an odd number of electrons, the atom can have less than a stable octet. An atom found in the third row or lower in the Periodic Table may have an **expanded octet** with more than eight electrons.

 ASSESSMENT ACTIVITIES

1 **Matching terms with descriptions**

a	stable arrangement of electrons with more than eight around an atom	dative bond
b	acid with H bonded to O	single bond
c	both electrons in bond pair supplied by one atom	double bond
d	formed when atoms share two pairs of electrons	oxyacid
e	formed when atoms share one pair of electrons	expanded octet

ISBN: 9780170355544   PHOTOCOPYING OF THIS PAGE IS RESTRICTED UNDER LAW.

2 **Drawing dative covalent bonds**

In both of the molecules CO and HCN, there is a triple bond. One bond in the triple bond is a dative bond. Draw Lewis structures for these molecules.

3 **Drawing Lewis structures for oxyacids**

For each molecule, draw a Lewis structure and predict the bond angle H–O–X and the one about the central atom.

    a HNO_3 **b** HNO_2 **c** H_2SO_4

    d H_2SO_3 **e** H_3PO_4 **f** $HClO$

    g $HClO_4$ **h** H_2CO_3

4 **Drawing Lewis structures with incomplete octets**

    a BH_3 **b** NO **c** H_3BO_3

5 **Finding bond angles**

    a Draw the Lewis structure for ethane, CH_3CH_3, and determine the approximate size of the bond angles.

    b Draw the Lewis structure for ethanoic acid, CH_3COOH, and determine the approximate sizes of the bond angles H–C–H, C–O–H, C–C=O.

    c Draw the Lewis structure for ethanol, C_2H_5OH, and determine the approximate sizes of the bond angles H–C–H, H–C–C and C–O–H.

6 **Comparing deviations from the tetrahedral angle**

Draw Lewis structures for CH_3OH (methanol) and CH_3NH_2 (aminomethane) and compare the bond angles C–O–H in methanol with C–N–H in aminomethane.

7 **Drawing Lewis structures for expanded octets**

Draw the possible Lewis structures and describe the shape of the molecule for each of the following:

    a PCl_5 **b** SO_2

    c SCl_6 **d** ClF_3

    e SCl_4 **f** Explain why NCl_5 and OCl_4 do not exist.

8 **Predicting molecular shape**

XeF_4 is one of the few compounds formed by noble gases.

    a Why does F react with Xe? (See electronegativities page 16.)

    b Draw a possible Lewis structure for XeF_4 and predict the shape of the molecule.

    c Why is it not possible for HeF_4 to form?

Revision One

1 Atoms and ions

a Copy and complete the following table.

	Symbol	Ground state electron configuration in *s*, *p*, *d* notation; [Ar] may be used as part of your answer
i	Na	
ii	Fe	
iii		$[Ar]3d^5$

b Which is larger, the sodium atom (Na) or the sodium ion (Na$^+$)? Justify your answer.

c **i** Write an equation for the reaction for which the enthalpy change is equal to the first ionisation enthalpy of sodium.

   ii Explain why the second ionisation enthalpy for sodium is much greater than the first.

d Which of the two atoms, sulfur or fluorine, is more electronegative? Explain your answer.

2 Molecules

a Copy and complete the following table by:
   - drawing Lewis structures for the two molecules
   - giving the name of the shape of each molecule.

Molecule	SF_2	SF_6
Lewis structure		
Name of shape		

b Describe the polarity of one S–F bond. Give your reasoning.

c Describe the polarity of the SF_2 molecule. Give your reasoning.

ISBN: 9780170355544 PHOTOCOPYING OF THIS PAGE IS RESTRICTED UNDER LAW.

Question II

1 **Atoms and ions**

    a Write the ground state electron configuration of the following atoms and ions in
s, p, d notation. You may use [Ar] as part of your answer.

      i   Cl   **ii** Cu^+ **iii** Mn^{2+}

    b Write the formula of an ion with the same electron configuration as Cl^-.

    c Which is larger, the Cl atom or the Cl^- ion? Explain your answer.

    d Write an equation for the reaction for which the enthalpy change is equal to the first
ionisation enthalpy of calcium.

    e Give reasons why the third ionisation enthalpy for calcium is much greater than the first and
second ionisation enthalpies.

    f Explain why fluorine has a higher electronegativity than oxygen.

2 **Molecules**

    a Copy and complete the following table by:

       • drawing Lewis structures for the two molecules

       • giving the name of the shape of each molecule.

Molecule	BF_3	NF_3
Lewis structure		
Name of shape		

    b Show the polarity of one N–F bond. Give your reasoning.

    c The polarities of the BF_3 and NF_3 are different. Discuss the reasons for this.

Question III

1 Atoms and ions

a Copy and complete the following table.

	Symbol	Ground state electron configuration in *s, p, d* notation; [Ar] may be used as part of your answer
i	As	
ii	Cr	
iii	Co^{3+}	

b Write the symbol for any atom or ion with the same electron configuration as Co^{3+}.

c Discuss the data given in the following table.

Atom/Ion	Radius (pm)
Cr	128
Co^{3+}	55

d Write an equation for the reaction for which the enthalpy change is equal to the first ionisation enthalpy of iron.

e Explain why the second ionisation enthalpy for iron is greater than the first.

f Ionisation enthalpies are always positive. Give the reason for this.

g State which has the larger electronegativity, P or As. Justify your answer.

2 Molecules

a Copy and complete the following table by:
- drawing Lewis structures for the two molecules
- giving the name of the shape of each molecule.

Molecule	NCl_3	PCl_5
Lewis structure		
Name of shape		

b Explain why the P–Cl bond is polar.

c Describe the polarity of the PCl_5 molecule. Give your reasoning.

d P forms PCl_3 and PCl_5 while the only compound of nitrogen and chlorine is NCl_3. Explain why this is so.

e N and Cl have similar electronegativity values. Discuss the reasons for this.

ISBN: 9780170355544 PHOTOCOPYING OF THIS PAGE IS RESTRICTED UNDER LAW.

Question IV

1 Atoms and ions

a Copy and complete the following table.

	Symbol	Ground state electron configuration in s, p, d notation; [Ar] may be used as part of your answer
i	K	
ii	Fe^{3+}	
iii	Cu	

b Write the symbol for any atom or ion with the same electron configuration as Fe^{3+}.

c Write these atoms and ions, K, Fe, Fe^{3+}, in order of increasing radius. Explain your answer.

d Write an equation for the reaction for which the enthalpy change is equal to the first ionisation enthalpy of potassium.

e Explain why the second ionisation enthalpy for potassium is much greater than the first ionisation enthalpy.

f Explain how the electronegativity of sulphur (S) differs from that of chlorine (Cl) and give the reason for this difference.

2 Molecules

a Copy and complete the following table by:
- drawing Lewis structures for the two molecules
- giving the name of the shape of each molecule.

Molecule	CF_4	SF_4
Lewis structure		
Name of shape		

b Describe the type and polarity of the bond between S and F in SF_4.

c Describe the polarity of (you may use diagrams as part of your answer):
  i the CF_4 molecule
  ii the SF_4 molecule.

Unit 5 | Bond enthalpy

Learning Outcomes — on completing this unit you should be able to:

- use bond enthalpy data as an indication of bond strength
- relate bond enthalpy to the nature of the bond involved.

Bonds are broken in the fuel and in oxygen. New bonds form and energy is released.

Bond enthalpy

- The **bond enthalpy** is the energy *required to break* one mole of isolated bonds to give isolated atoms or other uncharged fragments. The bonds must be away from the influence of any other particles. They must therefore be in the gas phase. Values given in tables of bond enthalpies are those measured at **constant pressure** (and usually at 25°C). *Bond enthalpy values are always positive*, as energy is always *needed* to break a bond. They are bond dissociation enthalpies.
- The bond enthalpy of the H–H bond is 436 kJ mol^{-1}.
  Therefore, $H_2(g) \longrightarrow 2H^\bullet(g)$ ($^\bullet$ represents an unpaired electron) $\Delta_r H$ = 436 kJ mol^{-1}.
  It takes 436 kJ to break 1 mol of H–H bonds.
- The bond enthalpy for the bond in NaCl is the energy required for the reaction:
  $NaCl(g) \longrightarrow Na^\bullet(g) + Cl^\bullet(g)$
- The bond enthalpy for the C–H bond is the energy required for the reaction:
  $CH_4(g) \longrightarrow {}^\bullet CH_3(g) + H^\bullet(g)$
  $^\bullet CH_3$ has an unpaired electron on C, which makes it very reactive. It is an example of a free radical.
- The bond enthalpy for the C–H bond can also be taken as quarter of the energy required for the reaction:
  $CH_4(g) \longrightarrow C^\bullet(g) + 4H^\bullet(g)$
- *Bond enthalpy* values:
  - *increase as the bonds become more polar* — more energy is needed to return to each atom an even share of electrons before the bonds break. The bond enthalpy of HCl > HBr > H$_2$.
  - *increase with bond order*. This means that a triple bond (C≡C, bond order 3) is more difficult to break than a double bond (C=C, bond order 2), which in turn is stronger than a single bond (C–C, bond order 1). Bond enthalpy values give information about the nature of bonds. The bond strength of a C=C bond is less than twice that of a C–C bond. One bond in the double bond is weaker than the other. This bond is broken in the addition reactions of alkenes. In a triple bond, one bond is stronger than the other two.
  - *vary a little from compound to compound*. For example, the bond enthalpy of the C–H bond in methane (CH$_4$) is not exactly the same as that of C–H bond in ethane (C$_2$H$_6$).

ISBN: 9780170355544 PHOTOCOPYING OF THIS PAGE IS RESTRICTED UNDER LAW.

ASSESSMENT ACTIVITIES

1 Matching terms with descriptions

a made up of two bonds of unequal strength	bond enthalpy
b energy required to break a bond	free radical
c a process which releases energy	bond order
d is 1 for a single bond, 2 for a double bond, and 3 for a triple bond	C=C double bond
e a species which is reactive because it has an unpaired electron	bond forming

2 Understanding the concept of bond energy

Write thermochemical equations to represent the following information. A thermochemical equation gives the reaction and the enthalpy change.

a The bond energy of the bond in H_2 is 436 kJ mol^{-1}.
b The bond energy of the H–Cl bond is 431 kJ mol^{-1}.
c The bond energy of the H–O bond in H_2O is 463 kJ mol^{-1}.
d The bond energy of the H–N bond in NH_3 is 391 kJ mol^{-1}.

3 Comparing bond energy values

Explain the trend in each of the following.

a The energy of the bond in H_2 is 436 kJ mol^{-1}, in O_2 is 498 kJ mol^{-1}, and in N_2 is 945 kJ mol^{-1}.
b The energy of the H–F bond is 567 kJ mol^{-1}, of the H–Cl bond is 431 kJ mol^{-1}, and of the H–I bond is 298 kJ mol^{-1}.
c The energy of the bond in H_2 is 436 kJ mol^{-1}, in Cl_2 is 242 kJ mol^{-1} and in I_2 is 151 kJ mol^{-1}.

4 Drawing Lewis structures

The following species have large bond enthalpies. Draw their Lewis structures to show what they have in common.

a Nitrogen (945 kJ mol^{-1})
b Cyanide ion (890 kJ mol^{-1})
c Carbon monoxide (1071 kJ mol^{-1})

5 Using bond enthalpy values

Copy and complete the table below right using the bond energy data in the table below left. The third column of the table below right lists some bonds. In the fourth column, calculate the energy required to break these bonds.

Bond	Average bond energy (kJ mol^{-1})
H–O	463
C–H	413
C–C	348
C=C	614
C–Cl	339
N–H	391

Molecule	Structural formula	Bonds broken	Energy required
H_2O	O⟋⟍ H H	all bonds	926 kJ mol^{-1}
CH_4		2 C–H bonds	
C_2H_6		all bonds	
C_2H_4		the C=C bond	
CH_3Cl		a C–Cl and a C–H bond	
NH_3		all bonds	

ISBN: 9780170355544   PHOTOCOPYING OF THIS PAGE IS RESTRICTED UNDER LAW.

Unit 6 | Intermolecular attractions

Learning Outcomes — on completing this unit you should be able to:

- describe the properties and structure of substances with different bond types
- describe the attractive forces between molecules
- explain the nature of hydrogen bonding and predict which compounds will form hydrogen bonds between molecules
- compare the strengths of different types of attractive forces
- relate the strength of intermolecular forces to the physical properties of substances.

Hydrogen bonds in wooden piles support big buildings.

Intermolecular attractions

- In Unit 3, we saw how atoms bond to form compounds. The bonds were either covalent bonds, in which electrons are shared, or ionic bonds, which are electrostatic attractions between oppositely charged ions.
- Intermolecular attractions are the attractive forces between molecules that hold them together in liquids and solids. In gases, these forces are weak or nonexistent. By looking at the nature of these intermolecular forces, we can explain the physical properties of substances. So first we will look at these properties.

Properties of substances

- The nature and the strength of forces give rise to the properties of molecular solids. They relate to the following:
  1 *Melting* and *sublimation temperatures* can be used to compare *strengths* of the *attractive forces* between particles in solids, provided they are measured at the same pressure.
  2 *Solubility* in polar and non-polar solvents gives information about the *polarity of particles and the ease with which they can be separated*. Polar molecules dissolve in polar solvents, non-polar molecules dissolve in non-polar solvents.
  3 The *hardness* of solids gives a measure of *bond strength and bond type*.
- The types of bonds which hold particles together in a solid are the same as those which hold them together in a liquid.

Types of intermolecular attraction

1 Instantaneous dipole-induced dipole

These forces are the result of orbiting electrons being at one end of the molecule for some instant.

For that instant, a dipole is set up. It can now induce a dipole on a neighbouring molecule.

These instantaneous dipole-induced dipole attractions are also called **dispersion forces (or London forces).** They exist between all particles because all particles have moving electrons!

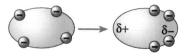

For an instant, the moving electrons in a non-polar molecule may be more concentrated in one region, causing it to become polar.

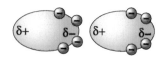

It then induces a dipole in a neighbouring molecule.

 ISBN: 9780170355544 PHOTOCOPYING OF THIS PAGE IS RESTRICTED UNDER LAW.

Particles in solids that are made up of *unbonded atoms, non-polar molecules* or *slightly polar molecule*s have only these forces between them. They are the weakest of all the forces which bind particles in a solid. Many substances with only these forces between the particles are liquids or gases at room temperature.

- Solids with only dispersion forces between the particles:
  — have low melting points and are soft
  — do not conduct electricity (the particles are uncharged)
  — tend to be soluble in non-polar solvents.
- Non-polar substances such as O_2 and CH_4 form solids at low temperatures. Large molecules have *large numbers of electrons*, e.g. CCl_4, and are more easily polarised. Such molecules have *stronger dispersion forces* than those with fewer electrons. Polythene is made up of very large non-polar molecules. It can be readily melted but the dispersion forces are strong enough for polythene to be used for shopping bags.

Butter is a solid with van der Waals forces between non-polar molecules.

- Shape also plays a part. Attractions between branched molecules are weaker than between chains as the molecules cannot be as close together.

2 Permanent dipole-induced dipole attraction between a polar molecule (permanent dipole) and a non-polar molecule

- A polar molecule can induce a dipole on a neighbouring molecule. Water is a polar molecule. It is a permanent dipole.
- Although O_2 is fairly insoluble in water, enough non-polar oxygen is dissolved in polar water to support a world of underwater life. The polar water molecule induces a dipole on the oxygen molecule, as shown in the diagram, and the two attract.

The negative end of a permanent dipole repels electrons around a non-polar molecule, thus inducing a dipole on it.

3 Dipole-dipole attraction between polar molecules

- The slightly negative end of a polar molecule, e.g. HCl, attracts the slightly positive end of another and a bond is formed between the two.
- At very low temperatures, HCl, HBr and HI form molecular solids.
- Polar molecules tend to dissolve in polar solvents such as water.

Permanent dipoles attract each other.

4 Ion-dipole attraction

- A cation (+ ion) is attracted to the negative end of a polar molar molecule, while an anion (– ion) is attracted to the positive end. Therefore ionic compounds, e.g. NaCl, will dissolve in polar solvents such as water.

5 Hydrogen bonds

- An important type of intermolecular forces are **hydrogen bonds**. These are relatively strong attractions and are stronger than those described earlier.
- Ice is found to have unexpectedly strong bonds between the H_2O molecules. The bonds in ice are examples of hydrogen bonds. *Hydrogen bonds occur between molecules in which the atoms N, O or F are covalently bonded to H, e.g. NH_3, H_2O and HF.* See positions of N, O and F on the Periodic Table.
- N, O and F are small, highly electronegative atoms. When bonded to H, the bonding electrons are attracted strongly away from the H atom towards the electronegative atom. This results in H's electron orbital being largely empty, so it can accommodate a share of a non-bonding electron pair from a neighbouring atom.

- The attraction of a *non-bonding electron pair* of an N, O or F atom in *one molecule* to a *hydrogen* atom in a *neighbouring molecule* is a hydrogen bond.

Hydrogen bonding.

Group 16 hydrides melting points

- *A hydrogen bond is relatively strong.* The compounds NH_3, H_2O and HF have unexpectedly high melting and boiling points for substances consisting of small molecules.
- With increasing molecular size, dispersion forces increase, for example the melting points of Group 16 hydrides (H_2O, H_2S, H_2Se and H_2Te) would be expected to increase. The melting temperatures are shown in the graph. The high melting point of H_2O is due to hydrogen bonding.
- Hydrogen bonds exist in water, alcohols and sugars (e.g. glucose) and cellulose-containing substances, such as wood.

The forces 1 to 5 described on pages 40–41 are examples of Van der Waals forces. They are not as strong as metallic bonds such as those in iron, ionic bonds such as those in sodium chloride and covalent bonds such as those in diamond. Types 2 and 4 are in mixtures only, the others are in pure substances and in mixtures.

Table of comparison of solids with van der Waals forces

Property	Solids with dispersion forces, e.g. CO_2 — dry ice	Solids with hydrogen bonding, e.g. H_2O — ice
Melting point	Low (in general less than 20 kJ needed to break 1 mole van der Waals bonds).	Higher than expected from the molar mass.
Electrical conductivity	Non-conducting — no charged particles.	Non-conducting — no charged particles.
Solubility of solids consisting of small molecules	More soluble in non-polar solvents.	Dissolve in polar solvents, unless the particles are too large. Particles are polar.
Hardness	May be cut with a knife.	Hard.

When dry ice is used for cooling it turns into a gas. No liquid puddle forms.

KEY POINTS SUMMARY

- Intermolecular forces are due to **dipole-dipole, dipole-induced dipole** or **instantaneous dipole-induced dipole** attractions and **hydrogen bonding** between neighbouring molecules. **Hydrogen bonding** is the attraction of a non-bonding electron pair of N, O or F in a molecule for a hydrogen atom in a neighbouring molecule.

 ISBN: 9780170355544 PHOTOCOPYING OF THIS PAGE IS RESTRICTED UNDER LAW.

ASSESSMENT ACTIVITIES

1 **Matching terms with descriptions**

a	attraction of a non-bonding electron pair of one molecule to a hydrogen atom of another		dispersion forces
b	an asymmetrically charged molecule		dipole
c	weakest of attractive forces between particles		hydrogen bond

2 **Understanding intermolecular attractions**
   a Explain how the forces in dry ice (solid CO_2) enable it to be in a solid state.
   b Explain the forces that exist between water and dissolved oxygen.
   c Explain why the forces in liquid helium (M = 4 g mol^{-1}) are weaker than those in liquid oxygen (M = 32 g mol^{-1}).
   d Explain why forces in liquid argon (M = 40 g mol^{-1}) are weaker than those in liquid hydrogen chloride (M = 36.5 g mol^{-1}).

3 **Understanding the role of hydrogen bonding**
   a The atoms C, N, O and F are in the same row of the Periodic Table. Boiling points of CH_4, NH_3, H_2O and HF are –162°C, –33°C, 100°C and 19°C, respectively.
   In terms of bond type and bond strength between the molecules, explain why CH_4 has a much lower boiling temperature.
   b The boiling points and molar masses of the hydrogen compounds of Group 15 are given in the table below.

Molecule	NH_3	PH_3	AsH_3	SbH_3
Boiling point	–33°C	–88°C	–57°C	–17°C
Molar mass (g mol^{-1})	17	34	78	125

   i Graph the boiling points of these compounds against the molar masses.
   ii Explain the unexpectedly high boiling point of NH_3.
   iii Explain the trend in boiling points for the compounds PH_3, AsH_3 and SbH_3.
   c In proteins there are hydrogen bonds between –NH groups of adjacent molecules. The molecules vary in shape, e.g. chains and globules. Give some examples of objects around (and in) us that are made up of protein molecules held together by hydrogen bonds.
   d The melting points of H_2O, H_2S and H_2Se are 0°C, –85°C and –68°C, respectively. Describe the types of intermolecular forces that exist for each substance, and explain this trend in their melting points.

e Account for the difference in the boiling points for the following pairs of compounds by comparing all the forces between the molecules.

	Boiling point (°C)	Molar mass (g mol⁻¹)
Compound A, CH_3CH_2OH	79	46.1
Compound B, CH_3CH_3	–88	30.1
Compound C, Cl_2	–34	70.9
Compound D, HCl	–85	36.5
Compound E, CH_3–CH_2–CH_2–OH	97	60.1

Compound F,
 	
    H
    |
CH3 – C – OH
    |
    CH_3
	83	60.1

f Explain the following in terms of the nature of the particles and the attractive forces between the particles.
   i Water is a liquid under standard conditions and petrol (mostly C_8H_{18} — it is non-polar) is also liquid.
   ii Butane, $CH_3CH_2CH_2CH_3$, boils at –0.5°C while methylpropane, $CH_3CHCH_3CH_3$, boils at –11.7°C.

g Account for the differences in boiling points of the three substances in the table below by comparing the nature of the particles, all the intermolecular forces between them and their molar masses.

Name	Structure	Boiling point (°C)	Molar mass (g mol⁻¹)
Ethanamine	$CH_3CH_2NH_2$	–81	45
Ammonia	NH_3	–33	17
Nitrogen	N_2	–196	28

h Wood is made up of cellulose with hydrogen bonds between –OH groups of nearby chain-like molecules. List uses of wood that indicate the strength of these hydrogen bonds.

4 Reviewing bond types

Explain the following in terms of the bonding between particles.
a Dry ice (solid CO_2) at –78°C dropped into a drink does not cool it as much as the same amount of ice at 0°C.
b Grease molecules (non-polar) dissolve in 1,1,1-trichloroethane.
c The solubility of oxygen is 0.04 g in 1 kg of water, while 500 g ammonia will dissolve in 1 kg of water.

ISBN: 9780170355544 PHOTOCOPYING OF THIS PAGE IS RESTRICTED UNDER LAW.

Unit 7 | Thermochemical principles

Learning Outcomes — on completing this unit you should be able to:

- use the vocabulary, symbols and conventions associated with thermochemistry
- understand the concepts of enthalpy of formation and enthalpy of combustion
- apply Hess's Law to calculations of enthalpy changes in reactions
- relate enthalpy and entropy changes to spontaneity in chemical reactions.

Exothermic and endothermic reactions

- Two energy/reaction progress graphs for chemical reactions are shown below.

- Graph 1 is for an **exothermic reaction**. *The products have less energy than the reactants.* Energy given out by the reaction is usually in the form of heat. If the heat change is measured at constant pressure, it is called the enthalpy change, ΔH, and is given a **negative sign**. If the reaction takes place in a test tube, the test tube feels hot. Energy may also be released in other forms, such as sound and light.

Graph 1: Energy/Reaction progress graph for an exothermic reaction

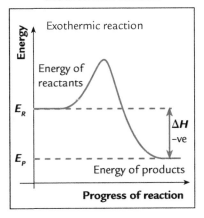

Graph 2: Energy/Reaction progress graph for an endothermic reaction

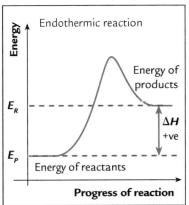

- Graph 2 is for an **endothermic reaction**. *The products have more energy than the reactants.* Energy is taken in from the surroundings by the reaction, usually in the form of heat. The enthalpy change, ΔH, is given a **positive sign**. If the reaction takes place in a test tube, the test tube feels cold. An endothermic reaction that takes in energy in the form of light is photosynthesis.

Standard enthalpy change for a chemical reaction

- The heat given out or taken in by a chemical reaction is often measured by a temperature change in a known mass of water. Two conditions are usually specified under which measurements are made. These are 298 K (25°C) and 101.3 kPa (1 atmosphere) pressure. They are referred to as **standard conditions** for measuring enthalpy changes. A substance in its state at 25°C and 101.3 kPa is in its **standard state**.

- A balanced chemical equation which also gives the value of ΔH for the molar quantities in the equation is a **thermochemical equation**. For example:

$$H_2(g) + \tfrac{1}{2}O_2(g) \longrightarrow H_2O(l) \quad \Delta_r H = -285 \text{ kJ mol}^{-1}$$

means that when 1 mole of hydrogen reacts with ½ mole of oxygen to form 1 mole of water, 285 kJ of energy is released.

$\Delta_r H$ is called the **enthalpy (heat) of reaction**. The subscript 'r' stands for 'reaction'.

For twice the amount of reactants, twice the amount of heat is given off:

$$2H_2(g) + O_2(g) \longrightarrow 2H_2O(l) \quad \Delta_r H = -570 \text{ kJ mol}^{-1}$$

If the reaction is reversed, the same amount of energy is taken in:

$$H_2O(l) \longrightarrow H_2(g) + \tfrac{1}{2}O_2(g) \quad \Delta_r H = +285 \text{ kJ mol}^{-1}$$

The *physical states* (gas, liquid, solid) of reactants and products must be given in a thermochemical equation. If, in the above reaction, $H_2O(g)$ was formed instead of $H_2O(l)$, less energy would be released. This is because when $H_2O(g)$ condenses to $H_2O(l)$, a further amount of energy is given out.

$$H_2(g) + \tfrac{1}{2}O_2(g) \longrightarrow H_2O(g) \quad \Delta_r H = -242 \text{ kJ mol}^{-1}$$

Standard enthalpy of formation and combustion

- The reaction $H_2(g) + \tfrac{1}{2}O_2(g) \longrightarrow H_2O(l) \quad \Delta_r H = -285 \text{ kJ mol}^{-1}$
  is the reaction for the formation of water from its elements under standard conditions.

  The **standard enthalpy of formation** of a compound is the enthalpy change when 1 mole of that compound in its standard state is formed from its *elements* in their standard states. It follows that the standard enthalpy of formation of any element is zero.

  Therefore, for water, we write: $\Delta_f H° \text{ (H}_2\text{O, } l\text{, 298 K)} = -285 \text{ kJ mol}^{-1}$. The meaning of each symbol is as follows:

Symbol	Meaning
ΔH	'the enthalpy change'
f	'for the formation of'
(H$_2$O, l, 298 K)	'liquid water at 298 K'
°	'under standard conditions'
−285 kJ	'285 kJ energy is given out'
mol^{-1}	'for every mole of water formed'

  With the symbol ° that means standard conditions, the state of water (l) and the temperature 298 K are implied. It is not necessary to have both.
  Enthalpies of formation can be *positive or negative*.

- The reaction $H_2(g) + \tfrac{1}{2}O_2(g) \longrightarrow H_2O(l) \quad \Delta_r H° = -285 \text{ kJ mol}^{-1}$
  is the reaction for the combustion of hydrogen under standard conditions.
  The **standard enthalpy of combustion** of a substance is the heat change when 1 mole of that substance in its standard state is *completely* burned to form products in their standard states.
  For hydrogen, we write: $\Delta_c H° \text{ (H}_2\text{, } g\text{, 298 K)} = -285 \text{ kJ mol}^{-1}$, where the subscript $_c$ stands for combustion. Enthalpies of combustion are *always negative*.

ISBN: 9780170355544 PHOTOCOPYING OF THIS PAGE IS RESTRICTED UNDER LAW.

- It follows that:

$\Delta_f H° (H_2O, l, 298K) = \Delta_c H° (H_2, g, 298K) = -285\,kJ\,mol^{-1}$

as $\Delta_f H°(H_2O, l, 298K)$ and $\Delta_c H°(H_2, g, 298K)$ are enthalpy changes for the same reaction.

Example

Write thermochemical equations for the following.

1. $\Delta_f H° (NO, g, 298K) = +90\,kJ\,mol^{-1}$

    NO is *formed*. It goes on the *product* side (right side) of the equation.

    $$\tfrac{1}{2}N_2(g) + \tfrac{1}{2}O_2(g) \longrightarrow NO(g) \quad \Delta_f H = +90\,kJ\,mol^{-1}$$

    The enthalpy change is for the formation of 1 mole of NO.

2. $\Delta_c H° (C_2H_6, g, 298K) = -1557\,kJ\,mol^{-1}$

    C_2H_6 is *burned*. It goes on the *reactant* side (left side) of the equation.

    $$C_2H_6(g) + 3\tfrac{1}{2}O_2(g) \longrightarrow 2CO_2(g) + 3H_2O(l) \quad \Delta_c H = -1557\,kJ\,mol^{-1}$$

    The enthalpy change is for the combustion of 1 mole of C_2H_6.

Hess's Law

- **Hess's Law** states that the enthalpy change for a chemical reaction is independent of the pathway.

    For example, if carbon is burned completely to carbon dioxide, the thermochemical equation is:

    $$C(s) + O_2(g) \longrightarrow CO_2(g) \quad \Delta_r H = -393\,kJ\,mol^{-1}$$

    If carbon is incompletely burned to carbon monoxide, and then the carbon monoxide reacted with oxygen to give carbon dioxide, the equations are as follows:

    $$C(s) + \tfrac{1}{2}O_2(g) \longrightarrow CO(g) \quad \Delta_r H = -282\,kJ\,mol^{-1}$$
    $$CO(g) + \tfrac{1}{2}O_2(g) \longrightarrow CO_2(g) \quad \Delta_r H = -111\,kJ\,mol^{-1}$$

    Adding the two reactions gives:

    $$C(s) + O_2(g) \longrightarrow CO_2(g) \quad \Delta_r H = -393\,kJ\,mol^{-1}$$

    In both cases, the formation of 1 mole of $CO_2(g)$ from 1 mole of $C(s)$ releases 393 kJ of energy.

SPECIAL SKILL 1: USING HESS'S LAW

A Using enthalpies of reactions, $\Delta_r H$

Example: A gaseous fuel can be more convenient than a solid fuel. For example, a gas transports itself once the gas lines are established. Water gas is a gaseous fuel consisting of $CO(g)$ and $H_2(g)$. It is produced by the reaction of steam on carbon.

$$C(s) + H_2O(g) \longrightarrow CO(g) + H_2(g)$$

Calculate the enthalpy change of the reaction for production of water gas, given these thermochemical equations:

$$C(s) + O_2(g) \longrightarrow CO_2(g) \quad \Delta_r H = -393\,kJ\,mol^{-1}$$

$$H_2(g) + \tfrac{1}{2}O_2(g) \longrightarrow H_2O(g) \quad \Delta_r H = -242\,kJ\,mol^{-1}$$

$$CO(g) + \tfrac{1}{2}O_2(g) \longrightarrow CO_2(g) \quad \Delta_r H = -282\,kJ\,mol^{-1}$$

»

Two methods are described below. Hess's Law states that the enthalpy change for the reaction is the same whether it is carried out in one step or in the three steps described below.

Method 1: Rearrange the thermochemical equations so that they can be added to give the required equation, which is:

$$C(s) + H_2O(g) \longrightarrow CO(g) + H_2(g)$$

a To get $C(s)$ on the left: $\quad C(s) + O_2(g) \longrightarrow CO_2(g) \quad \Delta_r H = -393 \text{ kJ mol}^{-1}$

b To get $H_2O(g)$ on the left: $\quad H_2O(g) \longrightarrow H_2(g) + \tfrac{1}{2}O_2(g) \quad \Delta_r H = +242 \text{ kJ mol}^{-1}$
This also gives $H_2(g)$ on the right.

c To get $CO(g)$ on the right: $\quad CO_2(g) \longrightarrow CO(g) + \tfrac{1}{2}O_2(g) \quad \Delta_r H = +282 \text{ kJ mol}^{-1}$

Combine equations and cancel: $\quad C(s) + O_2(g) + H_2O(g) + CO_2(g) \longrightarrow CO(g) + H_2(g)$
$+ \tfrac{1}{2}O_2(g) + CO(g) + \tfrac{1}{2}O_2(g)$

The thermochemical equation is: $\quad C(s) + H_2O(g) \longrightarrow CO(g) + H_2(g) \; \Delta_r H = +131 \text{ kJ mol}^{-1}$

Method 2: For the required reaction,
$C(s)$ is given by the first thermochemical equation, $H_2O(g)$ and $H_2(g)$ by **reversing** the second, $CO(g)$ by **reversing** the third.

$$\Delta_r H = -393 - (-242) - (-282) \text{ kJ mol}^{-1} = +131 \text{ kJ mol}^{-1}$$

B Using enthalpies of formation, $\Delta_f H°$

Because reactants are being un-formed and products are being formed:
$\Delta_r H° = \Sigma\Delta_f H° \textbf{ (products)} - \Sigma\Delta_f H° \textbf{ (reactants)}$
where $\Sigma\Delta_f H°$ (products) means the sum of the standard enthalpies of formation of the products, and $\Sigma\Delta_f H°$ (reactants) is the sum of the standard enthalpies of formation of the reactants.

Example: Solid sulfur is produced by the reaction of the two gases, sulfur dioxide and hydrogen sulfide:

$$SO_2(g) + 2H_2S(g) \longrightarrow 3S(s) + 2H_2O(l)$$

Calculate the standard enthalpy change for this reaction from the following standard enthalpies of formation:

$\Delta_f H° (H_2O, l, 298\,K) = -285 \text{ kJ mol}^{-1}$
$\Delta_f H° (SO_2, g, 298\,K) = -297 \text{ kJ mol}^{-1}$
$\Delta_f H° (H_2S, g, 298\,K) = -20 \text{ kJ mol}^{-1}$

Using the equation: $\Delta_r H° = \Sigma\Delta_f H°$ (products) $- \Sigma\Delta_f H°$ (reactants)

$$\Delta_r H° = [3 \times (0) + 2 \times (-285)] - [(-297) + 2 \times (-20)] = -233 \text{ kJ mol}^{-1}$$

Remember that the enthalpy of formation of any element is zero.

C Using enthalpies of formation, $\Delta_f H°$, and combustion, $\Delta_c H°$

Example: The gas H_2S is produced naturally in geothermal areas in New Zealand. It is also produced in large quantities at oil refineries. It burns in air to produce sulfur dioxide and water.

ISBN: 9780170355544 PHOTOCOPYING OF THIS PAGE IS RESTRICTED UNDER LAW.

Calculate the standard enthalpy of combustion of hydrogen sulfide from the following standard enthalpies of formation:

$\Delta_f H°$ (H_2O, l, 298 K) = –285 kJ mol^{-1}
$\Delta_f H°$ (SO_2, g, 298 K) = –297 kJ mol^{-1}
$\Delta_f H°$ (H_2S, g, 298 K) = –20 kJ mol^{-1}

$\Delta_c H°$ (H_2S, g, 298 K) is the standard enthalpy change for the reaction:

$$H_2S(g) + 1\tfrac{1}{2}O_2(g) \longrightarrow H_2O(l) + SO_2(g)$$

Using: $\Delta_r H° = \Sigma\Delta_f H°$ (products) $- \Sigma\Delta_f H°$ (reactants)

$$\Delta_c H° (H_2S, g, 298\,K) = [-285 + (-297)] - [-20 + (1\tfrac{1}{2} \times 0)] = -562\ kJ\ mol^{-1}$$

Specific heat capacity

- This is the amount of heat energy (J) needed to give a unit mass (1 g) of a substance a rise in temperature of one degree (K). It takes 4.184 J to raise the temperature of 1 g water by 1 K degree. So the specific heat capacity of water is 4.184 Jg^{-1}K^{-1}. By measuring the rise in temperature of a known mass of water caused by a known amount of reactant, the heat of reaction can be calculated. If a mass **m** g of water has a temperature rise of ΔT K, the enthalpy change is given by:

$$\Delta H = (4.184\ Jg^{-1}K^{-1} \times \textbf{m}\ g \times \Delta \textbf{T}\ K)\ joule$$

  A change of 1°C is the same as a change of 1K degree.
- In practice, heat changes are measured using a **calorimeter**. A bomb calorimeter allows transfer of the maximum amount of heat from the reaction to water. The loss of heat to containers and surrounding air is minimal. The heat capacity of a calorimeter is known and the information is supplied with the calorimeter.

SPECIAL SKILL 2: USING SPECIFIC HEAT CAPACITY TO CALCULATE ENTHALPY OF REACTION

Example: 2.000 g ethanol are burned in a calorimeter containing 500.0 g water. The temperature of the water increased from 21.00°C to 50.00°C. Calculate the enthalpy of combustion of ethanol. (M(ethanol) = 46.07 g mol^{-1}. The specific heat capacity of water is 4.184 Jg^{-1}K^{-1}.)

Heat released by burning ethanol = – heat absorbed by water

$$\Delta H = - \frac{4.184\ Jg^{-1}K^{-1}}{1000} \times 500\ g \times 29.0\ K = -60.67\ kJ$$

$$n\ (\text{ethanol}) = \frac{2.000}{46.07} = 0.04341\ mol$$

$$\Delta_c H(\text{ethanol}) = \frac{-60.67\ kJ}{0.04341\ mol} = -1398\ kJ\ mol^{-1}$$

- The enthalpy change for a reaction, e.g. $H_2(g) + \frac{1}{2}O_2(g) \longrightarrow H_2O(l)$ $\Delta H = -285$ kJ mol^{-1} measured under standard conditions is the **standard enthalpy of reaction** between hydrogen and oxygen; it is also the **standard enthalpy of formation** of water, and the **standard enthalpy of combustion** of hydrogen.
- Hess's Law states that the enthalpy change for a reaction is independent of the reaction pathway.
- For any reaction, $\Delta_r H° = \Sigma\Delta_f H°$ **(products)** $- \Sigma\Delta_f H°$ **(reactants)**
  i.e. the enthalpy of reaction equals the sum of the enthalpies of formation of the products minus the sum of the enthalpies of formation of the reactants.

Enthalpy and entropy

Enthalpy and entropy give measures of energy.

In chemistry, an **enthalpy** change is a change in heat energy measured at constant pressure. Enthalpy is a useful form of energy and can be converted to other forms. For example, heat from the sun gives kinetic energy to water particles so that they vaporise. Water vapour forms clouds that give rain, which fill hydroelectric dams and give us electrical energy. Electrical energy can be converted to many other useful forms of energy.

Hydroelectric power is the main source of electricity in New Zealand.

The water in the dam **spontaneously** flows downward, going to a **state of lower gravitational potential energy**.

The calculations you have been doing are associated with enthalpy changes.

Systems will spontaneously go to a state of minimum enthalpy.

Entropy is the energy of disorder that is brought about because particles (or objects) are moving. It is the useless form of energy and cannot be converted to other forms. Suppose you woke at 6 a.m. on Saturday morning. You have a box with a layer of red balls, then a layer of white ones and then a layer of black ones. You shake the box until 6 p.m. that night. The balls are now in a much disordered state. You have spent a lot of energy and are probably exhausted from all the shaking. The balls will not line up back in their layers and give you your energy back. Also, your energy has not been converted to any other useful form, except perhaps a little heat energy.

Ordered

Disorder is the result of moving particles

Moving particles **spontaneously go to a higher state of disorder**, a state of higher entropy.

ISBN: 9780170355544 PHOTOCOPYING OF THIS PAGE IS RESTRICTED UNDER LAW.

Systems with moving particles will spontaneously go a state of maximum entropy (maximum disorder). Particles that have more freedom of movement, such as gas particles, can become more disordered than those that have less, such as particles in a solid. Dissolved particles have more freedom of movement than solid particles.

An often quoted example is that contents of a teenager's room spontaneously go to a state of maximum disorder, but movement is an essential part of this. If objects are not moved, the disordered state will not happen.

Movement is needed to reach a state of disorder.

Application to chemistry

For a chemical reaction, the enthalpy change has the symbol $\Delta_r H$ and the entropy change has the symbol $\Delta_r S$.

The overall energy change for the reaction is $\Delta_r G$ given by

$$\Delta_r G = \Delta_r H - T\Delta_r S$$

T is the absolute temperature (K), H is measured in joules (J), and S is measured in joule/kelvin degree (JK^{-1}). A reaction is favoured by a negative value of $\Delta_r G$.

So a reaction is favoured by a drive to minimum enthalpy, so that $\Delta_r H$ is negative, and a drive to maximum entropy, so that $\Delta_r S$ is positive ($- T\Delta_r S$ is negative).

Consider the reaction $N_2(g) + 3H_2(g) \longrightarrow 2NH_3(g)$ $\Delta_r H = -90 \text{ kJ mol}^{-1}$

$\Delta_r H$ is negative. The reaction to the right is favoured by $\Delta_r H$. The drive to minimum enthalpy causes the reaction to form NH_3.

$\Delta_r S$ is also negative. The reaction to the left is favoured by $\Delta_r S$. There are four gas particles on the left (1 N_2 and 3 H_2) and two (2 NH_3) on the right. Four particles have more freedom of movement than two and can get more disordered. The drive to maximum entropy favours the formation N_2 and H_2 from NH_3.

When the drive to minimum enthalpy equals the drive to maximum entropy, the reaction reaches a state of equilibrium. At this stage, for every mole of NH_3 that forms, a mole of it breaks up into N_2 and H_2.

In the reaction, the $\Delta_r H$ is negative (favours formation of NH_3) and $\Delta_r S$ is also negative (does not favour formation of NH_3).

ASSESSMENT ACTIVITIES

1 **Explain fully the meaning of:**
   a $\Delta_f H^\circ (SO_2, g, 298K) = -297 \text{ kJ mol}^{-1}$
   b $\Delta_c H^\circ (C, s, 298K) = -393 \text{ kJ mol}^{-1}$

2 **Writing thermochemical equations from symbols**
   Write thermochemical equations to represent the following:
   a $\Delta_f H^\circ (CO_2, g, 298K) = -393 \text{ kJ mol}^{-1}$
   b $\Delta_c H^\circ (C, s, 298K) = -393 \text{ kJ mol}^{-1}$

c $\Delta_f H°(CH_4, g, 298\,K) = -75\ kJ\ mol^{-1}$
d $\Delta_c H°(CH_4, g, 298\,K) = -889\ kJ\ mol^{-1}$
e $\Delta_c H°(CO, g, 298\,K) = -282\ kJ\ mol^{-1}$
f $\Delta_f H°(NH_3, g, 298\,K) = -46\ kJ\ mol^{-1}$
g $\Delta_c H°(C_8H_{18}, l, 298\,K) = -5464\ kJ\ mol^{-1}$

3 Writing thermochemical equations from statements
Write thermochemical equations to represent the following:
a The standard enthalpy of formation of $HNO_3(l)$ is $-174\ kJ\ mol^{-1}$.
b The standard enthalpy of formation of $H_2SO_4(l)$ is $-814\ kJ\ mol^{-1}$.
c The standard enthalpy of formation of $NaCl(s)$ is $-411\ kJ\ mol^{-1}$.
d The standard enthalpy of formation of $NaBr(s)$ is $-360\ kJ\ mol^{-1}$.
e The standard enthalpy of formation of $NaI(s)$ is $-288\ kJ\ mol^{-1}$.
f The standard enthalpy of formation of $NaI.2H_2O(s)$ is $-883\ kJ\ mol^{-1}$.

The following questions relate to enthalpies of formation and Hess's Law.
$\Delta_r H° = \Sigma\Delta_f H°(\text{products}) - \Sigma\Delta_f H°(\text{reactants})$

4 Methane is used as a fuel.
Calculate the standard enthalpy change, $\Delta_r H°$, for the reaction:
$CH_4(g) + 2O_2(g) \longrightarrow CO_2(g) + 2H_2O(l)$
from the following standard enthalpies of formation:
$\Delta_f H°(CH_4, g, 298\ K) = -75\ kJ\ mol^{-1}$
$\Delta_f H°(CO_2, g, 298\ K) = -393\ kJ\ mol^{-1}$
$\Delta_f H°(H_2O, l, 298\ K) = -285\ kJ\ mol^{-1}$

5 Carbon monoxide and hydrogen are used to synthesise methanol at Methanex, Taranaki, New Zealand.
  a Calculate the standard enthalpy change, $\Delta_r H°$, for the reaction:
$$CO(g) + 2H_2(g) \longrightarrow CH_3OH(l)$$
given: $\Delta_f H°(CO, g, 298\ K) = -111\ kJ\ mol^{-1}$
         $\Delta_f H°(CH_3OH, l, 298\ K) = -239\ kJ\ mol^{-1}$
  b Is the reaction exothermic or endothermic? Give a reason for your answer.

6 Methanol can be dehydrated to give hydrocarbons:
$$2CH_3OH(l) \longrightarrow C_2H_4(g) + 2H_2O(l)$$
The hydrocarbons can then be polymerised and hydrogenated to give synthetic liquid fuels; octane C_8H_{18} is an important component of petrol:
$$4C_2H_4(g) + H_2(g) \longrightarrow C_8H_{18}(l)$$
  a Calculate the enthalpy change for each of the above two reactions.
$\Delta_f H°(CH_3OH, l, 298\ K) = -239\ kJ\ mol^{-1}$
$\Delta_f H°(C_2H_4, g, 298\ K) = +52\ kJ\ mol^{-1}$
$\Delta_f H°(H_2O, l, 298\ K) = -285\ kJ\ mol^{-1}$
$\Delta_f H°(C_8H_{18}, l, 298\ K) = -208\ kJ\ mol^{-1}$
  b What is the enthalpy change for the reaction that produces 1 mole of octane (C_8H_{18}) from 8 moles of methanol? Is the reaction exothermic or endothermic?

7 Ammonia is oxidised by the following reaction (ammonia is not used as a fuel):
$4NH_3(g) + 3O_2(g) \longrightarrow 2N_2(g) + 6H_2O(l)$
$\Delta_f H°(NH_3, g, 298\ K) = -46\ kJ\ mol^{-1}$
$\Delta_f H°(H_2O, l, 298\ K) = -285\ kJ\ mol^{-1}$
  a Calculate the enthalpy change for the reaction.
  b Suggest a reason why ammonia is not used as a fuel.

ISBN: 9780170355544 PHOTOCOPYING OF THIS PAGE IS RESTRICTED UNDER LAW.

8 Carbon monoxide is prepared by the dehydration of methanoic (formic) acid:

$HCOOH(l) \longrightarrow CO(g) + H_2O(l)$

$\Delta_fH°(HCOOH, l, 298\ K) = -409\ kJ\ mol^{-1}$

$\Delta_fH°(CO, g, 298\ K) = -111\ kJ\ mol^{-1}$

$\Delta_fH°(H_2O, l, 298\ K) = -285\ kJ\ mol^{-1}$

   a Calculate the enthalpy change for the dehydration of methanoic acid.

   b From the sign of your answer, is the reaction exothermic or endothermic? By looking at the nature of the reactants and products, suggest a reason why.

The following questions relate to enthalpies of combustion.

9 **Methane (in CNG), propane (in LPG) and octane (in petrol) are the most common motor fuels.**

Use the enthalpies of combustion below to calculate which fuel releases the most energy when completely burned:

   i on a mole basis

   ii on a gram basis ($M(H) = 1.0\ g\ mol^{-1}$, $M(C) = 12.0\ g\ mol^{-1}$).

$\Delta_cH°(CH_4, g, 298\ K) = -889\ kJ\ mol^{-1}$

$\Delta_cH°(C_3H_8, g, 298\ K) = -2217\ kJ\ mol^{-1}$

$\Delta_cH°(C_8H_{18}, l, 298\ K) = -5464\ kJ\ mol^{-1}$

10 **Methanol (CH_3OH) and ethanol (C_2H_5OH) are two fuels that are easily handled. Because of their other uses, which include synthesis of other compounds and as solvents, they are expensive. Petrol-containing compounds with oxygen in their molecules, such as ethanol, burn more cleanly; less carbon monoxide is formed. In countries with Clean Air Acts, the petrol may contain 10–20% ethanol.**

   a From these heats of formation, calculate the standard enthalpies of combustion of methanol and ethanol.

$\Delta_fH°(CH_3OH, g, 298\ K) = -239\ kJ\ mol^{-1}$

$\Delta_fH°(C_2H_5OH, g, 298\ K) = -276\ kJ\ mol^{-1}$

$\Delta_fH°(CO_2, g, 298\ K) = -393\ kJ\ mol^{-1}$

$\Delta_fH°(H_2O, l, 298\ K) = -285\ kJ\ mol^{-1}$

   b Which is the better fuel on a gram basis?

($M(H) = 1.0\ g\ mol^{-1}$, $M(C) = 12.0\ g\ mol^{-1}$, $M(O) = 16.0\ g\ mol^{-1}$)

11 **Methanol is a liquid fuel which can be synthesised from the gaseous fuels carbon monoxide and hydrogen:**

$$CO(g) + 2H_2(g) \longrightarrow CH_3OH(l)$$

The enthalpies of combustion of $CO(g)$, $H_2(g)$ and $CH_3OH(l)$ are:

$\Delta_cH°(CO, g, 298\ K) = -282\ kJ\ mol^{-1}$

$\Delta_cH°(H_2, g, 298\ K) = -285\ kJ\ mol^{-1}$

$\Delta_cH°(CH_3OH, l, 298\ K) = -727\ kJ\ mol^{-1}$

   a Calculate the enthalpy of reaction, Δ_rH, for $CO(g) + 2H_2(g) \longrightarrow CH_3OH(l)$.

   b Compare the reactants with the product as fuels.

   c What are the advantages of gaseous fuels over liquid fuels?

The following problems relate to using the heat capacity of water to measure the enthalpy of reactions.

12 A spirit burner containing ethanol weighed 53.8 g. It was used to heat 100 g water from 20.0 to 71°C. After it had cooled, the burner weighed 52.2 g. (M(ethanol) = 46.07 g mol⁻¹, specific heat capacity of water is 4.184 Jg⁻¹K⁻¹)

    a Calculate the enthalpy of combustion of ethanol from this information.

    b Explain why the value obtained is very different from –1364 kJ, the value given in a data book.

    c The burner's mass kept changing as the student weighed it. Describe how and why it would be changing.

13 A calorimeter is used to measure the enthalpy of reaction with minimal heat loss. The heat capacity of the calorimeter may be supplied with the calorimeter. For example, if the heat capacity of a calorimeter is 5.0 JK⁻¹, then 20.0 joules are used to raise its temperature by 4°C or 4 K (5.0 JK⁻¹ × 4 K).

    A calorimeter is used to measure the enthalpy of combustion of propane. 2.000 g propane are burned and the heat released is used to increase the temperature of 1000 g water. The initial temperature of the water is 20.20°C and the final temperature is 44.30°C.

    Specific heat capacity of water is 4.184 Jg⁻¹K⁻¹

    Heat capacity of the calorimeter is 5.200 JK⁻¹

    $M(C_3H_8)$ = 44.09 g mol⁻¹

    a Calculate the heat absorbed by the calorimeter.

    b Calculate the heat absorbed by 1000 g water.

    c Assuming all the heat formed by burning propane is transferred to the water and the calorimeter, calculate the heat released by burning 2.000 g propane.

    d Calculate the enthalpy of combustion of propane in kJ mol⁻¹.

The following questions relate to enthalpy and entropy changes.

14 For each of the following reactions, state whether the change in entropy has a positive sign, is zero, or has a negative sign.

    a $HCOOH(l) \longrightarrow CO(g) + H_2O(l)$ **b** $N_2(g) + O_2(g) \longrightarrow 2NO(g)$

    c $4NH_3(g) + 3O_2(g) \longrightarrow 2N_2(g) + 6H_2O(g)$ **d** $NH_3(g) + HCl(g) \longrightarrow NH_4Cl(s)$

    e $CH_4(g) + 2O_2(g) \longrightarrow CO_2(g) + 2H_2O(l)$ **f** $2NO(g) + O_2(g) \longrightarrow 2NO_2(g)$

15 For the following changes, give the sign (+ or −) of ΔH and ΔS.

    a water changes to steam

    b carbon dioxide gas forms dry ice

    c sugar dissolves in water (the temperature of the water drops).

16 For each of the following, compare the drive to minimum energy and the drive to maximum disorder for the change.

    a NaOH dissolves in water and the temperature of the water rises.

    b NH_4NO_3 dissolves in water and the temperature of the water drops.

ISBN: 9780170355544 PHOTOCOPYING OF THIS PAGE IS RESTRICTED UNDER LAW.

Revision Two

a Write balanced thermochemical equations for each of the following.

  i The reaction having an enthalpy change equal to the enthalpy of combustion of carbon, Δ_cH° (C, s, 298 K) = –393 kJ mol^{-1}

  ii The reaction having an enthalpy change equal to the enthalpy of formation of carbon dioxide, Δ_fH° (CO_2, g, 298 K) = –393 kJ mol^{-1}

  iii Explain why the enthalpy change is the same for the reactions in **i** and **ii**.

b Δ_fH° (CO , g, 298 K) is less negative than Δ_fH° (CO_2, g, 298 K). Explain why this is so.

c Use the following enthalpies of formation to calculate the enthalpy of combustion of methanol CH_3OH. The equation for the reaction is:

  $CH_3OH(l) + O_2(g) \longrightarrow CO_2(g) + 2H_2O(l)$

  Δ_fH° (H_2O, l, 298 K) = –285 kJ mol^{-1}

  Δ_fH° (CO_2, g, 298 K) = –393 kJ mol^{-1}

  Δ_fH° (CH_3OH, l, 298 K) = –239 kJ mol^{-1}

d Given the enthalpy of the reaction:

  $CH_4(g) + 2O_2(g) \longrightarrow CO_2(g) + 2H_2O(l)$

  is Δ_rH = –890 kJ mol^{-1}

  and

  Δ_fH (H_2O, l) = –286 kJ mol^{-1}, Δ_fH (CO_2, g) = –393 kJ mol^{-1},

  calculate the enthalpy of formation of $CH_4(g)$, Δ_fH (CH_4, g).

e 0.5200 g methane, CH_4, is burned in a calorimeter that contains 1.000 kg water. Assume that all the heat formed is transferred to the water. The temperature of the water increased from 15.40°C to 22.28°C.

  Calculate the enthalpy of combustion of methane, Δ_cH (CH_4).

  $M(CH_4)$ = 16.04 g mol^{-1}

  Specific heat capacity of water is 4.184 Jg^{-1} °C^{-1}

a Write balanced thermochemical equations for each of the following.

  i The reaction having an enthalpy change equal to the enthalpy of formation of methane, Δ_fH° (CH_4, g, 298 K) = –75 kJ mol^{-1}

  ii The reaction having an enthalpy change equal to the enthalpy of combustion of methane, Δ_cH° (CH_4, g, 298 K) = –882 kJ mol^{-1}

  iii Are these reactions exothermic or endothermic? Explain how you know.

b **i** Write an equation for the combustion of butane, $C_4H_{10}(g)$.

  ii Calculate the enthalpy of combustion of butane, Δ_cH°(C_4H_{10}, g).

    Δ_fH° (C_4H_{10}, g) = –126 kJ mol^{-1}

    Δ_fH° (CO_2, g) = –393 kJ mol^{-1}

    Δ_fH° (H_2O, l) = –286 kJ mol^{-1}

iii Explain why butane burns readily in oxygen. Your answer should describe both enthalpy and entropy changes.

iv The apparatus at right was used to determine the enthalpy of combustion of butane. Butane from a lighter was used to heat 80.0 g water in a beaker. When 0.600 g of butane was burned, the temperature of the water was found to increase from 19.5°C to 47.5°C. Calculate the experimental value of $\Delta_c H°$ (C_4H_{10}, g). Specific heat capacity of water = 4.18 $Jg^{-1}°C^{-1}$

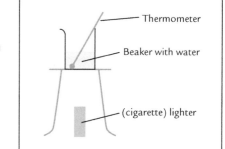

v The enthalpy of combustion determined by the experiment was only one-third of the value given in a data book. Give reasons for the difference AND suggest at least TWO ways this difference could be minimised.

c Methane can be used to produce iron metal from iron (II) oxide:
$$CH_4(g) + 3FeO(s) \longrightarrow 3Fe(s) + CO(g) + 2H_2O(g)$$

i Calculate the enthalpy change for this reaction using the following data:

$CO(g) + ½O_2(g) \longrightarrow CO_2(g)$ $\Delta_r H = -111$ kJ mol^{-1}

$CH_4(g) + 2O_2(g) \longrightarrow CO_2(g) + 2H_2O(l)$ $\Delta_r H = -882$ kJ mol^{-1}

$Fe(s) + ½O_2(g) \longrightarrow FeO(s)$ $\Delta_r H = -269$ kJ mol^{-1}

$H_2O(g) \longrightarrow H_2O(l)$ $\Delta_r H = -44$ kJ mol^{-1}

ii Explain whether entropy change in this reaction will favour the formation of products.

Question III

1 a Write balanced equations for each of the following.

i The reaction having an enthalpy change equal to the standard enthalpy of formation of water.

ii The reaction having an enthalpy change equal to the standard enthalpy of vaporisation of water.

b Give a reason why $\Delta_{vap} H°$ is always a positive value.

c Calculate a value for $\Delta_f H$ (H_2O, g) from the following data:

$\Delta_f H°$ (H_2O, l) = −286 kJ mol^{-1}

$\Delta_{vap} H°$ (H_2O, l) = +44 kJ mol^{-1}

2 a Explain why standard enthalpies of formation of elements are always zero.

b Calculate the enthalpy of combustion of ethanol, $\Delta_c H°$(C_2H_5OH, l), given:

$\Delta_f H°$ (C_2H_5OH, l) = −276 kJ mol^{-1}

$\Delta_f H°$ (CO_2, g) = −393 kJ mol^{-1}

$\Delta_f H°$ (H_2O, l) = −286 kJ mol^{-1}

c Propane gas is used in portable gas heaters. When it undergoes complete combustion, the energy released is −2217 kJ mol^{-1}.

$C_3H_8(g) + 5O_2(g) \longrightarrow 3CO_2(g) + 4H_2O(l)$ $\Delta_c H° = -2217$ kJ mol^{-1}

i Calculate the enthalpy of formation of propane, $\Delta_f H°$ (C_3H_8, g), given:

$\Delta_f H°$ (CO_2, g) = −393 kJ mol^{-1}

$\Delta_f H°$ (H_2O, l) = −286 kJ mol^{-1}

ii A Bunsen burner in a school laboratory uses propane gas. Heat from a Bunsen burner raised the temperature of 200.0 g water in a beaker from 22.5°C to 80.6°C. Given that the standard enthalpy of combustion of propane gas is −2217 kJ mol^{-1}, calculate the mass of propane gas that would be required for this temperature increase. Specific heat capacity of water = 4.184 $Jg^{-1}°C^{-1}$, $M(C_3H_8)$ = 44.1 g mol^{-1}

ISBN: 9780170355544 PHOTOCOPYING OF THIS PAGE IS RESTRICTED UNDER LAW.

iii The actual mass of propane gas used was more than twice the calculated amount. Give three reasons that could account for this.

d **i** Use the data below to find $\Delta_r H^\circ$ for the reaction:

$$4NH_3(g) + 5O_2(g) \longrightarrow 4NO(g) + 6H_2O(l)$$

Data: $N_2(g) + 3H_2(g) \longrightarrow 2NH_3(g)$ $\Delta_r H^\circ = -92.0 \text{ kJ mol}^{-1}$

$2H_2(g) + O_2(g) \longrightarrow 2H_2O(l)$ $\Delta_r H^\circ = -572 \text{ kJ mol}^{-1}$

$N_2(g) + O_2(g) \longrightarrow 2NO(g)$ $\Delta_r H^\circ = +180 \text{ kJ mol}^{-1}$

ii Give the sign of the entropy change for the following reaction. Explain your answer.

$$4NH_3(g) + 5O_2(g) \longrightarrow 4NO(g) + 6H_2O(l)$$

iii Explain how $\Delta_c H^\circ(NH_3)$ would differ from $\Delta_r H^\circ$ for the reaction:

$$4NH_3(g) + 5O_2(g) \longrightarrow 4NO(g) + 6H_2O(l)$$

e For the following changes, state whether the enthalpy (ΔH) is positive (+) or negative (−), and whether the entropy change (ΔS) is positive (+) or negative (−).

i Water changes to ice.

ii Propane burns in oxygen: $C_3H_8(g) + 5O_2(g) \longrightarrow 3CO_2(g) + 4H_2O(g)$

3.7

Demonstrate an understanding of oxidation-reduction processes

Internal assessment   **3 credits**

Unit	Content	Aim
8	**Redox reactions**	To be able to describe oxidation and reduction processes
9	**Electrochemical cells**	To be able to use the vocabulary, symbols and conventions associated with electrochemical cells To be able to calculate the cell voltage from standard electrode potentials
10	**Spontaneous reactions**	To calculate from standard electrode potentials whether a reaction is spontaneous under specified conditions
11	**Electrolytic cells and commercial electrical cells**	To understand the differences in conventions and purpose between electrolytic and electrochemical cells To be able to describe some commercial electrical cells

 ISBN: 9780170355544 PHOTOCOPYING OF THIS PAGE IS RESTRICTED UNDER LAW.

Unit 8 | Redox reactions

3.7

Learning Outcomes — on completing this unit you should be able to understand:

- oxidation and reduction in terms of gain or loss of O or H atoms
- oxidation and reduction in terms of loss or gain of electrons
- oxidation and reduction in terms of a change in oxidation number
- the difference between redox (oxidation-reduction) and other reactions
- how to write balanced equations for redox reactions in acidic, neutral and alkaline solutions.

Each year the US Government spends around US$23 billion fighting the oxidation of iron.

Oxidation and reduction

- Combustion reactions are examples of oxidation-reduction (redox) reactions. The fuel is oxidised. Oxygen is reduced.
- The earliest description (18th century) defined **oxidation** as the gain of oxygen or the loss of hydrogen. The concept of oxidation has been extended to a loss of electrons, or an increase in oxidation number.
- **Oxidation number** is the number of charges an atom would have in a molecule if electrons were transferred completely in the direction indicated by the electronegativities of the bonded atoms. Oxidation numbers will have a sign (+ or –) and a value between 1 and 7.
- **Reduction** is the opposite process to oxidation. Therefore, reduction is a loss of oxygen, or a gain of hydrogen, or a gain of electrons, or a decrease in oxidation number.
- All definitions are useful in recognising oxidation-reduction reactions.
- When methane burns in air, carbon dioxide and water form. The reaction is:
  $CH_4(g) + 2O_2(g) \longrightarrow CO_2(g) + 2H_2O(l)$
- Visually the reactions can be represented as:

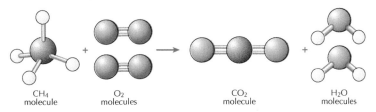

CH_4 molecule $\quad$ O_2 molecules $\quad$ CO_2 molecule $\quad$ H_2O molecules

- The C atom has lost 4 H atoms and gained 2 O atoms in forming CO_2; therefore it has been oxidised. H atoms have gained an O atom in forming H_2O; therefore they have been oxidised as well.
- Two of the O atoms have each gained 2 H atoms in forming H_2O molecules, therefore they have been reduced.
- When chlorine Cl_2 gas is bubbled through a colourless potassium iodide KI solution, iodine I_2 forms, resulting in a brown solution.
  $Cl_2(g) + 2I^-(aq) \longrightarrow 2Cl^-(aq) + I_2(aq)$
- The brown colour is due to $I_3^-(aq)$ formed by the subsequent reaction of I_2 molecules with I^- ions.

- Each I^- ion has lost an electron in becoming part of an I_2 molecule; therefore they have been oxidised. Cl atoms in the Cl_2 molecules gain electrons to become Cl^- ions; therefore they have been reduced.
- Another way of describing this reaction is to say that iodide ions have increased their oxidation number from –1 to 0; they have been oxidised. Chlorine atoms in elemental Cl_2 have decreased their oxidation number from 0 to –1; they have been reduced.
- If oxidation takes place in a reaction, then reduction must occur at the same time, and vice versa. Oxidation-reduction reactions, or **redox reactions**, are recognised by a transfer of oxygen or hydrogen, or by changes in oxidation numbers when the reaction occurs, or by gain or loss of electrons.

Redox equations

- One method of balancing redox equations is the **ion-electron method** (see the Special Skill boxes below). This involves breaking up the reaction into two half-reactions — one for the oxidation and one for the reduction. A half-equation is written for each half-reaction. The steps involved are shown for reactions in acidic solutions (Skill 1) and in neutral or basic solutions (Skill 2).

SPECIAL SKILL 1: BALANCING REDOX EQUATIONS IN ACIDIC CONDITIONS

Example: When a solution of potassium permanganate (the reacting ion is the purple ion MnO_4^-) is reacted with sodium sulfite (SO_3^{2-}) in acidic conditions, a colourless solution of manganous (Mn^{2+}) ions is formed. The sulfite ions form sulfate (SO_4^{2-}) ions.

Step 1	Write two half-equations for the reaction. Then balance each half-equation as follows. Note that the number of Mn is balanced, so is S.	$MnO_4^- \longrightarrow Mn^{2+}$	$SO_3^{2-} \longrightarrow SO_4^{2-}$
Step 2	Balance O with H_2O.	$MnO_4^- \longrightarrow Mn^{2+} + 4H_2O$	$SO_3^{2-} + H_2O \longrightarrow SO_4^{2-}$
Step 3	Balance H with H^+.	$8H^+ + MnO_4^- \longrightarrow$ $Mn^{2+} + 4H_2O$	$SO_3^{2-} + H_2O \longrightarrow$ $SO_4^{2-} + 2H^+$
Step 4	Balance the charges with electrons.	$8H^+ + MnO_4^- + 5e \longrightarrow$ $Mn^{2+} + 4H_2O$	$SO_3^{2-} + H_2O \longrightarrow$ $SO_4^{2-} + 2H^+ + 2e$
Step 5	Multiply the two half-equations by whole numbers so that electrons gained in one equals electrons given out by the other.	Multiplying the half-equation by 2 gives: $16H^+ + 2MnO_4 + 10e \longrightarrow$ $2Mn^{2+} + 8H_2O$	Multiplying the half-equation by 5 gives: $5SO_3^{2-} + 5H_2O \longrightarrow$ $5SO_4^{2-} + 10H^+ + 10e$
Step 6	Add the two half-reactions.	$16H^+ + 2MnO_4^- + 5SO_3^{2-} + 5H_2O \longrightarrow$ $2Mn^{2+} + 8H_2O + 5SO_4^{2-} + 10H^+$	

ISBN: 9780170355544
PHOTOCOPYING OF THIS PAGE IS RESTRICTED UNDER LAW.

Step 7	Subtract H⁺ and H₂O which occur on both sides.	$16H^+ + 2MnO_4^- + 5SO_3^{2-} + 5H_2O \longrightarrow 2Mn^{2+} + 8H_2O + 5SO_4^{2-} + 10H^+$ $-10H^+ \qquad\qquad\qquad -5H_2O \qquad\qquad -5H_2O \qquad\qquad -10H^+$ $6H^+(aq) + 2MnO_4^-(aq) + 5SO_3^{2-}(aq) \longrightarrow$ $2Mn^{2+}(aq) + 3H_2O(l) + 5SO_4^{2-}(aq)$

Sulfuric acid is used to supply H^+ ions for the reaction. It is not an oxidising or reducing agent under these conditions. Hydrochloric and nitric acids are not suitable. Hydrochloric acid is a reducing agent and can be oxidised to chlorine. Nitric acid is an oxidising agent and can be reduced to oxides of nitrogen such as NO or NO_2.

SPECIAL SKILL 2: BALANCING REDOX EQUATIONS FOR NEUTRAL OR ALKALINE CONDITIONS

Example 1: Balancing redox equations in neutral or slightly alkaline conditions

When a purple solution of potassium permanganate is reacted with sodium sulfite in neutral or slightly alkaline conditions, a brown precipitate of manganese dioxide (MnO_2) forms.

The procedure used for balancing equations in acidic conditions is used; then one OH^- is added for every H^+ in the equation.

Step 1: For the reaction of potassium permanganate with sodium sulfite in neutral or mildly alkaline conditions, the two half-reactions are:

$$MnO_4^- \longrightarrow MnO_2 \quad \text{and} \quad SO_3^{2-} \longrightarrow SO_4^{2-}$$

Steps 2–4: Following the procedure outlined in Special Skill 1 on page 60, the balanced half-equations are:

$$4H^+ + MnO_4^- + 3e \longrightarrow MnO_2 + 2H_2O \quad \text{and} \quad SO_3^{2-} + H_2O \longrightarrow SO_4^{2-} + 2H^+ + 2e$$

Steps 5–7: Multiplying the first equation by 2, the second by 3, then adding the two together and simplifying gives:

$$2H^+ + 2MnO_4^- + 3SO_3^{2-} \longrightarrow 2MnO_2 + H_2O + 3SO_4^{2-}$$

Step 8: Now add OH^- to convert any H^+ to H_2O. Any OH^- added to one side of the equation must also be added to the other side.

$$2H^+ + 2MnO_4^- + 3SO_3^{2-} \longrightarrow 2MnO_2 + H_2O + 3SO_4^{2-}$$
$$+ 2OH^- \qquad\qquad\qquad\qquad + 2OH^-$$

$$2H_2O(l) + 2MnO_4^-(aq) + 3SO_3^{2-}(aq) \longrightarrow 2MnO_2(s) + H_2O(l) + 3SO_4^{2-}(aq) + 2OH^-(aq)$$

which can be simplified by cancelling to:

$$H_2O(l) + 2MnO_4^-(aq) + 3SO_3^{2-}(aq) \longrightarrow 2MnO_2(s) + 3SO_4^{2-}(aq) + 2OH^-(aq)$$

Alternatively, step 8 can be carried out after step 4.

The addition of OH^- to convert H^+ to H_2O can be carried out earlier. OH^- can be added to the balanced half-equations.

$$4H^+ + MnO_4^- + 3e \longrightarrow MnO_2 + 2H_2O \quad \text{and} \quad SO_3^{2-} + H_2O \longrightarrow SO_4^{2-} + 2H^+ + 2e$$
$$+ 4OH^- \qquad\qquad\qquad + 4OH^- \qquad\qquad\qquad + 2OH^- \qquad\qquad\qquad + 2OH^-$$

$$4H_2O + MnO_4^- + 3e \longrightarrow MnO_2 + 2H_2O + 4OH^- \quad \text{and} \quad SO_3^{2-} + H_2O + 2OH^- \longrightarrow SO_4^{2-} + 2H_2O + 2e$$

Then steps 5–7 are completed.

Multiplying the first half-equation by 2, the second by 3, adding and simplifying gives the overall equation:

$$H_2O(l) + 2MnO_4^-(aq) + 3SO_3^{2-}(aq) \longrightarrow 2MnO_2(s) + 3SO_4^{2-}(aq) + 2OH^-(aq)$$

Example 2: Balancing redox equations in strongly alkaline conditions

When a purple solution of potassium permanganate is reacted with sodium sulfite in strongly alkaline conditions, a green solution of manganate ions (MnO_4^{2-}) forms.

Step 1: For the reaction of potassium permanganate with sodium sulfite in strongly alkaline conditions, the two half-reactions are

$$MnO_4^- \longrightarrow MnO_4^{2-} \qquad \text{and} \qquad SO_3^{2-} \longrightarrow SO_4^{2-}$$

Steps 2–8: Following the same process of balancing the equation in acid conditions, then adding OH^- for every H^+ results in:

$$2MnO_4^-(aq) + SO_3^{2-}(aq) + 2OH^-(aq) \longrightarrow 2MnO_4^{2-}(aq) + SO_4^{2-}(aq) + H_2O(l)$$
$$\text{purple} \qquad\qquad\qquad\qquad\qquad\qquad\qquad \text{green}$$

In some cases, it is easier to use OH^- right at the beginning. For example, in an alkaline battery, one half-reaction is

$$Cd \longrightarrow Cd(OH)_2$$

The easiest way to balance this is to add $2OH^-$ to the left side.

$$Cd + 2OH^- \longrightarrow Cd(OH)_2 + 2e$$

Oxidants and reductants

- An **oxidant** oxidises the other reactant in a reaction. It is also called the **oxidising agent**, and is itself reduced.
- An **reductant** reduces the other reactant in a reaction. It is also called the **reducing agent**, and is itself oxidised.
- The permanganate ion (in potassium permanganate) is an oxidant. The sulfite ion (in sodium sulfite) is a reductant.
- A compound that is both an oxidant and a reductant is the colourless solution hydrogen peroxide, H_2O_2.
- Hydrogen peroxide will oxidise sulfide S^{2-} to sulfate SO_4^{2-}. It will reduce acidified permanganate MnO_4^- to manganous Mn^{2+}.
- The half-reactions for hydrogen peroxide are:
  $$2H^+(aq) + H_2O_2(aq) + 2e \longrightarrow 2H_2O(l)$$
  H_2O_2 is acting an oxidant.
  $$H_2O_2(aq) \longrightarrow O_2(g) + 2H^+(aq) + 2e$$
  H_2O_2 is acting a reductant.

ISBN: 9780170355544 PHOTOCOPYING OF THIS PAGE IS RESTRICTED UNDER LAW.

On storage, hydrogen peroxide slowly decomposes:

$$2H_2O_2(aq) \longrightarrow 2H_2O(l) + O_2(g)$$

This reaction is obtained by adding the two half-equations above. It is an example of **auto-oxidation-reduction**.

- Manganese dioxide, MnO_2, a transition metal oxide, catalyses this reaction. Addition of MnO_2 causes bubbles of gas to form rapidly. This gas ignites a glowing piece of wood.
- An enzyme in blood, peroxidase, is also a catalyst for this reaction. A forensic test for blood on a piece of broken glass is to drop some H_2O_2 on it. If the red stain is blood, gas bubbles will form.

KEY POINTS SUMMARY

- **Oxidation** is a gain of oxygen, or a loss of hydrogen, or a loss of electrons, which results in an increase in **oxidation number**.
- **Reduction** is the opposite of oxidation. Reduction is a loss of oxygen, or a gain of hydrogen, or a gain of electrons, which results in a decrease in oxidation number.
- To **balance redox equations**:
  a balance all atoms other than O and H
  b balance O with H_2O
  c balance H with H^+
  d in neutral or alkaline solution, use OH^- to change all H^+ ions to H_2O.

ASSESSMENT ACTIVITIES

1 **Recognising oxidation and reduction**
State with reasons whether each of the following changes involve oxidation, reduction or neither.

a $CO(g) \longrightarrow CH_3OH(l)$
Methanol is formed from carbon monoxide at Methanex in Taranaki, New Zealand.

b $S_2O_3^{2-}(aq)$ [thiosulfate] $\longrightarrow S_4O_6^{2-}(aq)$ [tetrathionate]
This reaction is important in analytical chemistry for determining amounts of a variety of reagents.

c $Fe^{3+}(aq) + Fe(CN)_6^{4-}(aq) \longrightarrow Fe[Fe(CN)_6]^-(aq)$
The ion formed is the pigment Prussian blue.

d $AsO_4^{3-}(aq) \longrightarrow AsO_2^-(aq)$
Typical toxic arsenic compounds contain ions such as these.

e $Cr_2O_7^{2-}(aq) \longrightarrow CrO_4^{2-}(aq)$
This reaction is promoted by pH changes.

f $HNO_3(aq) \longrightarrow NO_2(g)$
Nitrogen dioxide gas must be prepared in a fume hood.

2 Balancing half-equations

Some antibacterial compounds are oxidising agents. These include chlorine, which is used in swimming pools; hydrogen peroxide, which is used on cuts; ozone, which is used to disinfect water supplies; and potassium peroxydisulfate, which is used for cleaning dentures. Use the ion-electron method to balance the following half-equations involving these oxidising agents:

a $Cl_2(aq) \longrightarrow Cl^-(aq)$

b $H_2O_2(aq) \longrightarrow H_2O(l)$ (acid solution)

c $O_3(g) \longrightarrow O_2(g)$ (acid solution)

d $S_2O_8^{2-}(aq) \longrightarrow SO_4^{2-}(aq)$

3 Balancing redox equations

For each of the following, balance two half-equations using the ion-electron method. Then combine them to write a balanced overall equation.

For each half equation, state whether oxidation or reduction has occurred.

a $H^+(aq) + Cr_2O_7^{2-}(aq) + Fe^{2+}(aq) \longrightarrow Cr^{3+}(aq) + Fe^{3+}(aq) + H_2O(l)$

b $H^+(aq) + IO_3^-(aq) + I^-(aq) \longrightarrow I_2(aq) + H_2O(l)$

c $H^+(aq) + Cr_2O_7^{2-}(aq) + C_2H_5OH(aq) \longrightarrow Cr^{3+}(aq) + CH_3COOH(aq) + H_2O(l)$

d $H^+(aq) + BrO_3^-(aq) + I^-(aq) \longrightarrow Br^-(aq) + I_2(aq) + H_2O(l)$

e $H^+(aq) + MnO_4^-(aq) + Fe^{2+}(aq) \longrightarrow Mn^{2+}(aq) + Fe^{3+}(aq) + H_2O(l)$

4 More redox equations

a Wine can be oxidised to vinegar by oxygen in the air. Ethanol is replaced by sour-tasting ethanoic (acetic) acid.

Write half-equations for the oxidation and reduction for the following reaction in acidic solution, then write the balanced overall equation:

$$C_2H_5OH(aq) + O_2(g) \longrightarrow CH_3COOH(aq) + H_2O(l)$$

b Chlorine gas can be prepared by the reaction of manganese dioxide with concentrated hydrochloric acid.

  i Balance the following equation for the reaction:

$$HCl(aq) + MnO_2(s) \longrightarrow Cl_2(g) + Mn^{2+}(aq)$$

  The write the balanced overall equation.

  ii What drives this reaction to the right?

  iii Describe precautions which must be taken when carrying out this reaction.

c One way of cleaning tarnished silver is to place it with a piece of aluminium foil into a solution of sodium bicarbonate and heat gently. Silver tarnishes when it forms silver sulfide on contact with sulfur-containing foods such as egg yolk. Silver sulfide is reduced to silver metal and is restored to the silver object. Aluminium is oxidised.

Balance the following equation for the reaction in alkaline solution:

$$Ag_2S(s) + Al(s) \longrightarrow Al(OH)_4^-(aq) + Ag(s) + S^{2-}(aq)$$

ISBN: 9780170355544   PHOTOCOPYING OF THIS PAGE IS RESTRICTED UNDER LAW.

d Iron is readily oxidised to ferric ions (Fe^{3+}) in alkaline solution. The low solubility of $Fe(OH)_3$ promotes the reaction $Fe \longrightarrow Fe^{3+}$.

Balance the following equation for the reaction in alkaline solution:

$$Fe(s) + O_2(g) + H_2O(l) \longrightarrow Fe(OH)_3(s)$$

e Products of redox reactions can sometimes be dependent on the pH of the solution.

Sulfur dioxide (SO_2) is oxidised to sulfate (SO_4^{2-}) ions. Write a balanced equation for the reaction of potassium permanganate solution with sulfur dioxide (a pungent-smelling gas formed by burning sulfur) in:

i an acid solution ($MnO_4^-(aq) \longrightarrow Mn^{2+}(aq)$)

ii a slightly basic solution ($MnO_4^-(aq) \longrightarrow MnO_2(s)$)

f Apple seeds contain cyanide bound in a compound so that it is harmless. However, small amounts of cyanide may be released by hydrolysis with warm stomach acid. The body contains small quantities of thiosulfate ions, which convert cyanide, CN^-, to relatively harmless thiocyanate, SCN^-.

The reaction is:

$$CN^-(aq) + S_2O_3^{2-}(aq) \longrightarrow SCN^-(aq) + SO_3^{2-}(aq)$$

i What is the change in oxidation number of the sulfur atom in the formation of SO_3^{2-}?

ii Has sulfur been oxidised or reduced in the formation of SO_3^{2-}?

iii In forming SCN^-, carbon (oxidation number +2) and sulfur have not changed oxidation numbers. What is the change in the oxidation number of N?

ISBN: 9780170355544   PHOTOCOPYING OF THIS PAGE IS RESTRICTED UNDER LAW.

Unit 9 | Electrochemical cells

Learning Outcomes — on completing this unit you should be able to:

- understand how an electrochemical cell can use a redox reaction to give an electric current
- use the correct vocabulary, symbols and conventions for an electrochemical cell
- write a cell in correct IUPAC convention
- write the electrochemical cell reaction
- calculate the electrochemical cell voltage.

Electrochemical cells and redox reactions

- In a **redox reaction**, the oxidising agent and the reducing agent collide and electron transfer takes place. The energy change for the reaction is given out as heat.
- It is possible to physically separate the oxidant and the reductant and make the electron transfer take place along an external circuit.
- The energy given out is partly electrical energy, which can immediately be converted to other useful energy forms such as light. Such a system is called an **electrochemical cell** or battery.
- For example, when a bar of zinc is placed in copper sulfate solution, the zinc goes into solution, a brown precipitate of copper forms and the blue colour of the copper sulfate solution fades.
- The reactions are as follows:

Oxidation	$Zn(s) \longrightarrow Zn^{2+}(aq) + 2e$
Reduction	$Cu^{2+}(aq) + 2e \longrightarrow Cu(s)$

Overall reaction $Cu^{2+}(aq) + Zn(s) \longrightarrow Cu(s) + Zn^{2+}(aq)$

- If the zinc bar is physically separated from the solution of copper ions, the transfer of electrons from zinc to the copper solution can be made to take place along an external circuit.
- The cell in the diagram below shows a **Daniell cell**. The copper solution and the zinc metal are placed in different containers.
- A **voltmeter** placed in the external circuit measures the electrical energy produced by the electrochemical cell. When the voltmeter is removed,

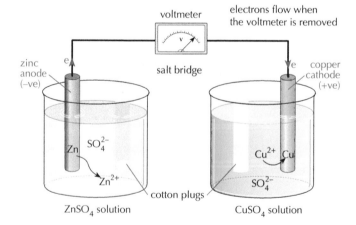

electrons travel in the direction shown. A voltmeter has a high internal resistance and there is only a very small flow of current when it is in the circuit.

ISBN: 9780170355544
PHOTOCOPYING OF THIS PAGE IS RESTRICTED UNDER LAW.

Daniell cell notation

- This particular cell can be represented as:

$$Zn(s) / Zn^{2+}(aq) // Cu^{2+}(aq) / Cu(s)$$

/ represents a surface that exists between Zn and Zn^{2+}
// indicates the two sides are in separate compartments
$Zn(s) / Zn^{2+}(aq)$ represents a **half-cell**, the zinc half-cell
$Cu^{2+}(aq) / Cu(s)$ represents the other half-cell, the copper half-cell.

- The cell is written so that the **reduction** reaction is on the **right**-hand side. Reduction takes place in the copper half-cell.

- At the copper electrode, the **reduction**

$$Cu^{2+}(aq) + 2e \longrightarrow Cu(s)$$

occurs. The copper bar is called the **cathode**.

- At the zinc electrode, the **oxidation**

$$Zn(s) \longrightarrow Zn^{2+}(aq) + 2e$$

occurs. The zinc bar is the called the **anode**.

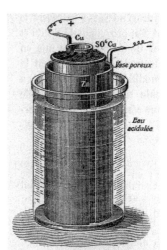

Fig. 284. — Élément Daniell.

Daniell cells were used for telegraphy from 1836 until they were replaced by Leclanché cells 30 years later.

- Electrons leave the zinc bar and move along the external circuit to the copper bar.
- Electrons enter the copper solution through the copper electrode (this electrode need not be made of copper; it could be any conducting, but otherwise unreactive, material).
- The most important feature of an electrochemical cell is the **production of an electric current by a spontaneous redox reaction**. Electrodes are labelled positive and negative by the direction of this current. Negative electrons leave the zinc electrode. This electrode is therefore negative.
- In an electrochemical cell, oxidation takes place at the anode, which is negative.
- Electrons are attracted to the copper electrode. This electrode is therefore positive.
- In an electrochemical cell, reduction takes place at the cathode, which is positive.
- The negative sign given to the anode and positive sign for the cathode is for electrochemical cells only. In electrolytic cells, the discharge of ions is the most important feature. Positive cations are reduced at the cathode; therefore the cathode is negative. The anode is positive.
- As Zn^{2+} cations enter the solution at the anode, the concentration of cations builds up in this half-cell but the number of anions is the same. As Cu^{2+} ions form Cu at the cathode, the concentration of cations drops. Anions are needed at the anode but are in surplus at the cathode.
- The **salt bridge** provides a path for movement of ions between the half-cells to complete the circuit. A salt bridge can be an inverted U-tube containing a solution of a strong electrolyte. This usually is a salt that does not react chemically with any species in the cell. For example, NaCl cannot be used in a salt bridge if Ag^+ is in any half-cell.
- An electrochemical cell can be based on any spontaneous redox reaction. It must have:
  — the oxidant in one compartment and the reductant in another
  — an external pathway for electrons
  — a device such as a salt bridge to complete the circuit.

$E°$ values — standard electrode potentials

- In the Daniell cell, reduction takes place spontaneously in the copper half-cell, not in the zinc half-cell. Therefore reduction of Cu^{2+} must take place more readily than the reduction of Zn^{2+}.
- How easily reduction occurs in a half-cell is measured by its **standard electrode potential**. This is an experimentally determined value and is given the symbol **$E°$**.

- To measure the $E°$ value, the half-cell is connected to a hydrogen half-cell:

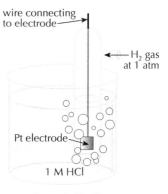

$$Pt(s) \:/\: H_2(g),\: H^+(aq)$$

under standard conditions and the voltage is measured.
- Standard conditions for measuring $E°$ values are:
  - concentrations of all aqueous reactants are 1 mol L^{-1}
  - metals are pure metals
  - gas pressure is 101.3 kPa (1 atmosphere)
  - temperature is 298 K (25°C).

Hydrogen half-cell

- When neither reactant in a half-cell is in the solid state (it cannot be an electrode), a non-reactive electrode is used. Often it is a graphite or platinum electrode.
- To measure the **standard electrode potential** of the zinc half-cell, the cell:

$$Pt(s) \:/\: H_2(g,\: 101.3\text{ kPa}),\: H^+(aq,\: 1\text{ mol L}^{-1}) \:// \: Zn^{2+}(aq,\: 1\text{ mol L}^{-1}) \:/\: Zn(s)$$

is set up and the **voltage** measured. The voltage is found to be 0.76 volts.
Because oxidation will occur in the zinc half-cell, the standard electrode potential for the zinc half-cell is recorded as –0.76 V. It is written as:

$$E°_{(Zn^{2+}/Zn)} = -0.76\text{ V}$$

- An electric current flows from the zinc half-cell to the hydrogen half-cell when the voltmeter is removed and the circuit reconnected without it. Standard electrode potentials are **reduction potentials**. The reduction is:

$$Zn^{2+} + 2e \longrightarrow Zn$$

Thus $E°$ for the zinc half-cell is given the symbol:

$$E°_{(Zn^{2+}/Zn)}$$

that is, $E°_{(oxidised\ form/reduced\ form)}$

- The negative sign indicates that reduction takes place less readily in that half-cell than in the hydrogen half-cell. The value gives a measure of the energy released by the reaction in the half-cell.
- Obviously, a hydrogen half-cell connected to another hydrogen half-cell would not produce electrical energy. Therefore:

$$E°_{(H^+/H_2)} = 0\text{ V}$$

Finding the overall cell voltage $E°_{cell}$

- For two half-cells connected to form an electrochemical cell, the cell voltage or emf $E°_{cell}$ is calculated from standard electrode potentials by:

  $E°_{cell} = E°_{red} - E°_{ox}$
  where $E°_{red}$ is the $E°$ value for the half-cell in which reduction takes place,
  and $E°_{ox}$ is the $E°$ value for the half-cell in which oxidation takes place.

　　ISBN: 9780170355544　　PHOTOCOPYING OF THIS PAGE IS RESTRICTED UNDER LAW.

Convention

- The IUPAC (International Union of Physics and Applied Chemistry) convention for writing a cell diagram puts the half-cell in which reduction takes place on the right-hand side.
- The hydrogen-zinc cell would be written as:

$$Zn^{2+}(aq, 1 \text{ mol } L^{-1}) / Zn(s) // H_2(g, 101.3 \text{ kPa}), H^+(aq, 1 \text{ mol } L^{-1}) / Pt(s)$$

and the cell voltage would be calculated by:

$$E^{\circ}_{cell} = E^{\circ}_{RHS} - E^{\circ}_{LHS}$$

where E°_{RHS} is the E° value for the half-cell written on the right-hand side,
and E°_{LHS} is the E° value for the half-cell on the left-hand side.

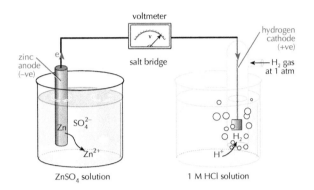

Using E° values to find E°_{cell} and the cell reaction

For any two half-cells connected together:

a reduction takes place in the half-cell with the higher E° (reduction potential) value
b the electrode in the half-cell with the higher (more positive) E° value is the positive electrode
c electrons will flow to the positive electrode.
  When writing the cell diagram in the conventional way, the half-cell with the higher E° value is written on the right side. This is the IUPAC convention.

Example: Answer the following for a zinc-iron cell, given
  $E^{\circ}_{(Zn^{2+}/Zn)} = -0.76$ V and $E^{\circ}_{(Fe^{2+}/Fe)} = -0.44$ V

1 State which electrode is the positive electrode.	Fe (this half-cell has the more positive value)
2 Give the half-cell where reduction takes place.	Fe²⁺/Fe (this half-cell has the greater reduction potential) Note that in **1**, an electrode is the answer. In this question, the answer is a half-cell.
3 Write the cell diagram in the IUPAC convention.	$Zn(s) / Zn^{2+}(aq) // Fe(s) / Fe^{2+}(aq)$ Reduction occurs on the right side
4 Calculate the voltage of the cell.	$E^{\circ}_{cell} = -0.44 - (-0.76) = 0.32$ V Note: voltage of cell is always positive
5 Write the cell reaction.	Reduction takes place at the positive electrode $Fe^{2+} + 2e \longrightarrow Fe$ $Zn \longrightarrow Zn^{2+} + 2e$ $Zn(s) + Fe^{2+}(aq) \longrightarrow Zn^{2+}(aq) + Fe(s)$
6 State the direction of flow of electrons in the external circuit.	Electrons flow from the Zn electrode (−) to the Fe electrode (+)

KEY POINTS SUMMARY

- An **electrochemical cell** uses a **spontaneous redox reaction** to produce an electric current.
- It consists of two **half-cells**, one in which oxidation takes place, the other in which reduction takes place.
- The ease of reduction in a half-cell is measured by its $E°$ value. $E°$ values are **standard electrode potentials**; they are standard **reduction** potentials.
- Reduction takes place in the half-cell with the higher $E°$ value.
- The overall cell voltage can be calculated using this formula:

$$E°_{cell} = E°_{reduction} - E°_{oxidation}$$

$E°_{oxidation}$ is the standard electrode potential for the oxidation half-reaction.

ASSESSMENT ACTIVITIES

Analysing electrochemical cells

1 Combining half-cells to make electrochemical cells

Electrochemical cells: reduction will take place in the half-cell with the more positive $E°$ value. This fact allows all the questions about cells to be answered.

Use this list of $E°$ values:

$E°_{(Zn^{2+}/Zn)} = -0.76$ V

$E°_{(Ni^+/Ni)} = -0.25$ V

$E°_{(Ag^+/Ag)} = +0.80$ V

$E°_{(Fe^{2+}/Fe)} = -0.44$ V

$E°_{(Cu^{2+}/Cu)} = +0.34$ V

For the following cells:

- zinc-copper
- copper-silver
- nickel-silver
- silver-iron
- zinc-nickel
- iron-zinc,

answer these questions:

a In which half-cell does reduction take place?

b At which electrode will oxidation take place?

c Calculate the cell voltage $E°_{cell}$.

d In which direction will electrons flow in the external circuit?

e Represent the cell by the notation:

A / A oxidised // B / B reduced

f Write the cell reaction.

2 Electrochemical cell with high $E°$

Use the $E°$ values given for the five electrode systems in question **1** to answer this question.

a Which two electrodes would be combined to produce a cell with the greatest voltage?

b Calculate $E°_{cell}$ for this electrochemical cell.

ISBN: 9780170355544 PHOTOCOPYING OF THIS PAGE IS RESTRICTED UNDER LAW.

3 Electrochemical cell with permanganate as the oxidant

Consider the cell:

$$Pt(s) / Cl^-(aq), Cl_2(aq) // MnO_4^-(aq, acid), Mn^{2+}(aq) / Pt(s)$$

$E^\circ_{(Cl_2/Cl^-)}$ = +1.40 V
$E^\circ_{(MnO_4^-/Mn^{2+})}$ = +1.51 V

a Write an equation for the reduction half-reaction.

b Write an equation for the oxidation half-reaction.

c Write the cell reaction.

d In which direction do electrons flow in the external circuit?

e Calculate the value of E°_{cell}.

f What would you observe as the cell delivers a current?

4 Electrochemical cell with nitrate as the oxidant

The following cell is set up:

$$Pt(s) / Fe^{2+}(aq), Fe^{3+}(aq) // NO_3^-(aq), NO_2(g) / Pt(s)$$

$E^\circ_{(Fe^{3+}/Fe^{2+})}$ = +0.77 V
$E^\circ_{(NO_3^-/NO_2)}$ = +0.80 V
$E^\circ_{(Cl_2/Cl^-)}$ = +1.40 V

a Write equations for the oxidation and reduction half-reactions. Write the overall equation for the cell reaction.

b Under standard conditions, what is the reading on the voltmeter?

c Explain why it is not possible to prepare the salt $Fe(NO_3)_2$.

d Is it possible to prepare the salt $FeCl_3$?

5 Electrochemical cell with dichromate as the oxidant

Consider the cell:

$$Pt(s) / SO_3^{2-}(aq), SO_4^{2-}(aq) // Cr_2O_7^{2-}(aq), Cr^{3+}(aq) / Pt(s)$$

$E^\circ_{(Cr_2O_7^{2-}/Cr^{3+})}$ = +1.33 V

$E^\circ_{(SO_4^{2-}/SO_3^{2-})}$ = +0.17 V

a Draw a schematic diagram of the cell.

b Calculate the value of E°_{cell}.

c Explain why the solution in the dichromate half-cell must be acidic.

d Write a balanced equation for the reduction half-reaction.

e Write a balanced equation for the oxidation half-reaction.

f Write a balanced equation for the cell reaction.

g Describe what you would observe as current is drawn from the cell.

6 Electrochemical cell involving iodine and iodide

Consider the cell shown in the diagram at right and the following E° values:

$E^\circ(I_2/I_3^-)$ = +0.62 V

$E^\circ(MnO_4^-/ Mn^{2+}$, acid solution $)$ = +1.51 V

a Write a balanced equation for the half-reaction that takes place at electrode A.

b Write a balanced equation for the half-reaction that takes place at electrode B.

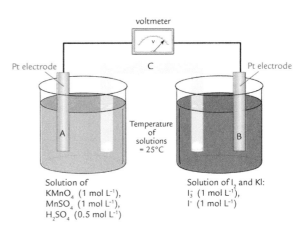

c   Explain the purpose of C. What name is given to it and what does it contain?

d   Write an equation for the cell reaction.

e   What would you observe as current is drawn from the cell?

f   In which direction do electrons flow in the external circuit?

g   Write the cell using the convention:
    X / X oxidised // Y / Y reduced

7   Matching terms with descriptions

a electrode where reduction takes place	electrochemical cell
b contains a solution of a strong electrolyte to provide a pathway for the flow of ions	half-cell
c system that uses an oxidation-reduction reaction to produce an electric current	cathode
d $E^\circ_{red} - E^\circ_{ox}$	anode
e gives a measure of how readily reduction will occur in a half-cell	salt bridge
f electrode where oxidation takes place	standard electrode potential
g part of an electrochemical cell in which either oxidation or reduction occurs	cell voltage, E°_{cell}

ISBN: 9780170355544   PHOTOCOPYING OF THIS PAGE IS RESTRICTED UNDER LAW.

Unit 10 | Spontaneous reactions

Learning Outcomes — on completing this unit you should be able to:

- associate a positive $E°_{cell}$ value with a spontaneous redox reaction under standard conditions
- use $E°$ values to distinguish redox reactions that can happen from those that can never happen.

Spontaneous reactions and $E°$ values

- Standard electrode potentials, $E°$ values, can be used to predict whether a redox reaction can occur under standard conditions. They do not give information about the rate, so while $E°$ values tell us that a reaction can occur, a small number are too slow to be observed.
- However, if $E°$ values tell us that a reaction will not occur, then the reaction will never occur under standard conditions.
- To predict whether a reaction will occur, the $E°$ value for a cell using the reaction is calculated.
- If the calculated value is positive, the reaction will occur spontaneously. A cell set up using the reaction will deliver a current as the reaction proceeds.
- If the calculated value is negative, the reaction will not occur. The reverse reaction will be spontaneous.
- If $E° = 0$, the reaction will be in equilibrium.
- To calculate the $E°$ value for a reaction, use the formula:

$$E°_{cell} = E°_{for\ the\ reduction} - E°_{for\ the\ oxidation} \text{ or } E°_{cell} = E°_{red} - E°_{ox}$$

where $E°_{red}$ is the standard electrode potential for the reduction half-reaction, and $E°_{ox}$ is the standard electrode potential for the oxidation half-reaction.

Oxidising strength

- $E°$ values are standard reduction potentials. The higher the value, the greater the ease of reduction.
- The list given earlier is:

$E°_{(Mg^{2+}/Mg)} = -2.37$ V   Strongest reductant
$E°_{(Al^{3+}/Al)} = -1.66$ V
$E°_{(H^+/H_2)} = 0.00$ V
$E°_{(Cu^{2+}/Cu)} = 0.34$ V
$E°_{(Fe^{3+}/Fe^{2+})} = 0.77$ V
$E°_{(NO_3^-, H^+/NO_2)} = 0.79$ V
         Strongest oxidant

- The highest $E°$ value (reduction potential) in the above list is:

$$E°_{(NO_3^-, H^+/NO_2)} = 0.79 \text{ V}$$

- This means that NO_3^- is readily reduced to NO_2 in acid solution. The oxidising strength of NO_3^- is great, the reducing strength of NO_2 is small.
- Therefore NO_3^- is the strongest oxidant and NO_2 is the weakest reductant in the list.

- Also from the above list, the lowest $E°$ value (reduction potential) is:

$$E°_{(Mg^{2+}/Mg)} = -2.37 \text{ V}$$

- This means Mg^{2+} is the most difficult to reduce. The weakest oxidant in the list is Mg^{2+} and the strongest reductant is Mg.
- While NO_3^- is readily reduced, Mg is readily oxidised.
- The two half-cells can be combined into the reaction:

$$Mg(s) + 4H^+(aq) + 2NO_3^-(aq) \longrightarrow Mg^{2+}(aq) + 2NO_2(g) + 2H_2O(l)$$

and $E°_{cell}$ would be 0.79 – (–2.37) = 3.16 V

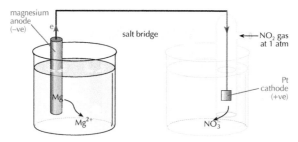

Electrochemical cell for
Mg(s) / Mg^{2+}(aq) // NO$_3^-$(aq), NO$_2$(g) / Pt(s)

 KEY POINTS SUMMARY

- A reaction will occur spontaneously under standard conditions if, for a cell set up using the reaction,

$$E°_{cell} = E°_{\text{for the reduction reaction}} - E°_{\text{for the oxidation reaction}}$$ has a positive value.

- A **strong oxidant** is the oxidised species associated with a large $E°$ value.
- A **strong reductant** is the reduced species associated with a small $E°$ value.

Example: Relating $E°_{cell}$ values to spontaneous redox reactions

The following list of standard electrode potentials is used to illustrate the worked problems **1** to **5** that follow:

$E°_{(Mg^{2+}/Mg)} = -2.37 \text{ V}$ $\qquad$ $E°_{(Al^{3+}/Al)} = -1.66 \text{ V}$ $\qquad$ $E°_{(H^+/H_2)} = 0.00 \text{ V}$
$E°_{(Cu^{2+}/Cu)} = 0.34 \text{ V}$ $\qquad$ $E°_{(Fe^{3+}/Fe^{2+})} = 0.77 \text{ V}$ $\qquad$ $E°_{(NO_3^-, H^+/NO_2)} = 0.79 \text{ V}$

1 **Predict whether the following reaction will occur:**
   $2Al(s) + 3Mg^{2+}(aq) \longrightarrow 2Al^{3+}(aq) + 3Mg(s)$

   The reduction reaction is:
   $Mg^{2+} + 2e \longrightarrow Mg$
   $E°_{cell} = E°_{red} - E°_{ox} = -2.37 \text{ V} - (-1.66 \text{ V}) = -0.71 < 0$
   Therefore the reaction will not occur.

2 **Determine whether copper will react with the nitric acid.**

   Nitric acid contains H^+ and NO_3^-. Two reactions need to be considered: the reactions of Cu with H^+ and NO_3^-. In both cases, Cu would be oxidised to Cu^{2+}.
   With H^+: $\quad E°_{cell} = E°_{red} - E°_{ox} = 0.00 \text{ V} - 0.34 \text{ V} = -0.34 \text{ V} < 0$
   Copper does not react with nitric acid to give H_2.
   With NO_3^-: $\quad E°_{cell} = E°_{red} - E°_{ox} = 0.79 \text{ V} - 0.34 \text{ V} = 0.45 \text{ V} > 0$
   Copper can react with nitric acid to give NO_2.

ISBN: 9780170355544
PHOTOCOPYING OF THIS PAGE IS RESTRICTED UNDER LAW.

3 **State what you would observe when a strip of magnesium ribbon is placed in copper sulfate solution.**

Reactants are Mg and Cu^{2+}. If a reaction occurred, the reduction is:
$Cu^{2+} + 2e \longrightarrow Cu$
$E°_{cell} = E°_{red} - E°_{ox} = 0.34\ V - (-2.37\ V) = 2.71\ V > 0$
Therefore the reaction will occur.
The magnesium ribbon goes into solution. A brown precipitate of copper metal forms and the blue colour of the solution fades.

4 **Explain why $Fe(NO_3)_2$ does not exist.**

If Fe^{2+} and NO_3^- are present, Fe^{2+} could be oxidised to Fe^{3+} and NO_3^- reduced to NO_2.
For this reaction:
$E°_{cell} = E°_{red} - E°_{ox} = 0.79\ V - 0.77\ V = 0.02\ V > 0$
NO_3^- can oxidise Fe^{2+} to Fe^{3+} and $Fe(NO_3)_2$ does not form.

5 **When ferric sulfate solution is added to potassium iodide solution, the solution turns brown. Addition of a drop of starch solution gives a blue-black colour. What can be said about the value of the standard electrode potential, $E°_{(I_2/I^-)}$?**

Iodide has been oxidised to iodine. Ferric ions must therefore have been reduced.
$E°_{cell} = E°_{red} - E°_{ox}$

Therefore: $0.77\ V - E°(I_2/I^-) > 0$

Therefore: $E°_{(I_2/I^-)} < 0.77\ V$

ASSESSMENT ACTIVITIES

1 **Matching terms with descriptions**

a the reaction can be used to make an electrochemical cell	spontaneous redox reaction
b readily oxidises another species and is itself readily reduced	strong oxidant
c readily reduces another species and is itself readily oxidised	strong reductant

2 **Determining whether a reaction is spontaneous**
Spontaneous reactions are associated with positive $E°_{cell}$ values, and:

$$E°_{cell} = E°_{for\ the\ reduction} - E°_{for\ the\ oxidation}$$

Some aqueous bleach solution ($Cl_2(aq)$) is shaken with potassium iodide solution in a test tube. A brown-coloured solution forms from the two colourless solutions.
a Write a balanced half-equation for the reduction reaction.
b Write a balanced half-equation for the oxidation reaction.
c Write a balanced equation for the reaction.

d Use these electrode potentials to explain why the reaction occurs.

$$E°_{(I_2/I^-)} = +0.62 \text{ V} \qquad\qquad E°_{(Cl_2/Cl^-)} = +1.40 \text{ V}$$

e What would be observed if some starch indicator was added to the test tube?

3 Looking at more reactions under standard conditions

Use this list of $E°$ values to answer the questions below for reactions under standard conditions.

$E°_{(Mg^{2+}/Mg)} = -2.36 \text{ V}$ $\qquad\qquad E°_{(Zn^{2+}/Zn)} = -0.76 \text{ V}$

$E°_{(Fe^{2+}/Fe)} = -0.41 \text{ V}$ $\qquad\qquad E°_{(Pb^{2+}/Pb)} = -0.13 \text{ V}$

$E°_{(Cu^{2+}/Cu)} = +0.34 \text{ V}$ $\qquad\qquad E°_{(I_2/I^-)} = +0.62 \text{ V}$

$E°_{(Fe^{3+}/Fe^{2+})} = +0.77 \text{ V}$ $\qquad\qquad E°_{(Ag^+/Ag)} = +0.80 \text{ V}$

$E°_{(Cl^2/Cl^-)} = +1.40 \text{ V}$ $\qquad\qquad E°_{(H_2O_2/H_2O)} = +1.77 \text{ V}$

For questions **a** to **g**, identify the reduction reaction and the oxidation reaction. Then use the formula:

$$E°_{cell} = E°_{for\ the\ reduction} - E°_{for\ the\ oxidation}$$

to determine whether the reaction occurs under standard conditions.

a Deduce whether the following reaction is possible:

$$Cl_2(aq) + 2Fe^{2+}(aq) \longrightarrow 2Cl^-(aq) + 2Fe^{3+}(aq)$$

b Is the reaction $2Ag^+(aq) + 2Cl^-(aq) \longrightarrow 2Ag(s) + Cl_2(g)$ possible?

c Predict whether a solution of hydrogen peroxide can be reduced by ferrous sulfate solution.

d The reaction occurs spontaneously.

$$Cl_2(aq) + 2Br^-(aq) \longrightarrow 2Cl^-(aq) + Br_2(aq)$$

What can be deduced about the value of the cell voltage $E°_{(Br2/Br^-)}$?

e Write the equation for the reaction of copper metal with chlorine gas.

f Calculate the cell voltage $E°_{cell}$ for the reaction in question **e**.

g Explain why the salt FeI_3 does not exist.

h State the metals in the list that can reduce lead ions to lead.

i List all the species that can react with aqueous acid to produce hydrogen gas.

4 Considering standard and non-standard conditions

In the laboratory, chlorine gas can be prepared by the reaction of concentrated hydrochloric acid with manganese dioxide.

a Write a balanced half-equation for the oxidation reaction.

b Write a balanced half-equation for the reduction reaction.

c Given:

$E°_{(Cl_2/Cl^-)} = +1.40V$

$E°_{(MnO_2/Mn^{2+})} = +1.23V$

calculate the cell voltage $E°_{cell}$ for the reaction.

d Explain why it is possible to prepare chlorine from manganese dioxide and concentrated hydrochloric acid.

5 Comparing the strengths of oxidants and reductants

Choose from the species given in the list of standard electrode potentials in question **3** to answer the following questions.

a Which metal ion is the strongest oxidant? **b** Which metal ion is the weakest oxidant?

c Which metal is the strongest reductant? **d** Which metal is the weakest reductant?

e Which species is the strongest oxidant?

ISBN: 9780170355544   PHOTOCOPYING OF THIS PAGE IS RESTRICTED UNDER LAW.

Unit 11 | Electrolytic cells and commercial electrical cells

Learning Outcomes — on completing this unit you should be able to:

- explain how an electrolytic cell uses an electric current to force a non-spontaneous redox reaction to occur
- understand the difference between an electrolytic cell and an electrochemical cell which uses a spontaneous redox reaction to produce an electric current
- describe commercial cells and how they are suited to their purpose.

An electrolytic cell reverses spontaneous reactions, but none has been developed to change CO_2 and water back to petrol.

Electrochemical and electrolytic cells

- **Electrochemical cells** are also known as **voltaic cells**, **galvanic cells** and **batteries**. They use a spontaneous redox reaction to produce an electric current.
- There are a large number of such reactions to choose from and one is chosen to suit the purpose of the cell. Factors considered include the voltage required, the cost, the size and the safety of the chemicals.
- Everyday batteries range from lead-acid batteries in cars to small batteries in hearing aids and pacemakers that regulate heartbeat.
- **Electrolytic cells** use an electric current to reverse a spontaneous redox reaction. Commercially, these cells are more commonly used in industrial processes. However, a common household electrolytic cell is a battery charger. It uses an electric current from the main power supply to reverse the spontaneous reaction that has taken place in a battery.

Electrolytic cells

- If we were not able to reverse spontaneous reactions, we would end up with a lot of useless products. Fortunately, methods can be found to reverse some reactions, and these methods are both practical and economical. For example, the reduction of metal oxides to metals. Other reactions are not economical or practical to reverse at present. For example, the conversion of carbon dioxide and water back to petrol. For these reactions, the sources of reactants must be monitored.
- Using an electric current to reverse a spontaneous reaction is called **electrolysis**. The system in which it is carried out is called an **electrolytic cell**.

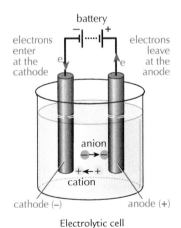

Electrolytic cell

Terminology and sign convention

- An electrochemical cell uses a spontaneous redox reaction to produce an electric current. An electrolytic cell uses an electric current to force a non-spontaneous redox reaction to occur.
- In both types of cells, oxidation takes place at the anode (electrons are released), reduction takes place at the cathode (electrons are used up).

- The important feature of an **electrochemical cell** is the current of electrons produced in the external circuit, and the signs of the electrodes relate to this. Negatively charged electrons leave the anode (–) where oxidation (loss of electrons) has occurred. **O**xidation takes place at the **a**node, the **negative** electrode.
- The negative electrons are attracted to the cathode (+). Electrons entering the cathode caused reduction to take place. **R**eduction takes place at the **c**athode, the **positive** electrode.
- The important feature of an **electrolytic cell** is the discharge of ions, and the signs relate to this. Cations (+ ions) are attracted to the cathode (–) where they are reduced. **R**eduction takes place at the **c**athode, the **negative** electrode.
- Anions (– ions) are attracted to the anode (+) where they are oxidised. That is, **o**xidation takes place at the **a**node, the **positive** electrode.

Commercial electrochemical cells

- The **Daniell cell** described in Unit 9 is an example of an electrochemical cell or battery. It uses a spontaneous redox reaction to produce an electric current. Because of the nature of the reactants, this cell has limited use. As it contains two aqueous solutions, it is not convenient to carry it around in a torch or a hearing aid. Electrochemical cells are designed to fit a purpose.
- **Torch batteries** have dry reactants and products. They are light and relatively inexpensive. They can be thrown out when the amounts of reactants are insufficient to deliver the required current.
- A **car battery** is heavy and cumbersome as it consists of lead plates. It contains sulfuric acid solution and must be kept upright. However, the spontaneous chemical reaction occurring is readily reversed and the battery can be 'recharged'. A current produced by the car's alternator will reverse the spontaneous reaction. This is an important advantage. Because the battery is carried around in a vehicle, the weight and the need to be kept upright are not serious disadvantages.
- **Fuel cells** consist of a typical fuel such as methane or hydrogen and an oxidant such as oxygen. Combustion of fuels typically produces large amounts of energy. Combustion of hydrogen in oxygen is an explosive reaction. These combustion reactions are also **redox reactions**. If the fuel can be physically separated from oxygen, then the reaction can be used to deliver relatively large amounts of electrical energy. Fuel cells have developed rapidly as a result of spacecraft research. Hydrogen and oxygen are light to carry. The water formed by the reaction can be used in cooling systems or for drinking.

Non-rechargeable portable batteries

- Some of the batteries we use today are thrown away when they no longer provide the required voltage. They are inexpensive to replace and are usually readily available.

Dry cell or Leclanché cell

- The common torch battery is called a **dry cell** or **Leclanché cell**.
- Oxidation occurs at the anode:

  $Zn \longrightarrow Zn^{2+} + 2e$

- Reduction occurs at the cathode:

  $MnO_2 + 4H^+ + 2e \longrightarrow Mn^{2+} + 2H_2O$

- Mn^{3+} may also form:

  $MnO_2 + H_2O + e \longrightarrow MnO(OH) + OH^-$

- H^+ is supplied by NH_4^+. NH_4Cl is in a paste with MnO_2.

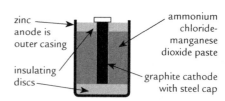

ISBN: 9780170355544 PHOTOCOPYING OF THIS PAGE IS RESTRICTED UNDER LAW.

- The reduction reaction can be written as:

$$MnO_2 + 4NH_4^+ + 2e \longrightarrow Mn^{2+} + 4NH_3 + 2H_2O$$

- The overall reaction is:

$$Zn + MnO_2 + 4NH_4^+ \longrightarrow Zn^{2+} + Mn^{2+} + 4NH_3 + 2H_2O$$

- The NH_3 resulting from the reaction forms complex ions with Zn^{2+}:

$$Zn^{2+} + 4NH_3 \longrightarrow Zn(NH_3)_4^{2+}$$

- The reaction is therefore:

$$Zn + MnO_2 + 4NH_4^+ \longrightarrow Zn(NH_3)_4^{2+} + Mn^{2+} + 2H_2O$$

Alkaline batteries

- **Alkaline batteries** rely on a similar reaction. In place of ammonium chloride, which is acidic, they contain potassium hydroxide, a strong base. Manganese dioxide is the oxidant (and so is reduced) and zinc is the reductant (so zinc is oxidised).
- Oxidation occurs at the zinc anode and the cathode is graphite. No gases are produced in the reactions and this prolongs the battery life.

Mercury batteries

- **Mercury batteries** also rely on the oxidation of zinc as the source of electrons. Mercury oxide, HgO, is reduced to mercury. These batteries are very small and portable and are used in cameras and electronic equipment.
- However, mercury and its compounds are poisonous and the battery has to be tightly sealed. This in itself becomes another hazard; the tightly sealed batteries explode when heated.
- Mercury batteries are being phased out. When buying equipment that relies on batteries, make sure the battery is a common, readily available type.

Lithium batteries

- **Lithium batteries** have lithium instead of zinc as the reductant. Lithium is much lighter and a much stronger reducing agent than zinc. The density of zinc is 7.1 gcm^{-3} while the density of lithium is 0.53 gcm^{-3}. In some lithium batteries, manganese dioxide is the oxidant, but this varies. The battery must have alkaline conditions as acid causes a violent exothermic reaction with lithium.
- These light batteries are used in pacemakers and other equipment.

Rechargeable portable batteries

- Many batteries can be **recharged** by having their spontaneous reaction reversed. These include batteries in mobile phones, electronic books and portable computers.
- A lead-acid battery in a car provides electrical energy for the starter motor, lights, horn, windscreen wipers, radio, stereo equipment, burglar alarm, clock, automatic door locks, water pump and air-conditioning to mention a few uses! The **alternator** uses kinetic energy provided by the moving car to constantly recharge the battery.
- Many other batteries can be charged when they are 'flat'. An electric current from the power supply or another battery (electrochemical cell) reverses the reaction that has taken place in these batteries.
- While a battery is delivering current, it is an electrochemical cell. While a battery is being recharged, it behaves as an electrolytic cell.

Lead-acid battery

- A typical **lead-acid battery** is shown in the photo at right. It has plates of Pb and PbO_2 in sulfuric acid.
- When the cell (battery) is delivering a current, Pb^{2+} is formed in both the oxidation and the reduction reactions.
- Oxidation occurs at the anode (negative terminal):

$$Pb(s) + SO_4^{2-}(aq) \longrightarrow PbSO_4(s) + 2e$$

- Reduction occurs at the cathode (positive terminal):

$$PbO_2(s) + 4H^+(aq) + SO_4^{2-}(aq) + 2e \longrightarrow PbSO_4(s) + 2H_2O(l)$$

- The overall reaction is:

$$Pb(s) + PbO_2(s) + 4H^+(aq) + 2SO_4^{2-}(aq) \longrightarrow 2PbSO_4(s) + 2H_2O(l)$$

- This reaction produces electrical energy. The reaction can be reversed by supplying electrical energy. Although a lead-acid battery is heavy and cumbersome, no replacement has been found that is as efficient and inexpensive.

Nickel-cadmium battery

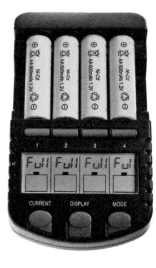

- **Nickel-cadmium** (Ni-cad) batteries are lightweight rechargeable batteries used in cordless appliances, mobile phones and other equipment.
- When the cell is delivering a current, cadmium metal (density 8.7 g cm^{-3}) is oxidised to cadmium hydroxide, and nickel (IV) oxide is reduced to nickel (I) hydroxide. These cells produce a constant voltage while they function. The products that form do not flow in the cell and stay close to the electrodes so that the reaction can be readily reversed. Cadmium, like mercury, is toxic.
- New batteries are being developed that use polymers to bind the chemicals, to render them safer and to hold them close to the electrodes so that the batteries can be recharged.

Fuel cells

- The space programme is largely responsible for the development of **fuel cells**. These electrochemical cells enable the large amount of heat energy released when a fuel burns to be released as electrical energy instead. The cells are highly efficient for producing large amounts of energy for long periods of time. Prototype fuel cell-operated cars are being produced.
- A **hydrogen-oxygen** or a **hydrogen-air** fuel cell relies on the high-energy-producing reaction:

$$2H_2(g) + O_2(g) \longrightarrow 2H_2O(l)$$

for the electric current. H_2 is a lightweight fuel and the product, water, is non-polluting.
- The anode and cathode are porous graphite.
At the anode, this reaction occurs:

$$H_2(g) + 2OH^-(aq) \longrightarrow 2H_2O(l) + 2e$$

- At the cathode, this reaction occurs:

$$O_2(g) + 2H_2O(l) + 4e \longrightarrow 4OH^-(aq)$$

 ISBN: 9780170355544 PHOTOCOPYING OF THIS PAGE IS RESTRICTED UNDER LAW.

- Unfortunately, hydrogen does not occur naturally. It has to be prepared by the electrolysis of water or other chemical means. Energy has to be expended to provide this new source of energy.
- The supply of gas must also be constant. Having cylinders of gas under pressure makes the fuel cell less portable than other batteries.

Prototype fuel cell vehicle.

KEY POINTS SUMMARY

- **Electrochemical cells** use a spontaneous chemical reaction to deliver an electrical current. They are also known as **voltaic cells** and **batteries**.
- **Electrolytic cells** use an electrical current to reverse a spontaneous redox reaction. This current may be provided by an electrochemical cell.
- **Non-rechargeable batteries** in common use include dry cells, alkaline batteries, mercury batteries and lithium batteries. The type of battery used for equipment is chosen to suit the purpose.
- **Rechargeable batteries** in common use include lead-acid and Ni-cad.
- **Fuel cells** convert the energy released from combustion of a fuel into electricity.

ASSESSMENT ACTIVITIES

1 **Dry cell**

A dry cell has a carbon electrode in a paste of MnO_2 and NH_4Cl enclosed in a zinc case.
   a Describe the anode of a dry cell and write a half-equation for the anode reaction which occurs when the cell is delivering a current.
   b Write a half-equation for the reaction that takes place at the cathode.
   c Combine the equations in **a** and **b** to write the overall cell reaction.
   d Ammonia is formed when a dry cell is in use. Why is the smell of ammonia not obvious?

2 **Alkaline battery**

At the graphite electrode of an alkaline battery the reaction is described by the unbalanced half-equation:

$$MnO_2(s) + H_2O(l) \longrightarrow Mn(OH)_2(s)$$

The other electrode is zinc.
   a What has been gained by MnO_2? Is the reaction oxidation or reduction?
   b Write the overall cell reaction for an alkaline battery.
   c Which electrode is the anode? What sign is given to the anode?
   d What is the advantage of an alkaline battery over a Leclanché dry cell?

3 Mercury battery

The diagram shows a mercury battery.
Conditions in the cell are strongly alkaline.
Zn and HgO are reactants.
ZnO and Hg are formed in the reaction.

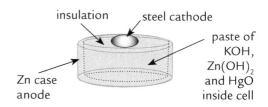

a Write a balanced half-equation for the oxidation.
b Write a balanced equation for the reduction.
c Write the overall cell reaction.
d Which electrode is the anode?
e Which electrode is the negative electrode?

4 Li-Mn battery

Consider a lithium-manganese dioxide battery. Conditions must be alkaline, as acid will give a violent reaction with lithium.

a Write a balanced half-equation for the oxidation.
b What is the oxidising agent (oxidant)?
c Write a balanced equation for the reduction reaction.
d Which electrode is negative?
e Describe two advantages of this battery over the mercury battery.

5 Lead-acid battery

When a lead-acid battery is delivering a current, the reaction is:

$$Pb(s) + PbO_2(s) + 4H^+(aq) + 2SO_4^{2-}(aq) \longrightarrow 2PbSO_4(s) + 2H_2O(l)$$

a Describe the electrode that is the anode.
b Write a balanced half-equation for the reaction at the anode.
c What happens to the concentration of sulfuric acid as current is drawn from the cell?
d Sulfuric acid is more dense than water. What happens to the density as current is drawn from the cell?
e The lead-acid battery can be recharged by applying an electric current to reverse the flow of electrons. In a car, this current can be generated by the kinetic energy of the moving vehicle. While the battery is delivering a current, it is an electrochemical cell. When an external supply of energy is recharging the battery, it is an electrolytic cell.
   Write an equation for the reaction that occurs when the battery is recharging.
f Water has evaporated from a battery of a car that has been in storage. Will this affect the performance of the battery?
g Describe two advantages and two disadvantages of a lead-acid battery.

6 Ni-cad battery

a Like a lead-acid battery, a Ni-cad battery can be recharged. In the oxidation reaction of a Ni-cad battery, cadmium metal forms cadmium hydroxide, $Cd(OH)_2$.
   Write a balanced half-equation for the reaction.
b The reduction reaction involves the formation of nickel (I) hydroxide (NiOH) from nickel (IV) oxide (Ni_2O_4).
   Write a balanced half-equation for the reaction.
c Write a balanced equation for the reaction that occurs when current is being drawn from the cell.
d Write a balanced equation for the reaction which occurs when the cell is being 'recharged'.

ISBN: 9780170355544 PHOTOCOPYING OF THIS PAGE IS RESTRICTED UNDER LAW.

7 Hydrogen-oxygen fuel cells

At present, hydrogen-oxygen fuel cells function best with pure hydrogen.

a For a hydrogen-oxygen fuel cell, write a balanced half-equation for the reaction that would occur at the negative electrode.

b Write an equation for the overall reaction that occurs when the cell is delivering a current.

c What features of a hydrogen-oxygen fuel cell make it suitable for use on spacecraft?

Fuel cells that run on less pure hydrogen and other fuels are being developed. If a carbon-air fuel cell was developed:

d Name the substance that would be oxidised.

e What product would be formed by the cell?

f Write an equation for the reaction which would occur when a current was drawn from the cell.

g Describe two advantages and two disadvantages of this cell compared with a hydrogen-oxygen fuel cell.

8 Designing a battery

a Use these $E°$ values to design a battery that will provide a voltage of 1.5 V under standard conditions.

With a diagram, describe the cathode, the anode and the overall design.

$E°_{(Mg^{2+}/Mg)} = -2.36$ V

$E°_{(Fe^{2+}/Fe)} = -0.41$ V

$E°_{(Cu^{2+}/Cu)} = +0.34$ V

$E°_{(Fe^{3+}/Fe^{2+})} = +0.77$ V

$E°_{(Zn^{2+}/Zn)} = -0.76$ V

$E°_{(Pb^{2+}/Pb)} = -0.13$ V

$E°_{(I_2/I^-)} = +0.62$ V

b Describe a possible use for your battery.

9 Matching terms with descriptions

a	uses a reaction of a fuel with oxygen to produce an electric current	electrochemical cell
b	produces an electric current from the reaction of zinc with manganese dioxide paste	electrolytic cell
c	uses an electric current to force a non-spontaneous redox reaction to occur	dry cell
d	uses a spontaneous redox reaction to produce an electric current	lead-acid cell
e	a special type of cell that can both deliver a current and be recharged by applying an electric current	fuel cell

ISBN: 9780170355544

PHOTOCOPYING OF THIS PAGE IS RESTRICTED UNDER LAW.

3.7

Revision Three

Question I

1 Redox reactions

When ferrous ions (Fe^{2+}) are reacted with potassium permanganate, the permanganate ion (MnO_4^-) is reduced to Mn^{2+} ions.

a Write a balanced half-equation showing the reduction of MnO_4^- to Mn^{2+} ions.

b Write a balanced half-equation for the oxidation of Fe^{2+} ions.

c Combine the two half-reactions in **a** and **b** to write the equation for the reaction.

d Describe what you would observe in the reaction of acidified potassium permanganate with ferrous sulphate solution.

e Write the name or formula of an acid that could be used to acidify the solution for the reaction to take place.

Potassium dichromate solution will oxidise sulfite (SO_3^{2-}) ions in acid conditions. $Cr_2O_7^{2-}$ is changed to Cr^{3+}.

f Write a balanced half-equation showing the change of $Cr_2O_7^{2-}$ to Cr^{3+}.

g Write a balanced half-equation showing the change of SO_3^{2-} to SO_4^{2-} in acid conditions.

h Combine the two half-equations that you have written in **f** and **g** to show the reaction of sulfite ions with dichromate ions in acidic solution.

i Describe what you would observe in the above reaction of dichromate with sulfite.

Copper (II) ions will oxidise iodide ions. Copper (I) iodide is formed.

j Write a balanced equation for the reaction of the Cu^{2+} ions.

k Write a balanced equation for the reaction of the I^- ions.

l Write a balanced equation for the overall reaction.

m Describe what you would observe in the reaction of Cu^{2+} ions with I^- ions.

2 Oxidation numbers

Write the oxidation number of:

a Cr in CrO_4^{2-}

b Cr in $Cr_2(SO_4)_3$

c Cr in $Cr_2O_7^{2-}$

3 Electrochemical cells

Use the following standard electrode potentials to answer the questions below.

$E^{\circ}_{Ag^+/Ag} = +0.80$ V $E^{\circ}_{Cu^{2+}/Cu} = +0.34$ V

An electrochemical cell (battery) that spontaneously delivers a current is made by combining a zinc half-cell with a silver half-cell.

a Write the standard cell diagram for this cell (using IUPAC convention).

b In which half-cell does reduction occur?

c Write a balanced half-equation for the reaction at the negative electrode.

d Write an equation to represent the cell reaction.

e Calculate E°_{cell} for this cell.

f In which direction do electrons flow in the external circuit when the cell is delivering a current?

ISBN: 9780170355544 PHOTOCOPYING OF THIS PAGE IS RESTRICTED UNDER LAW.

g One electrode is found to weigh less after the cell had been operating for 30 minutes. Which electrode was this? Justify your answer.

h In the cell, a salt bridge joined the two half-cells. Name a compound that could be used in the salt bridge and state the purpose of this salt bridge.

4 Chemical reaction

Answer the following questions with reference to ions and atoms given with these standard electrode potentials.

$E^\circ_{S_4O_6^{2-}/S_2O_3^{2-}}$ = +0.08 V (acid conditions) $E^\circ_{Cu^{2+}/Cu}$ = +0.34 V

$E^\circ_{I_2/I^-}$ = +0.54 V $E^\circ_{Br_2/Br^-}$ = +1.09 V

a Which species is the strongest oxidant?

b Which species is the strongest reductant?

c i Determine whether I^- can be oxidised to I_2 with $S_4O_6^{2-}$ ions under standard conditions. Give a reason for your answer.

  ii What would you observe if small pieces of copper were placed in bromine solution under standard conditions?

d A cell is set up with I_2 and I^- in one half-cell and Br^- and Br_2 in the other half.

  i Write the cell reaction.

  ii Calculate E°_{cell}.

  iii The cell is delivering a current. What would you observe after a while?

5 Commercial cells

I Dry cell (Leclanché cell)

In a dry cell, a simplified possible reaction is:

$$Zn(s) + 2NH_4^+(aq) + 2MnO_2(s) \longrightarrow Zn^{2+}(aq) + 2NH_3(aq) + Mn_2O_3(s) + H_2O(l)$$

a Write the half-equation for the reaction at:

  i the positive electrode

  ii the negative electrode.

b What material is used for:

  i the positive electrode?

  ii the negative electrode?

c i Give an advantage of the dry cell battery.

  ii Give a disadvantage of the dry cell battery.

II Mercury battery

The mercury battery delivers a more constant voltage than a dry cell and also has a longer life. They are used in heart pacemakers and hearing aids. In a mercury battery the conditions are alkaline (basic). Zinc is oxidised to zinc oxide (ZnO) and mercury oxide (HgO) is reduced to mercury.

a Write the half-equation for the reaction at:

  i the positive electrode

  ii the negative electrode.

b This battery is contained in a small stainless steel cylinder. Explain why this is necessary.

III Lead-acid batteries

These are used to supply electrical energy for short periods of time in cars. Plates of Pb and PbO_2 are immersed in sulfuric acid.

The equation for the reaction that occurs when the cell is delivering a current is:

$$Pb(s) + PbO_2(s) + 4H^+(aq) + 2SO_4^{2-}(aq) \longrightarrow 2PbSO_4(s) + 2H_2O(l)$$

a The lead-acid battery is able to be recharged. Write the equation for the reaction that occurs when the battery is recharged.

b Name the electrolyte in the lead-acid battery.

c When the battery fails to deliver a current, the density of the solution inside the battery is low. Give a reason for this.

d i Give an advantage of the lead-acid battery.

ii Give a disadvantage of the lead-acid battery.

IV A fuel cell

A fuel cell contains methane (CH_4) in one half-cell and oxygen at the other. The electrodes are made of conducting materials that are also catalysts for the reaction.

a What is contained at the negative electrode: methane or oxygen?

b Write the overall cell reaction (the products are carbon dioxide and water).

c An advantage of a fuel cell is that it efficiently converts chemical energy to electrical energy. What is a disadvantage of a fuel cell?

6 Electrolysis

An electric current is passed through a dilute solution of sodium chloride. Oxygen gas is formed at one electrode and hydrogen gas at the other.

a Write a balanced half-equation for the reaction that occurs at the positive electrode.

b Write a balanced half-equation for the reaction that occurs at the negative electrode.

c Write the overall equation for the reaction.

d Pieces of red and blue litmus paper are placed under the negative electrode. What change is observed as electrolysis proceeds?

Question II

1 An electrochemical cell

An electrochemical cell was constructed with acidified dichromate ($Cr_2O_7^{2-}$) and chromic (Cr^{3+}) ions in one half-cell. Iron (II) (Fe^{2+}) and iron (III) (Fe^{3+}) ions are in the other half-cell.

$E°_{Cr_2O_7^{2-}(aqueous\ acid)/Cr^{3+}}$ = +1.33 V $E°_{Fe^{3+}/Fe^{2+}}$ = +0.77 V

a Write the names or formulae of possible substances for each of the solutions 1, 2 and 3, and electrode 1.

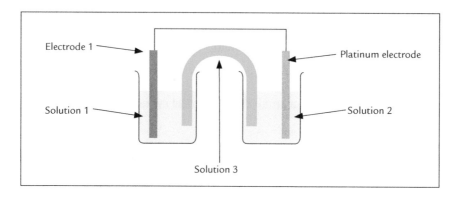

b Copy and complete the standard cell diagram for the cell.

/	//	/

c Calculate the cell potential under standard conditions.
d In which direction do electrons flow in the external circuit?
e Which half-cell contains the cathode? Give a reason for your answer.

2 Copper (II) + iodide

A solution of copper sulfate is reacted with potassium iodide in a test tube. A white precipitate of CuI and a brown solution of I_2 form.

a Write a balanced half-equation for the oxidation reaction.
b Write a balanced half-equation for the reduction reaction.
c Write a balanced equation for the overall reaction.
d Sodium thiosulfate solution is added to the mixture in the test tube. Explain why the brown colour disappears. The thiosulfate ion ($S_2O_3^{2-}$) forms the tetrathionate ion ($S_4O_6^{2-}$).

3 Using $E°$ values

Use the following data to answer the questions below.

Redox couple	$E°$/V at pH 1
Zn^{2+}/Zn	−0.76 V
$MnO_2/Mn^{2+}_{(acid)}$	1.23 V
Al^{3+}/Al	−1.66 V
$NO_3^-/NO_{2(acid)}$	0.80 V
Cu^{2+}/Cu	0.34 V
I_2/I^-	0.54 V

a Which species is the strongest oxidising agent?
b Which species is the strongest reducing agent?
c Zinc and MnO_2 are the reactants in a dry cell. Use the data above to calculate the expected cell potential.
d Which two half-cells would give the largest cell potential when they are combined? Calculate this cell potential.
e State whether aluminium ions (Al^{3+}) can be reduced by iodide (I^-) ions. Give a reason for your answer.

Question III

1 Electrochemical cells

Use the following standard electrode potentials to answer the questions below.

$E°_{Zn^{2+}/Zn} = -0.76V$ $E°_{Fe^{3+}/Fe^{2+}} = 0.77 V$

An electrochemical cell (battery) that spontaneously delivers a current is made by combining a zinc half-cell with a half-cell containing dissolve Fe^{2+} and Fe^{3+} ions.

   a Write this cell using standard notation (IUPAC notation).

   b In which half-cell does reduction occur?

   c Which electrode is the negative electrode?

   d Write an equation to represent the cell reaction.

   e Calculate E°_{cell} for this cell.

   f In which direction do electrons flow in the external circuit?

   g A salt bridge was placed between the two half-cells.

      i Give the formula of a compound that could be used in the salt bridge.

      ii Give two properties that this compound should have.

     iii Describe the purpose of the salt bridge.

2 Using E° values

Answer the following questions with reference to ions and atoms given with these standard electrode potentials.

$$E^\circ_{SO_4^{2-}/SO_3^{2-}} = +0.17 \text{ V} \qquad\qquad E^\circ_{Fe^{3+}/Fe^{2+}} = +0.77 \text{ V}$$

$$E^\circ_{Ni^{2+}/Ni} = -0.23 \text{ V} \qquad\qquad E^\circ_{I_2/I^-} = +0.54 \text{ V}$$

   a Which species is the strongest oxidising agent (oxidant)?

   b Which species is the strongest reducing agent (reductant)?

   c Determine whether sulfite ions can be oxidised by Fe^{3+} ions under standard conditions. Give a reason for your answer.

   d Using the E° values given above, write the cell that would give the greatest possible voltage and calculate this voltage. Write the cell in IUPAC notation.

   e For the cell that you have written in **d**, name a material that could be used for:

      i the left-hand electrode

     ii the right-hand electrode.

   f **i** Given the following standard electrode potentials:

$$E^\circ_{Zn^{2+}/Zn} = -0.76\text{V} \qquad\qquad E^\circ_{NO_3^-/NO_2} = 0.80\text{V}$$

      describe possible reactions of zinc with nitric acid (H^+ and NO_3^-).
      Describe the reactions by giving the reactants and products in each case.
      Explain why the reactions are possible.

     ii Given further that:

$$E^\circ_{Cu^{2+}/Cu} = 0.34\text{V}$$

      compare the reactions of zinc with the reactions of copper with nitric acid.
      Explain your answers.

3 Commercial cells

Dry cells (Leclanché cells) consist of a zinc case and a carbon rod which is surrounded by a paste of MnO_2 and NH_4Cl. When the cell is delivering current, the zinc case reacts and MnO_2 becomes Mn^{2+}.

   a Is the zinc case the cathode or the anode?

   b Write equations for the anode and cathode reactions, and the overall reaction.

ISBN: 9780170355544 PHOTOCOPYING OF THIS PAGE IS RESTRICTED UNDER LAW.

Question IV

1 Oxidation numbers

Write the oxidation number of:

a Mn in MnO_4^-

b Mn in MnO_2

c Mn in MnO_4^{2-}

d Mn in Mn_2O_3

2 Redox equations

A solution of acidified potassium permanganate is reacted with hot oxalic acid solution. The MnO_4^- ion forms Mn^{2+} and $(COOH)_2$ forms CO_2 gas. Heat is required for the reaction. It is too slow at room temperature.

a Write a balanced ion-electron equation for the reduction half-reaction.

b Write a balanced ion-electron equation for the oxidation half-reaction.

c Write a balanced ion-electron equation for the overall reaction.

d Describe what you would observe in the reaction.

3 Electrochemical cells

Use the following standard electrode potentials to answer the questions below.

$$E^\circ_{Cu^{2+}/Cu} = 0.34 \text{ V} \qquad\qquad E^\circ_{Fe^{2+}/Fe} = -0.44 \text{ V}$$

a Explain the meaning of $E^\circ_{Cu^{2+}/Cu} = 0.34$ V.

An electrochemical cell (battery) that spontaneously delivers a current is made by combining a copper half-cell with an iron half-cell.

b Write this cell using IUPAC notation.

c In which half-cell does reduction occur?

d Write a balanced half-equation for the reaction at the anode.

e Which electrode is the negative electrode?

f Write an equation to represent the cell reaction.

g Calculate E°_{cell} for this cell.

h In which direction do electrons flow in the external circuit?

i Describe the changes occurring in each half-cell while the cell is delivering a current.

j In the cell, a salt bridge was placed joining the two half-cells. Name a compound that could be used in the salt bridge and state the purpose of this salt bridge.

4 Chemical reaction

Answer the following questions with reference to ions and atoms given with these standard electrode potentials.

$$E^\circ_{S_4O_6^{2-}/S_2O_3^{2-}} = +0.08 \text{ V} \qquad\qquad E^\circ_{Cu^{2+}/Cu} = +0.34 \text{ V}$$

$$E^\circ_{I_2/I^-} = +0.54 \text{ V} \qquad\qquad E^\circ_{Br_2/Br^-} = +1.09 \text{ V}$$

$$E^\circ_{MnO_4^-/Mn^{2+}} = +1.51 \text{ V (acid conditions)} \qquad\qquad E^\circ_{Cl_2/Cl^-} = +1.36 \text{ V}$$

$$E^\circ_{NO_3^-/NO} = +0.96 \text{ V (acid conditions)} \qquad\qquad E^\circ_{Fe^{3+}/Fe^{2+}} = +0.77 \text{ V}$$

a Which species is the strongest oxidant?

b Which species is the strongest reductant?

c For each of the following, give a calculation using E° values as part of your answer.

  i Determine whether I^- can be oxidised to I_2 with $S_4O_6^{2-}$ ions under standard conditions. Give a reason for your answer.

  ii What would you observe if small pieces of copper were placed in bromine solution under standard conditions?

 d List all the species in the list that can be oxidised by acidified nitrate ions under standard conditions.

5 Electrolysis

An electric current is passed through a solution of sodium chloride. Chlorine gas is formed at one electrode and hydrogen gas at the other.

 a Write a balanced half-equation for the reaction that occurs at the positive electrode (anode).

 b Write a balanced half-equation for the reaction that occurs at the negative electrode (cathode).

 c Explain why this process is used for the commercial production of sodium hydroxide.

6 An electrochemical cell

Use the following electrode potentials to answer the questions.

$E^\circ_{Ag^+/Ag} = +0.80$ V $E^\circ_{Zn^{2+}/Zn} = -0.76$ V

 a Write the standard cell diagram for a cell made by combining a silver half-cell with a zinc half-cell in this form.

 b Below is the schematic diagram for the cell. Give possible species for solutions 1, 2 and 3, and electrode 1.

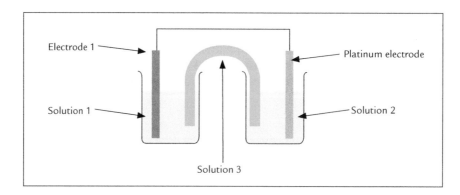

 c Which electrode is the positive electrode?

 d When the cell is delivering a current, is the reaction at the positive electrode oxidation or reduction?

 e Write the cell reaction.

 f Calculate E°_{cell}.

 g Explain your choice for solution 3 in part **b** of this question.

 h In which direction would the electrons flow in the external circuit if a current flows?

 i The salt bridge allows for the movement of anions and cations between the half-cells. In which direction do anions flow in the salt bridge?

ISBN: 9780170355544 PHOTOCOPYING OF THIS PAGE IS RESTRICTED UNDER LAW.

3.6

Demonstrate an understanding of equilibrium principles in aqueous systems

External assessment   **5 credits**

Unit	Content	Aim
12	**Systems in equilibrium**	To understand the concept of **chemical equilibria in an aqueous system**
13	**Solubility of ionic solids**	To apply equilibrium principles in **aqueous equilibria** involving **ionic solids** qualitatively and quantitatively
14	**Acids**	To distinguish between **strong** and **weak acids** by the **way they ionise**, their **pH** and **conductivity**
15	**Bases**	To distinguish between **strong** and **weak bases** by the way they **react with water**, their **pH** and **conductivity**
16	**Titration curves**	To relate the shapes of **titration curves** to the **strengths** of the **acids** and **bases** titrated, where the curves show changes in **pH** and **conductivity**.
17	**pH of buffers and salt solutions**	To understand that a buffer solution must contain a **weak acid** and its **conjugate base**, and to relate the action of a buffer solution to the presence of the weak acid and its conjugate base

In this section, a hydrogen ion in solution is written both as $H^+(aq)$ and $H_3O^+(aq)$. Both represent a proton surrounded with water molecules. In reality, a proton in solution is attracted to several water molecules and not just one. The '(aq)' denotes this — note that this can also be written as a subscript.

Unit 12 | Systems in equilibrium

Learning Outcomes — on completing this unit you should be able to:

- understand the concept that a chemical equilibrium will exist when two factors oppose
- understand that these two factors are the drive to minimum energy and the drive to maximum entropy
- understand that an equilibrium exists when these two factors are such that the rates of forward reaction and backward reactions are the same
- understand that equilibria can exist only in closed systems.

An aquarium with its finite volume of water is a closed system.

Factors that drive a reaction

- Two factors determine whether a reaction is likely to occur. If the reaction does occur, these factors determine the extent to which it occurs. These factors are also discussed in Unit 7. They are as follows.
  1. The drive to achieve a state of minimum energy
     - Objects fall down spontaneously losing gravitational potential energy.
     - A compressed spring spontaneously releases its elastic potential energy.
     - Energy changes in a chemical reaction can be measured as **enthalpy** changes, $\Delta_r H$ (heat changes occurring at a constant pressure). The drive to achieve minimum energy favours exothermic reactions, those for which $\Delta_r H$ is negative.
  2. The drive to achieve a state of maximum entropy (maximum disorder)
     - Moving particles attempt to achieve a state of maximum disorder.
     - In a Lotto draw, the numbered balls are tumbled in an attempt to achieve a state of maximum disorder. The energy expended in achieving disorder cannot be recovered. When the Lotto balls are returned to numerical order, energy is not released.
- The higher the temperature, the greater the kinetic energy of particles and the greater the drive to become more disordered. A measure of a system's disorder is called its **entropy**.
- In a chemical reaction, the drive to maximum disorder favours the side where particles have the greatest freedom of movement. Particles in the gas or dilute aqueous phases have greater freedom of movement than those in the liquid phase. Particles in a liquid have more freedom of movement than particles in a solid. *If both factors favour a reaction, or one factor is dominant, then the reaction goes to completion.* Otherwise, the reaction reaches a state of equilibrium.

Moving Lotto balls increase their state of disorder. The entropy of the balls has increased.

   ISBN: 9780170355544   PHOTOCOPYING OF THIS PAGE IS RESTRICTED UNDER LAW.

Equilibria

- When the rate of forming products (called the forward reaction) equals the rate of decomposition of products (called the backward reaction), the system is said to have reached **equilibrium**. At this stage, the equilibrium is affected only by changes in **concentrations** of any reactant or product or changes in its **temperature**.

- For example, the ionisation of CH_3COOH is endothermic:

$$CH_3COOH\ (aq) \rightleftharpoons CH_3COO^-\ (aq) + H^+\ (aq)$$

The drive to achieve minimum energy favours the exothermic formation of CH_3COOH but the drive to achieve maximum disorder favours the formation of the CH_3COO^- and H^+. Two dissolved ions on the right have greater freedom of movement than one dissolved molecule on the left.

If the two factors oppose and one factor is not much greater than the other, then the reaction achieves a state of equilibrium.

Dissolved molecules of CH_3COOH are constantly forming ions, but the ions $CH_3COO^- + H^+$ ions are forming CH_3COOH molecules at the same rate. A dynamic **equilibrium** exists between the dissolved molecules and ions.

At this stage, concentrations of reactants and products remain essentially constant.

- The expression $\frac{[CH_3COO^-][H^+]}{[CH_3COOH]}$, where $[CH_3COO^-]$ is the concentration of CH_3COO^- in mol L^{-1}, etc. is called the **reaction quotient**, Q. When the system is in equilibrium, the value of the reaction quotient has a constant value. This value is K_c and is called the **equilibrium concentration constant** or, more simply, the **equilibrium constant**. It has a **constant** value provided the **temperature remains constant**.

- For a reaction $aA(aq) + bB(aq) \rightleftharpoons cC(aq) + dD(aq)$

$$Q = \frac{[C]^c[D]^d}{[A]^a[B]^b}$$

- When the system is in equilibrium, $Q = K_c$ and the **rate of forward reaction equals the rate of backward reaction.**

Equilibria can exist only in closed systems

- Substances dissolved in water are confined to the volume of water. They are in a **closed system**.

- In the equilibrium above, $CH_3COOH(aq) \rightleftharpoons CH_3COO^-(aq) + H^+(aq)$, only a small number of CH_3COOH molecules are ionised. The concentrations of each species in the dynamic equilibrium remain essentially the same.

- If nothing is added to the system, the only factor that can change the number of molecules and ions in this equilibrium is a change in temperature.

- *Equilibria can be established only in closed systems.* Reactants and products must both be present so that they can collide with one another, and the rates of the two reactions **Reactants ⟶ Products** and **Products ⟶ Reactants** become equal.

Water in the puddle is open to the air. Water and water vapour cannot reach a state of equilibrium. The system is not closed.

Factors that affect an equilibrium

When changes are made to the **concentration** of a species or the **temperature** of a system in chemical equilibrium, the system will adjust to relieve that change. This is a version of **Le Chatelier's Principle**.

- An increase in the **concentration** of a substance favours the reaction that will use up that substance and reduce its concentration. A decrease in the concentration of a substance favours the reaction that forms it.
- An increase in **temperature** favours the reaction that uses up heat — the endothermic reaction. A decrease in temperature favours the reaction that produces heat — the exothermic reaction.
- Only substances whose concentrations can be changed appear in the **reaction quotient**. These are substances in the **gas** or **aqueous** phases. In these phases, substances can be concentrated or diluted. Substances in liquid and solid phases do not appear in the equilibrium expression.
- Reacting gases are in a closed system if they are in a fixed-volume container. It is important to note that the pressure in a gas phase equilibrium can be increased either by a decrease in volume or by adding an unreactive gas to the system. **Decreasing the volume** increases the concentration so **the equilibrium will adjust** to form fewer gas particles. **Adding an unreactive gas** does not change the concentration of reactants or products. Increasing the pressure in this way **has no effect** on the equilibrium. If there were 2 moles of hydrogen in a 20 L container, adding a mole of helium does not change the fact of that there are 2 moles of hydrogen in 20 L.

Example

Natural gas wells contain methane and carbon dioxide. The mixture can be reacted with steam to produce 'synthesis gas', a mixture of carbon monoxide and hydrogen. A number of products can be produced from synthesis gas, which is also a good fuel. In a closed system, this equilibrium reaction can be established:

$$3CH_4(g) + CO_2(g) + 2H_2O(g) \longrightarrow 4CO(g) + 8H_2(g) \quad \Delta_r H = +658 \text{ kJ mol}^{-1}$$

The equilibrium constant $K_c = \dfrac{[CO]^4[H_2]^8}{[CH_4]^3[CO_2][H_2O]^2}$

The following table summarises the effects of changes made on this equilibrium reaction.

Change made to system	Initial effect of change	Shift in equilibrium*	Final [CO₂]**	Final [H₂O]**	Final [H₂]**	Effect on K_c
Add CO_2	concentration of CO_2 is increased	forward reaction favoured	increased; forward reaction does not use up all the added CO_2	decreased	increased	no change
Remove H_2	concentration of H_2 is decreased	forward reaction favoured	decreased	decreased	decreased; forward reaction does not restore the original amount	no change
Increase pressure by decreasing the volume	all concentrations increased	backward reaction favoured; fewer gas moles occupy less volume	increased by the decrease in volume and reaction	increased by the decrease in volume and reaction	increased by the decrease in volume	no change
Increase the pressure by adding an unreactive gas	no change in concentration of reactants or products, or temperature	no change	no change	no change	no change	no change
Heat the system	increase in temperature	forward (endothermic) reaction favoured	decrease	decrease	increase	increase
Add a catalyst	no change in concentration of reactants or products, or temperature	no change	no change	no change	no change	no change

* If the forward reaction is favoured, the forward reaction rate is temporarily faster than the backward reaction rate until equilibrium is restored.

** Final concentrations are those when equilibrium has been re-established. If a substance is added, the system will attempt to use it up, but the added amount will not be completely used up.

ISBN: 9780170355544 PHOTOCOPYING OF THIS PAGE IS RESTRICTED UNDER LAW.

KEY POINTS SUMMARY

- Some reactions go to completion. Others reach an **equilibrium** when they are contained in a closed system.
- In a **chemical equilibrium**, the rate at which products form equals the rate at which they decompose again. Concentrations of substances remain essentially constant, $Q = K_c$.
- Concentrations in an equilibrium system are used to calculate the **equilibrium constant**, K_c. The magnitude of K_c gives an indication of how complete the reaction is when it reaches equilibrium. Only species which are in the gas or aqueous phases appear in the reaction quotient, Q.
- Changes in concentrations or temperature will cause changes in a chemical equilibrium.
- Only a change in temperature will change K_c's value. Heating will decrease K_c's value for an exothermic reaction.

ASSESSMENT ACTIVITIES

1 Tendency to achieve maximum entropy

For each of the following, state whether the drive to maximum entropy will favour the formation of reactants (on the left) or products (on the right) or neither.

a $N_2(g) + O_2(g) \longrightarrow 2NO(g)$

b $N_2(g) + 3H_2(g) \longrightarrow 2NH_3(g)$

c $4NH_3(g) + 3O_2(g) \longrightarrow 2N_2(g) + 6H_2O(g)$

d $NH_3(g) + HCl(g) \longrightarrow NH_4Cl(s)$

e $C_2H_2(g) + H_2(g) \longrightarrow C_2H_4(g)$

2 Effect of changes to an equilibrium system

a Synthesis gas can be used to produce methanol in the reaction:

$$CO(g) + 2H_2(g) \longrightarrow CH_3OH(g) \qquad \Delta_r H = -90 \text{ kJ mol}^{-1}$$

Copy and complete the following table for this equilibrium reaction.

Change made to system	Initial effect of change	Shift in equilibrium	Final [CO]	Final [H₂]	Final [CH₃OH]	Effect on K_c
Add H₂	concentration of H₂ increased	forward reaction favoured	decreased	increased	increased	no change
Remove CO						
Increase pressure by decreasing the volume						
Increase pressure by adding an unreactive gas						
Heat the system						
Add a catalyst						

b Dimethyl ether (CH_3OCH_3) is a useful chemical that can be formed by the dehydration of methanol in the reaction:

$$2CH_3OH(g) \longrightarrow CH_3OCH_3(g) + H_2O(g) \qquad \Delta_r H = \text{-24 kJ mol}^{-1}$$

Copy and complete the following table for this equilibrium reaction.

Change made to system	Initial effect of change	Shift in equilibrium	Final [CH$_3$OH]	Final [CH$_3$OCH$_3$]	Final [H$_2$O]	Effect on K_c
Add CH_3OH						
Remove CH_3OCH_3						
Increase pressure by decreasing the volume						
Increase pressure by adding an unreactive gas						
Heat the system						
Add a catalyst						

ISBN: 9780170355544
PHOTOCOPYING OF THIS PAGE IS RESTRICTED UNDER LAW.

Unit 13 | Solubility of ionic solids

Learning Outcomes — on completing this unit you should be able to:

- demonstrate understanding of equilibria involving solubility of ionic solids in water
- calculate concentrations of ions in saturated aqueous solutions using K_s
- calculate the solubility of a substance in a solution already containing one of the ions (a common ion)
- predict whether precipitation will occur when two solutions are mixed.

Aqueous solutions

- When solids dissolve in water, the particles in the solid are attracted to water molecules. Attractive forces holding the particles together in the solid are overcome.
- When a sugar crystal is placed in water, the sucrose molecules become separated from one another and are surrounded with water molecules. When an ionic solid such as $MgCl_2$ dissolves in water, it completely breaks up into positive and negative ions. As the number of dissolved ions increases, the **electrical conductivity** of the solution increases. There are more mobile ions to carry charge in the solution. Conductivity of a solution is measured with a conductivity meter.

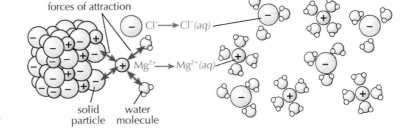

forces of attraction

$Cl^- \longrightarrow Cl^-(aq)$

$Mg^{2+} \longrightarrow Mg^{2+}(aq)$

solid particle   water molecule

- When $MgCl_2$ is dissolved in water, the magnesium and chloride ions separate.
- Ionic solids **ionise** completely when they dissolve.
- Soluble ionic solids can give solutions with high concentrations of Ions. These solutions are **good conductors of electricity**. Soluble ionic compounds are **strong electrolytes**.
- Sucrose does not form ions in solution and the solution will not conduct electricity. Sucrose is a **non electrolyte**. Ethanol is also a non electrolyte.
- The following graphs show the concentrations of species (components) in solution for 1 mol L^{-1} sucrose (left) and 1 mol L^{-1} $MgCl_2$ (right).

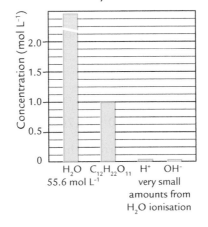

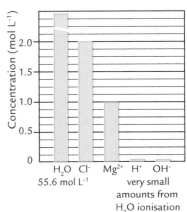

3.6

Solubility of ionic solids

- Most ionic solids will dissolve in water at first. After a while, a stage is reached where no more solid will dissolve. The conductivity of the solution remains constant as there are no additional ions entering the solution to carry charge. At this stage, the number of ions that go into solution in any given time equals the number that precipitate out.
- An **equilibrium** is reached between particles in the solid state and those in solution. The solution is said to be **saturated**.

In a saturated solution, as particles enter the solution the same number leave.

When calcium hydroxide dissolves in water it gives a colourless solution at first. Then the solution becomes cloudy, and a white solid settles at the bottom of the container. An equilibrium exists between solid calcium hydroxide and the dissolved ions.

$$Ca(OH)_2(s) \rightleftharpoons Ca^{2+}(aq) + 2OH^-(aq)$$

- Because the concentrations of ions stay the same, the reaction quotient Q_s (for dissolved ions it is also called the ion product) now has a constant value. This value is the **solubility constant or solubility product**, which has the symbol K_s.

    $Q_s = [Ca^{2+}][OH^-]^2$

    In a saturated solution, $Q_s = K_s = [Ca^{2+}][OH^-]^2$

    K_s has a constant value provided the temperature remains constant.
- Compounds with low solubility have a small value of K_s.
- For example, K_s for $Ca(OH)_2$ is 4×10^{-6}. K_s for $Mg(OH)_2$ is 1×10^{-11}. In both cases, $K_s = $ [metal ion] $[OH^-]^2$. So $Mg(OH)_2$ is much less soluble than $Ca(OH)_2$.
- The electrical conductivity of a saturated solution of $Mg(OH)_2$ is much lower than a saturated solution of $Ca(OH)_2$.

Relationship between K_s (solubility constant) and Q_s (solubility reaction quotient)

$Q_s = K_s$	Saturated solution, system is at equilibrium
$Q_s < K_s$	Unsaturated, precipitate will not form
$Q_s > K_s$	Supersaturated, sudden precipitation may occur when system is disturbed (shock, temperature changes, etc.). It is more usual that such a solution will not form. Solid will precipitate out until $Q_s = K_s$

Example: If a solution contains 4.2×10^{-6} mol L^{-1} silver ions and 4.2×10^{-6} mol L^{-1} chloride ions, does a precipitate form?

$K_s(AgCl) = 1.7 \times 10^{-10}$

Calculate Q_s:

$Q_s = [Ag^+] \times [Cl^-] = 4.2 \times 10^{-6} \times 4.2 \times 10^{-6} = 1.76 \times 10^{-11} < 1.7 \times 10^{-10}$

In this case $Q_s < K_s$ and a precipitate will NOT form — the solution is unsaturated.

ISBN: 9780170355544
PHOTOCOPYING OF THIS PAGE IS RESTRICTED UNDER LAW.

Example: Solving problems using K_s and the reaction quotient Q

Calculating K_s for a compound

A saturated solution of lead chloride, $PbCl_2$, contains 0.0162 mol L^{-1} lead ions.

a Write an equation for the equilibrium reaction of lead chloride in the saturated solution.

$$PbCl_2(s) \rightleftharpoons Pb^{2+}(aq) + 2Cl^-(aq)$$

b Write an expression for the solubility product K_s for lead chloride.

$$K_s = [Pb^{2+}][Cl^-]^2$$

c Calculate the concentration of chloride ions in the solution.

$n(Cl^-) = 2 \times n(Pb^{2+}) = 2 \times 0.0162$ mol $= 0.0324$ mol
This amount is in 1 L.
$c(Cl^-) = 0.0324$ mol L^{-1}

d Calculate the value of K_s.

$K_s = [Pb^{2+}][Cl^-]^2$
$= 0.0162 \times 0.0324^2$
$= 1.70 \times 10^5$

Using the reaction quotient to predict precipitation

A solution is made by mixing equal volumes 0.0100 mol L^{-1} lead nitrate, $Pb(NO_3)_2$, and 0.0100 mol L^{-1} sodium chloride solutions. By calculating the value of Q_s product for lead chloride, deduce whether a precipitate of lead chloride will form.

After dilution (when the solutions are mixed), $[Pb^{2+}] = 0.01/2 = 0.00500$
Similarly, $[Cl^-] = 0.00500$
$Q_s = [Pb^{2+}][Cl^-]^2$
$= 0.00500 \times 0.00500^2$
$= 1.25 \times 10^{-7}$
$< K_s$
No precipitate of lead chloride will form.

Calculating the amount of solute dissolved in a saturated solution

A solution is saturated with silver sulphate, Ag_2SO_4, at 25°C. Calculate the concentration of Ag^+ ions in the solution.

$K_s(Ag_2SO_4) = 1.22 \times 10^{-5}$ at 25°C
$M(Ag) = 108$ g mol^{-1}

a Write an equation for the equilibrium reaction of silver sulfate in the saturated solution.

$$Ag_2SO_4(s) \rightleftharpoons 2Ag^+(aq) + SO_4^{2-}(aq)$$

b Write an expression for the solubility product K_s for silver sulfate.

$$K_s = [Ag^+]^2[SO_4^{2-}]$$

c Calculate the concentration (mol L^{-1}) of silver ions in the saturated solution.

If x moles Ag_2SO_4 dissolve in 1 L,
$[Ag^+] = 2x$ and $[SO_4^{2-}] = x$
Substituting in the expression for K_s:
$K_s = 1.22 \times 10^{-5} = [2x]^2[x] = 4x^3$
$\therefore x = 1.45 \times 10^{-2}$ mol L^{-1}
$[Ag^+] = 2[x] = 2.90 \times 10^{-2}$ mol L^{-1}
$c(Ag^+) = 2.90 \times 10^{-2}$ mol L^{-1}

Common ion effect

- In the solution that contained 4.2×10^{-6} mol L^{-1} silver ions and 4.2×10^{-6} mol L^{-1} chloride ions, a precipitate did not form.

$$Q_s = 1.76 \times 10^{-11} \qquad K_s(AgCl) = 1.7 \times 10^{-10} \qquad Q_s < K_s$$

- What if we added more chloride ions to the solution of Ag^+ and Cl^-? If NaCl was added until $[Cl^-]$ was so large that $Q_s = K_s$, the solution would be saturated. If even more NaCl is added, a precipitate would form.
- AgCl and NaCl have a common ion, Cl^-. AgCl and $AgNO_3$ also have a common ion, Ag^+.
- In a saturated solution of AgCl, the following equilibrium exists:

$$AgCl(s) \rightleftharpoons Ag^+(aq) + Cl^-(aq)$$

- If a common ion, for example Cl^- in the form of an NaCl solution, is added, the equilibrium will attempt to reduce the concentration of the common ion. The equilibrium shifts to the left and AgCl(s) forms. AgCl(s) is precipitated out by the addition of a common ion.
- AgCl is less soluble in a solution that contains Cl^-. Addition of $AgNO_3$ solution would have the same effect but would be expensive.
- **A solid is less soluble in a solution that contains a common ion.**
- A saturated solution of silver chloride, AgCl, in water contains only 0.000013 (1.3×10^{-5}) mol L^{-1} or 0.0014 g Ag^+. Even this small amount of silver can be removed from the solution by adding a common ion.

ISBN: 9780170355544 PHOTOCOPYING OF THIS PAGE IS RESTRICTED UNDER LAW.

Example

Calculate the maximum concentration of silver ions in a 1 mol L^{-1} sodium chloride solution, given that:

$$K_s(AgCl) = 1.7 \times 10^{-10}$$

If x mol L^{-1} AgCl dissolves, $[Ag^+] = x$ and $[Cl^-] = 1 + x \simeq 1$. (Two sources of chloride ions are NaCl and AgCl. From NaCl, there is 1 mol L^{-1}. From AgCl, there is x mol L^{-1}, a very small and insignificant amount* compared with the contribution from NaCl.)

$$1.7 \times 10^{-10} = [Ag^+][Cl^-]$$
$$\longrightarrow 1.7 \times 10^{-10} = [Ag^+] \times 1$$
$$\longrightarrow [Ag^+] = 1.7 \times 10^{-10}$$

The concentration of Ag$^+$ is now only 0.00000000017 mol L^{-1}.
(* The chloride contribution from AgCl is also 1.7×10^{-10} mol L^{-1}.)
So most of the silver ions in a saturated solution of AgCl can be precipitated out by making the solution 1 mol L^{-1} in NaCl. This process is an example of **'salting out'**.
Addition of an ion to a saturated solution that already contains that ion can cause precipitation. This is the **common ion effect**.

KEY POINTS SUMMARY

- For any solution of an ionic solid, a **reaction quotient** Q_s or **ion product** can be calculated.
- For a **saturated solution**, the reaction quotient has a constant value K_s and is called the **solubility constant**.
- In a saturated solution, an **equilibrium** exists between the solid and the dissolved particles.
- K_s is an **equilibrium constant**.
- The addition of a **common ion** to a saturated solution results in precipitation.

ASSESSMENT ACTIVITIES

1 Matching terms with descriptions

a product of concentrations of ions in mol L^{-1} with each concentration raised to the power of the stoichiometric number in the formula		ionise
b a solution in which no more solute will dissolve		reaction quotient
c the value of the reaction quotient for a saturated solution		saturated solution
d further addition of an ion already in solution causes precipitation		solubility constant
e break up to give positive and negative ions		common ion effect

2 Writing reaction quotients (ion products)

In any aqueous solution of dissolved ions, a reaction quotient, Q_s, can be written. For each of the following in aqueous solution, write an expression for Q_s.

a $NaCl$

b $CaSO_4$

c Na_2SO_4

d $Ca(OH)_2$

e $PbCl_2$

f $Fe(OH)_3$

g Li_2CO_3

h $Fe_2(SO_4)_3$

i $PbSO_4$

j Ag_2SO_4

k $CaCO_3$

l $MgBr_2$

m $PbCrO_4$

n PbI_2

o $Cu(OH)_2$

3 Listing species present in saturated solutions

List all species present in the following saturated solutions in order of decreasing concentrations. Some ions may be in equal concentrations.

a $AgCl$

b $CaCO_3$

c $PbCl_2$

d $AgSO_4$

4 Calculating solubility constants K_s

In any saturated solution of ions, $Q_s = K_s$.

Information given in this question applies to solutions at 25°C.

a The concentration of silver chloride in a saturated solution 1.40×10^{5} mol L^{-1}. Calculate K_s for silver chloride.

b The concentration of calcium hydroxide in a saturated solution is 1.00×10^{-2} mol L^{-1}. Calculate K_s for calcium hydroxide.

c A saturated solution of calcium carbonate contains 0.00710 g L^{-1}. Calculate K_s for calcium carbonate. $M(CaCO_3) = 100$ g mol^{-1}

d Calcium fluoride is found in teeth. A saturated solution of calcium fluoride contains 0.0150 g L^{-1}. Calculate K_s for calcium fluoride. $M(Ca) = 40.08$ g mol^{-1}, $M(F) = 19.00$ g mol^{-1}

5 Calculating the amount that will dissolve

Values of solubility constants given in this question are measured at 25°C.

a The solubility constant of barium sulfate is 1.00×10^{-10}. Calculate the concentration of barium sulfate in mol L^{-1} in a saturated solution.

b The solubility constant of magnesium hydroxide is 1.006×10^{-11}. Calculate the concentration of magnesium hydroxide in mol L^{-1} in a saturated solution. What are the concentrations of magnesium ions and hydroxide ions?

c Magnesium carbonate has a solubility constant 1×10^{-5}. Calculate the mass in grams of magnesium carbonate dissolved in 1 L of saturated solution. $M(Mg) = 24.31$ g mol^{-1}, $M(C) = 12.01$ g mol^{-1}, $M(O) = 16.00$ g mol^{-1}

d i Lead chloride is soluble in hot water but at 25°C its solubility constant is 2×10^{-5}. Calculate the mass of lead chloride in a litre of saturated solution at 25°C. $M(Pb) = 207.2$ g mol^{-1}, $M(Cl) = 35.45$ g mol^{-1}

   ii What is the mass of lead ions in 1 L of this solution?

e The solubility constant of calcium fluoride is 3×10^{-11}. Sodium fluoride is readily soluble in water.

   i Calculate the maximum amount of dissolved fluoride in a calcium fluoride solution in mol L^{-1} and g L^{-1}. $M(F) = 19.00$ g mol^{-1}

   ii Why do fluoride tablets from the chemist contain sodium fluoride and not calcium fluoride?

   iii Why is it an advantage to have calcium fluoride in teeth enamel?

6 Predicting whether precipitation will occur

In any solution the maximum value a solubility reaction quotient Q_s can have is the value of the solubility constant K_s. If the concentrations are such that Q_s is larger than K_s, precipitation occurs until concentrations are such that Q_s equals K_s.

ISBN: 9780170355544
PHOTOCOPYING OF THIS PAGE IS RESTRICTED UNDER LAW.

For a to d, cations and anions are from the same compound.

Calculate Q_s and compare it with the solubility constant to determine whether it is possible to make each of the following solutions.

a   0.001 mol L^{-1} MgCO$_3$ K_s(MgCO$_3$) = 1 × 10^{-5}

b   0.01 mol L^{-1} Ca(OH)$_2$ K_s(Ca(OH)$_2$) = 4 × 10^{-6}

c   0.001 mol L^{-1} PbBr$_2$ K_s(PbBr$_2$) = 9 × 10^{-6}

d   0.02 mol L^{-1} Ba(OH)$_2$ K_s(Ba(OH)$_2$) = 5 × 10^{-3}

For e to h, the cations and anions are from different sources.

Calculate the appropriate Q_s for each of the following solutions and determine whether precipitation will occur by comparing your value with the given K_s .

e   To a litre of solution containing 1 × 10^{-10} mol Ag$^+$(aq) ions, 1 mole of solid sodium chloride is added. Determine whether silver chloride will be precipitated. K_s(AgCl) = 2 × 10^{-10}

f   Tap water that contains 3 × 10^4 mol L^{-1} lead ions is chlorinated. The chloride concentration becomes 2 × 10^{-3} mol L^{-1}. Will lead chloride be precipitated? K_s(PbCl$_2$) = 2 × 10^{-5}

g   A solution contains 5.3 × 10^{-8} mol L^{-1} Ag$^+$ ions. Calculate the mass of sodium chloride that must be added to cause precipitation of silver chloride. K_s(AgCl) = 1.7 × 10^{-10}, M(Na) = 23.0 g mol^{-1}, M(Cl) = 35.5 g mol^{-1}

h   Copy and complete the following table. Calculate the value of the reaction quotients Q_s when the following solutions are mixed. Compare Q_s with K_s to predict whether a precipitate will form.

Solution 1	Solution 2	Q_s	K_s	Will a precipitate form? Y/N
50 mL 0.0010 mol L^{-1} Ag$^+$	50 mL 0.0010 mol L^{-1} Cl^{-1}		K_s(AgCl) = 1.8 × 10^{-10}	
20 mL 0.010 mol L^{-1} Pb^{2+}	20 mL 0.00010 mol L^{-1} I$^-$		K_s(PbI$_2$) = 7.1 × 10^{-9}	
10 mL 0.00010 mol L^{-1} Ag$^+$	40 mL 0.0010 mol L^{-1} CrO$_4^{2-}$		K_s(Ag$_2$CrO$_4$) = 1.1 × 10^{-12}	
20 mL 0.010 mol L^{-1} Mg^{2+}	20 mL 0.0010 mol L^{-1} OH$^-$		K_s(Mg(OH)$_2$) = 1.8 × 10^{-11}	

Complete the following: In a saturated solution, Q is _____ K_s.

When no precipitate forms, Q is _____ K_s.

7   **Using the common ion effect**

In a solution of a compound that has a low solubility, addition of a common ion can cause precipitation of that compound.

a   Given that K_s(AgCl) = 1.7 × 10^{-10}, calculate the solubility of silver chloride in mol L^{-1} in

    i   water

    ii   a 1.0 mol L^{-1} sodium chloride solution.

    iii   Explain why the answers to i and ii are different.

b   When a solution of sulfuric acid is added to a saturated solution of silver sulfate, a precipitate forms. K_s(Ag$_2$SO$_4$) = 2.00 × 10^{-5}

    i   Calculate the solubility of silver sulfate in water and in a 1 mol L^{-1} solution of sulfuric acid.

    ii   In your calculation of the solubility of silver sulfate in 1 mol L^{-1} acid, the contribution of anions from silver sulfate need not be taken into account. Why must it be taken into account when determining the solubility of silver sulfate in 1 × 10^{-3} mol L^{-1} solution of the acid?

3.6

c Copy and complete the following table.

Solubility constant	Equilibrium reaction	Expression for K_s	Solubility in water	Solubility in
$K_s(BaSO_4) = 1.1 \times 10^{-10}$				0.10 mol L^{-1} Na$_2$SO$_4$
$K_s(CaF_2) = 5.3 \times 10^{-9}$				0.10 mol L^{-1} NaF
$K_s(FeS) = 6.0 \times 10^{-19}$				0.10 mol L^{-1} FeSO$_4$
$K_s(Fe(OH)_2) = 8.0 \times 10^{-16}$				0.10 mol L^{-1} NaOH
$K_s(Ag_2SO_4) = 1.4 \times 10^{-5}$				0.10 mol L^{-1} Na$_2$SO$_4$

8 Summarising aqueous solutions

Solubility problem 1

Lead iodide (PbI_2) is a bright-yellow solid with a limited solubility in water.

a Write an equation for the equilibrium that exists in a saturated solution of PbI_2.

b Write an expression for the solubility product, K_s.

$K_s(PbI_2)$ at 25°C is 8.0×10^{-9}, $M(PbI_2) = 461$ g mol^{-1}

c Calculate the solubility of lead iodide in water at 25°C in

  i   mol L^{-1}

  ii   g L^{-1} (use a Periodic Table to find molar masses).

d Calculate the solubility of lead iodide in 0.10 mol L^{-1} potassium iodide solution.

e Explain why your answer in **d** differs from your answer in **c i**.

f 1.0×10^{-3} mol KI are added to 1 L 0.010 mol L^{-1} Pb(NO$_3$)$_2$ solution. Explain, with a calculation, whether precipitation will occur.

Solubility problem 2

$Zn(OH)_2$ is a white solid with a limited solubility in water.

a Write an equation for the equilibrium that exists in a saturated solution of $Zn(OH)_2$.

b Write an expression for the solubility product, K_s.

$K_s(Zn(OH)_2)$ at 25°C is 1.20×10^{-17}, $M(Zn(OH)_2) = 99.4$ g mol^{-1}

c Calculate the solubility of zinc hydroxide in water at 25°C in

  i   mol L^{-1}

  ii   g L^{-1}.

d Calculate the solubility of zinc hydroxide in 0.0100 mol L^{-1} sodium hydroxide solution.

e Explain how the solubility of zinc hydroxide changes as

  i   the pH is increased (concentration of OH$^-$ is increased)

  ii   the pH is decreased (concentration of H$^+$ is increased).

Solubility problem 3

Silver carbonate is a white solid with a low solubility in water.

$K_s(Ag_2CO_3)$ at 25°C is 8.00×10^{-12}

a Write an expression for the solubility product, K_s, for Ag$_2$CO$_3$.

b Calculate the solubility in mol L^{-1} of Ag$_2$CO$_3$ in water at 25°C.

c Calculate the solubility in mol L^{-1} of Ag$_2$CO$_3$ in 0.100 mol L^{-1} sodium carbonate solution.

d Explain the differences in your answers to parts **b** and **c**.

e A saturated solution of silver carbonate is in equilibrium with solid silver carbonate. Describe what you would observe when

  i   nitric acid is added

  ii   ammonia is added.

Write equations to support your answers.

ISBN: 9780170355544   PHOTOCOPYING OF THIS PAGE IS RESTRICTED UNDER LAW.

Unit 14 | Acids

Learning Outcomes — on completing this unit you should be able to:

- describe the difference between a strong acid and a weak acid by the way it ionises in solution
- list and write the concentrations of species present in strong and weak acids
- relate the concentrations of hydrogen ions to pH
- relate the concentrations of total ions to conductivity
- perform calculations involving pH and concentrations of solutions of strong and weak acids.

Fruit contains mamy different weak acids.

3.6

Aqueous solutions of acids

Strong acids

- A **strong acid** is completely ionised in solution. Reaction of the acid with water molecules causes the ions to separate. The ions are surrounded by water molecules and remain separated.
- For example, hydrochloric acid is completely ionised (see Graph 1). There are no HCl molecules in solution, only H^+ and Cl^- ions.
- A solution of a strong acid has a high conductivity. The conductivity of the $H^+(aq)$ ion is five times, and that of the $OH^-(aq)$ ion is three times, that of any other singly charged ion.
- Strong acids are **strong electrolytes**.

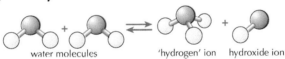

water molecules   'hydrogen' ion   hydroxide ion

$$HCl(g) + H_2O(l) \longrightarrow H_3O^+(aq) + Cl^-(aq) \text{ or } HCl(g) \longrightarrow H^+(aq) + Cl^-(aq)$$

- A hydrogen ion bonded to water can be written H_3O^+ (which is $H_2O + H^+$), $H_5O_2^+$ (which is $2H_2O + H^+$), and so on, or simply $H^+(aq)$. The chloride ion in solution is written $Cl^-(aq)$.
- As a strong acid becomes more concentrated, the ionisation is incomplete. In concentrated HCl (12.1 mol L^{-1}), HCl gas can be detected when the bottle is opened. For example, damp blue litmus paper will turn pink or ammonia gas will react with the HCl gas to form a white powder of NH_4Cl.
- Strong acids include HNO_3, HCl, H_2SO_4.

Graph 1: Species present in 1 mol L^{-1} hydrochloric acid solution

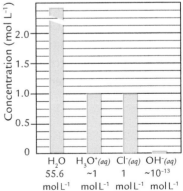

	H_2O	$H_3O^+(aq)$	$Cl^-(aq)$	$OH^-(aq)$
	55.6	~1	1	~10^{-13}
	mol L^{-1}	mol L^{-1}	mol L^{-1}	mol L^{-1}

Weak acids

- A **weak acid** is partly ionised in solution. Only a few particles react with water to give hydrogen ions, $H^+(aq)$, and anions.
- For example, an ethanoic acid solution has a few hydrogen ions, $H^+(aq)$, and ethanoate ions $CH_3COO^-(aq)$, but has mainly ethanoic acid molecules, $CH_3COOH(aq)$.
- When the number of ions in solution is small, the solution has a low conductivity.
- Ethanoic acid is a **weak electrolyte**.

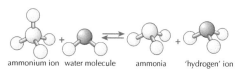

ethanoic acid   water molecule CH_3COO^- hydrogen ion

$$CH_3COOH(aq) + H_2O(l) \rightleftharpoons CH_3COO^-(aq) + H_3O^+(aq)$$

- Acid molecules, $CH_3COOH(aq)$, are in equilibrium with the ions $CH_3COO(aq)$ and $H^+(aq)$.
- An acid need not be molecular. Ions can be acids. For example, ammonium ions, $NH_4^+(aq)$, partially ionise in solution to release hydrogen Ions. Ammonium ions form when a salt such as ammonium chloride NH_4Cl dissolves in water.
- $NH_4Cl(s) \longrightarrow NH_4^+(aq) + Cl^-(aq)$ (The salt is completely ionised.)
  Then $NH_4^+(aq) \rightleftharpoons NH_3(aq) + H_3O^+(aq)$
- The ammonium ion is therefore a weak acid.
- The solution of $NH_4^+(aq)$ and $Cl^-(aq)$ contains a high concentration of ions. The solution has a high conductivity because NH_4Cl is a strong electrolyte.

ammonium ion   water molecule   ammonia   'hydrogen' ion

- Weak acids include CH_3COOH, $HCOOH$, NH_4^+, $CH_3NH_3^+$.

Describing acids

- A molecule or ion that ionises to give one H^+ ion is a **monoprotic** acid, for example HCl, CH_3COOH and NH_4^+. One that gives two hydrogen ions is **diprotic**, for example H_2SO_4. H_2SO_4 gives one hydrogen ion readily and the second a little less readily. An example of a triprotic acid is H_3PO_4, which is a weak acid with respect to all three H^+ ions.
- A **concentrated acid** solution has a large amount of acid dissolved in solution. **Concentrated** hydrochloric acid has 12.1 mol L^{-1} HCl. Concentrated ethanoic (acetic) acid contains 17.4 mol L^{-1} CH_3COOH. The concentration of an acid is **not** related to the acid strength.
- A **dilute acid** solution is taken to have a concentration of 2 mol L^{-1} or less. When the concentration becomes as low as 10^{-6} mol L^{-1}, the contribution of hydrogen ions from the ionisation of water becomes significant.

Ionisation of water

- In any aqueous solution, a small number of water molecules are ionised to give 'hydrogen' ions and hydroxide ions.

$$H_2O(l) \rightleftharpoons H^+(aq) + OH^-(aq)$$

and the equilibrium constant $K_w = 1 \times 10^{-14}$ at 25°C.

- In pure water, hydrogen ions and hydroxide ions are formed from water only. Their concentrations are equal. In a solution of acid, the concentration of hydrogen ions is greater than the concentration of hydroxide ions.

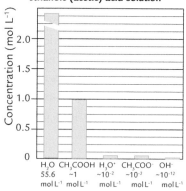

Graph 2: Species present in 1 mol L^{-1} ethanoic (acetic) acid solution

	H_2O	CH_3COOH	H_3O^+	CH_3COO^-	OH^-
	55.6	~1	~10^{-2}	~10^{-2}	~10^{-12}
	mol L^{-1}	mol L^{-1}	mol L^{-1}	mol L^{-1}	mol L^{-1}

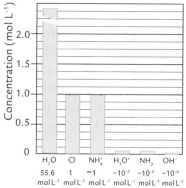

Graph 3: Species present in 1 mol L^{-1} ammonium chloride solution

	H_2O	Cl^-	NH_4^+	H_3O^+	NH_3	OH^-
	55.6	1	~1	~10^{-5}	~10^{-5}	~10^{-9}
	mol L^{-1}	mol L^{-1}	mol L^{-1}	mol L^{-1}	mol L^{-1}	mol L^{-1}

ISBN: 9780170355544 PHOTOCOPYING OF THIS PAGE IS RESTRICTED UNDER LAW.

pH calculations

- By definition:
  $pH = -\log[H_3O^+]$ or $-\log[H^+]$
  Therefore if the $[H^+]$ is known, the pH can be calculated.
- For pure water at 25°C
  $[H^+][OH^-] = 10^{-14}$ and $[H^+] = [OH^-]$
  Therefore: $[H^+] = 10^{-7}$
  and $pH = -\log[H^+] = 7$

pH of a strong acid

- For solutions less than 3 mol L⁻¹ of strong monoprotic acids, the acid can be regarded as completely ionised and the hydrogen ion concentration equals the concentration of acid.

Example: Calculating the pH of a solution of a strong acid

Calculate the pH of 0.150 mol L⁻¹ HCl solution.

$$pH = -\log 0.150 = 0.824$$

(On a calculator, press –, log, 0.150, then = .)
In the ionisation equation below, the concentrations are written under each species.

$$HCl(g) \longrightarrow H^+(aq) + Cl^-(aq)$$

0 mol L⁻¹   0.150 mol L⁻¹   0.150 mol L⁻¹

pH values of different concentrations of HCl are given in the following table; use your calculator to check the pH values.

Concentration of HCl	[H⁺]	pH
2 mol L⁻¹	2	-0.3
1 mol L⁻¹	1	0
0.01 (= 1 x 10⁻²) mol L⁻¹	1 x 10⁻²	2
0.00352 mol L⁻¹	0.00352	2.45
2.3 x 10⁻⁵ mol L⁻¹	2.3 x 10⁻⁵	4.6
1.00 x 10⁻⁸ mol L⁻¹	1.00 × 10⁻⁷ + 1.00 x 10⁻⁸ = 1.10 × 10⁻⁷	6.96*

* In any aqueous solution, there are H⁺ and OH⁻ ions formed by the ionisation of water. In pure water $[H^+] = [OH^-] = 10^{-7}$. In acid solutions, the contribution from water is even less than this (common ion effect) and is negligible compared to $[H^+]$ from the acid. However, when the $[H^+]$ from an acid is less than 10^{-6}, the contribution from water becomes significant. In 1 x 10⁻⁸ mol L⁻¹ HCl, $[H^+]$ from water is approximately 10^{-7} and $[H^+]$ from HCl is 10^{-8} (or 0.1 x 10⁻⁷).
Total $[H^+] = 1.1 \times 10^{-7}$ and pH = 6.96

pH of a weak acid

- Weak acids are only partially ionised.
- In solutions of weak acids, an equilibrium exists between the dissolved acid and the ions that it forms in solution.
- The equilibrium constant is called the **acid ionisation constant**, and is given the symbol K_a. Tables often give pK_a values. pK_a is equal to $-\log K_a$.
- For ethanoic acid (acetic acid):

$$CH_3COOH(aq) + H_2O(l) \rightleftharpoons CH_3COO^-(aq) + H_3O^+(aq)$$

or $CH_3COOH(aq) \rightleftharpoons CH_3COO^-(aq) + H^+(aq)$

$$K_a = \frac{[CH_3COO^-][H_3O^+]}{[CH_3COOH]} \quad \text{or} \quad K_a = \frac{[H^+][CH_3COO^-]}{[CH_3COOH]}$$

At 25°C, $K(CH_3COOH) = 1.74 \times 10^{-5} = \dfrac{1.74}{100,000}$

- This means that the denominator (100,000) is very large compared to the numerator (1.74). Most of the acid is in the form of CH_3COOH molecules that are not ionised. This solution has a low conductivity.

Examples: Calculating the pH of a solution of a weak acid

Example 1: Calculate the pH of 0.150 mol L^{-1} CH$_3$COOH solution.

$$K_a(CH_3COOH) = 1.74 \times 10^{-5}$$

First find $[H_3O^+]$, then use the definition pH = $-\log[H_3O^+]$.
For a 0.150 mol L^{-1} solution containing the equilibrium shown below, the concentrations are written underneath the equation.

$$CH_3COOH(aq) + H_2O(l) \rightleftharpoons CH_3COO^-(aq) + H_3O^+(aq)$$
$$(0.15 - x) = 0.15 \text{ mol L}^{-1} \qquad x \text{ mol L}^{-1} \qquad x \text{ mol L}^{-1}$$

where x mol L^{-1} of acid is ionised. The assumption $(0.150 - x) = 0.150$ is valid if K_a is small (as a general rule 10^{-4} or less). Because K_a is small, x is small compared with 0.150.

Substituting the values in:

$$K_a = \frac{[CH_3COO^-][H_3O^+]}{[CH_3COOH]} \quad \text{gives} \quad 1.74 \times 10^{-5} = \frac{[x] \cdot [x]}{0.150} = \frac{x^2}{0.150}$$

$$\therefore x^2 = 2.61 \times 10^{-6} \longrightarrow x = 1.62 \times 10^{-3} = [H_3O^+]$$

$$pH = -\log[H_3O^+] = -\log 1.62 \times 10^{-3} = 2.79$$

Compare this value with 0.150 mol L^{-1} for a strong acid on page 107. Its pH is 0.824. If the assumption is not made, then the equation:

$$1.74 \times 10^{-5} = \frac{[x] \cdot [x]}{(0.150 - x)} = \frac{x^2}{(0.150 - x)} \text{ must be solved.}$$

ISBN: 9780170355544   PHOTOCOPYING OF THIS PAGE IS RESTRICTED UNDER LAW.

Rearranging the equation gives:

⟶ $1.74 \times 10^{-5} \times (0.150 - x) = x^2$

⟶ $x^2 + 1.74 \times 10^{-5}x - 2.61 \times 10^{-6} = 0$

which can be solved using the quadratic formula:

$$ax^2 + bx + c = 0, \quad x = \frac{-b \pm \sqrt{(b^2 - 4ac)}}{2a}$$

The solution for x results in a pH of 2.794 compared with 2.792 obtained by making the assumption.

Example 2: Calculate the pH of a 0.150 mol L^{-1} ammonium chloride solution. $K_a(NH_4^+) = 5.75 \times 10^{-10}$

A 0.150 mol L^{-1} ammonium chloride solution contains 0.150 mol L^{-1} NH_4^+ ions and 0.150 mol L^{-1} chloride ions.

NH_4^+ is a weak acid and the value of K_a is given for it.

$$NH_4^+(aq) + H_2O(l) \rightleftharpoons NH_3(aq) + H_3O^+(aq)$$
$$(0.15 - x) = 0.15 \text{ mol L}^{-1} \quad x \text{ mol L}^{-1} \quad x \text{ mol L}^{-1}$$

where x mol L^{-1} of the weak acid is ionised. The assumption that $(0.150 - x)$ is approximately equal to 0.150 is valid as K_a is extremely small ($<< 10^{-4}$).

Substituting in:

$$K_a = \frac{[NH_3][H_3O^+]}{[NH_4^+]}$$

gives $5.75 \times 10^{-10} = \dfrac{x^2}{0.150}$ so $x = 9.29 \times 10^{-6} = [H_3O^+]$

pH $= -\log[H_3O^+] = -\log 9.29 \times 10^{-6} = 5.03$

Comparing strong and weak acids

- From the calculations above, the pH of a 0.150 mol L^{-1} solution of the strong acid HCl is 0.824, the pH for a 0.150 mol L^{-1} solution of the weak acid CH_3COOH is 2.79, and the pH of a 0.150 mol L^{-1} solution of the very weak acid NH_4^+ is 5.03.
- The greater the concentration of hydrogen ions in solution, the lower the pH.
- In pH calculations, the acid constant K_a must be given for a weak acid. The smaller the K_a value, the larger the value of pK_a and the weaker the acid. For a strong acid, $[H_3O^+]$ is taken to be the concentration of the acid unless information is given otherwise.
- A strong acid is completely ionised in solution. For a weak acid, the proportion of acid that is ionised increases as the acid becomes more dilute.
- Glacial (concentrated) ethanoic acid is practically all in the molecular form. As water is added, more molecules ionise.
- Ethanoic acid (CH_3COOH) has a pK_a value of 4.76. In solutions with pH less than 4.76, the molecular form (CH_3COOH) of the acid is dominant. At higher pH values, the ionised form (CH_3COO^-) dominates.

3.6

Conjugate acid-base pairs

- Two species related by a hydrogen ion, H^+, are defined as a **conjugate acid-base pair**.
- Examples in this unit are:
  - HCl / Cl^-
  - CH_3COOH / CH_3COO^-
  - NH_4^+ / NH_3
  - H_3O^+ / H_2O
  - H_2O / OH^-
- The conjugate acid written on the left loses one H^+ to give the base on the right.
- Species that can both accept or donate protons are called **amphiprotic**, for example H_2O donates a proton to become OH^- and accepts a proton to become H_3O^+.

Example: Calculating hydrogen ion concentration from the pH

If the pH of a solution is less than 7, it is acidic and the hydrogen ion concentration will be greater than 10^{-7} mol L^{-1}.
Because pH = $-\log[H_3O^+]$,
$[H_3O^+]$ = antilog$(-pH)$ or 10^{-pH}

The pH of a solution is 4.2. Calculate the hydrogen ion concentration.
On a calculator, press 10^x, $-$, 4.2, then =.
Concentration of H_3O^+ = 6.31×10^{-5} mol L^{-1}
Use your calculator to check out the values in the following table:

pH	$[H_3O^+]$
0.024	0.95
9	10^{-9}
13.6	2.51×10^{-14}

KEY POINTS SUMMARY

- A **strong acid** is completely ionised in solution.
- A **weak acid** is partially ionised in solution.
- A **concentrated acid** has a large amount of acid in solution.
- A **dilute acid** has less than 2 mol L^{-1} acid in solution.
- **pH** is equal to $-\log[H^+]$.
- The **acid dissociation constant** for a weak acid is given the symbol K_a.
- pK_a is equal to $-\log K_a$.
- If the value of K_a is small, the value of pK_a is big and only a small amount of the acid is ionised. The acid is a very weak acid.
- **Conjugate acid-base pairs** are related by a proton.

 ISBN: 9780170355544 PHOTOCOPYING OF THIS PAGE IS RESTRICTED UNDER LAW.

ASSESSMENT ACTIVITIES

All values of K_a and pK_a given in this section are those measured at 25°C.

1 Distinguishing between terms

Write statements and give examples to explain the difference between:

a a strong acid and a concentrated acid

b a weak acid and a dilute acid

c a monoprotic and a diprotic acid

d a molecular acid and an ionic acid

e an acid and its conjugate base.

2 Writing ionisation equations

Acids ionise in solution and are proton donors. Write equations to represent the ionisation of the following acids in aqueous solution.

Strong acids:

a HCl

b H_2SO_4 (donates one proton readily)

Weak acids:

c CH_3COOH

d NH_4^+

e HSO_4^-

f H_2CO_3 (a weak acid with respect to both protons)

3 Species in solution

List all the species present in the following aqueous solutions and give their relative concentrations; omit water.

Strong acids:

a 1 mol L^{-1} HCl

b 1 mol L^{-1} H_2SO_4

Weak acids:

c 1 mol L^{-1} CH_3COOH, pH $\approx$ 2

d 1 mol L^{-1} HCN, pH $\approx$ 4

e 1 mol L^{-1} NH_4^+, pH $\approx$ 5

4 pH calculations

In any aqueous solution, pH = $-\log[H^+]$ and K_w = 1 x 10^{-14}.

I Calculate the pH of the following aqueous solutions of strong acids.

  a 0.01 mol L^{-1} HCl

  b 0.035 mol L^{-1} HCl

  c 0.000526 mol L^{-1} H_2SO_4 (assume complete ionisation of both protons at this concentration)

  d 3.54 x 10^{-4} mol L^{-1} HNO_3

  e 1.000 x 10^{-9} mol L^{-1} HCl

  f 3.0 mol L^{-1} HCl

II Calculate the pH of the following aqueous solutions of weak acids stating any assumptions made in questions **g** to **l** in your calculations.

  g 0.01 mol L^{-1} CH_3COOH (K_a(CH_3COOH) = 1.74 x 10^{-5})

  h 0.035 mol L^{-1} CH_3COOH (K_a(CH_3COOH) = 1.74 x 10^{-5})

  i 1.0 mol L^{-1} NH_4Cl (K_a(NH_4^+) = 5.75 x 10^{-10})

  j 0.025 mol L^{-1} NH_4Cl (K_a(NH_4^+) = 5.75 x 10^{-10})

3.6

k 1×10^{-3} mol L^{-1} HF (K_a(HF) = 6.76×10^{-4})

l 1.2 mol L^{-1} HF (K_a(HF) = 6.76×10^{-4})

m 0.01 mol L^{-1} CH_3COOH (K_a (CH_3COOH) = 1.74×10^{-5}), but do not make any assumptions.

5 Comparing strong and weak acids

The strength of an acid is related to how readily it donates hydrogen ions.

a Calculate the pH of

1.0 mol L^{-1} HCl

0.10 mol L^{-1} HCl

0.010 mol L^{-1} HCl.

b Calculate the pH of

1.0 mol L^{-1} CH_3COOH

0.10 mol L^{-1} CH_3COOH

0.010 mol L^{-1} CH_3COOH solutions.

(K_a(CH_3COOH) = 1.74×10^{-5})

State any assumptions that you make in your calculation.

c By comparing your answers in **a** and **b**, describe how you could distinguish between a solution of a strong acid from one of a weak acid.

d By considering your answers in **b**, state how the proportion of weak acid that is ionised in solution varies as the acid becomes more dilute.

6 To neutralise a spill of a strong alkali such as oven cleaner or sugar soap, the spill is washed with plenty of water, neutralised with dilute acetic acid (for example vinegar) and washed with plenty of water again.

a Why is hydrochloric acid (for example bricklayer's acid) not used?

b Why is the spill washed before application of the acid?

7 Finding the hydrogen ion concentration from pH

The pH of a solution is readily measured with a pH meter or, less accurately, with pH indicator papers and solutions. From the measured values, the concentration of hydrogen ions in solution can be calculated.

For solutions with the following pH values, calculate the hydrogen ion concentration.

a 1	**b** 3	**c** 5	**d** 9	**e** 0
f 4.50	**g** 2.80	**h** 0.250	**i** 10.4	**j** −0.0330

8 Comparing weak acids

a The value of pK_a for ethanoic (acetic) acid is 4.76. For methanoic (formic) acid, the pK_a is 3.75. In solution, which acid is more ionised?

b Equimolar amounts of methanoic acid and sodium ethanoate are dissolved in water. Consider the equilibrium:

methanoic acid + ethanoate ion $\rightleftharpoons$ ethanoic acid + methanoate ion

Is the ratio below greater than, equal to, or less than 1? Explain your answer.

$$\frac{[\text{methanoate}][\text{ethanoic acid}]}{[\text{ethanoate}][\text{methanoic acid}]}$$

9 Pyruvic acid ($CH_3COCOOH$) has a pK_a value of 2.49. Ethanoic acid has a pK_a value of 4.76. Consider a solution containing equal amounts of the two acids in which the pH has been adjusted with sodium hydroxide solution to 3.5.

Which ion is present in higher concentration, pyruvate or ethanoate? Give an explanation for your answer.

ISBN: 9780170355544   PHOTOCOPYING OF THIS PAGE IS RESTRICTED UNDER LAW.

10 Conjugate acid-base pairs

Conjugate acid-base pairs are related by one proton H^+.

Copy and complete the following table.

Acid	Conjugate base
HCl	Cl$^-$
H$_2$SO$_4$	
HSO$_4^-$	
	OH$^-$
H$_3$O$^+$	
	NH$_3$
H$_2$CO$_3$	
	CO$_3^{2-}$
CH$_3$COOH	
	HCOO$^-$

11 Amphiprotic species can both accept and donate protons.

Use your completed table in question 9 to write down three amphiprotic species.

12 Understanding diprotic and triprotic acids. This is not required for NCEA.

Phosphoric acid, H_3PO_4, is important in the manufacture of fertilisers. Many New Zealand soils are deficient of phosphorus, and fertilisers such as superphosphate need to be applied. However, when these fertilisers are washed into lakes and streams, established ecological systems can be upset.

For phosphoric acid, $pK_{a,1}$ = 2.13, $pK_{a,2}$ = 7.20 and $pK_{a,3}$ = 12.36.

Write the formula of the predominant phosphoric acid species present in a solution of pH:

a 1

b 5

c 14

d Explain why $pK_{a,1} < pK_{a,2} < pK_{a,3}$.
  (Why does H_3PO_4 lose a proton more readily than $H_2PO_4^-$, and so on?)

3.6

Unit 15 | Bases

Learning Outcomes — on completing this unit you should be able to:

- describe the difference between a strong base and a weak base by the way it reacts in solution
- list and write the concentrations of species present in strong and weak bases
- relate the concentrations of OH⁻ ions to pH
- relate the concentrations of total ions to conductivity
- perform calculations involving pH and concentrations of solutions of strong and weak bases.

Aqueous solutions of bases

- In all aqueous solutions there are hydrogen ions and hydroxide ions. In aqueous solutions of **bases**, there are more hydroxide ions than hydrogen ions.
- Hydroxides of Group 1 elements such as KOH and NaOH are **strong bases**. They are soluble and ionise completely in water. The hydroxide ions formed are attracted to water molecules and are written $OH^-(aq)$.
- Solutions of strong bases are **strong electrolytes**. The solutions have a high conductivity. Because the conductivity of the OH^- ion is less (about 0.6 times) than that of the $H^+(aq)$ ion, an NaOH solution has lower conductivity than an HCl solution of the same concentration.
- For example, sodium hydroxide in aqueous solution:

$$NaOH(s) \longrightarrow Na^+(aq) + OH^-(aq)$$

Sodium hydroxide is extremely soluble and is completely ionised in water.

Graphs 1 and 2: Species in solution for 0.01 mol L⁻¹ solutions of NaOH (left) and Ca(OH)₂ (right)

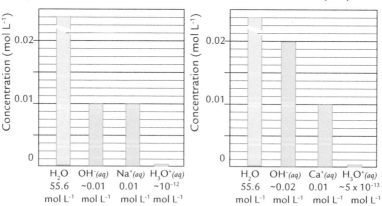

- $Ca(OH)_2$ is not very soluble in H_2O, but all that is dissolved is completely ionised.

$$Ca(OH)_2(s) \longrightarrow Ca^{2+}(aq) + 2OH^-(aq)$$

- Because 1 mol of NaOH releases 1 mol of OH^- ions in solution, it is said to be **monobasic**. $Ca(OH)_2$, which releases 2 mol of OH^- ions, is said to be **dibasic**.

ISBN: 9780170355544
PHOTOCOPYING OF THIS PAGE IS RESTRICTED UNDER LAW.

- Ammonia and a group of compounds related to ammonia called amines are **weak bases**. A small amount will react with water to give $OH^-(aq)$ ions. In the reaction, they attract a hydrogen ion from water.
- For example, ammonia in aqueous solution.

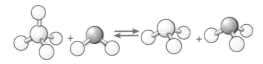

$$NH_3(aq) + H_2O(l) \longrightarrow NH_4^+(aq) + OH^-(aq)$$

- Weak bases are **weak electrolytes**.
- A base need not be molecular. Ions can be bases. For example, ethanoate ions, $CH_3COO^-(aq)$, react incompletely with water to give $OH^-(aq)$ ions. Ethanoate ions are formed when a salt such as sodium ethanoate, CH_3COONa, dissolves in water.
- $CH_3COONa(s) \longrightarrow CH_3COO^-(aq) + Na^+(aq)$. The salt is completely ionised.
- Then $CH_3COO^-(aq) + H_2O(l) \rightleftharpoons CH_3COOH(aq) + OH^-(aq)$. The reaction is not complete. It reaches an equilibrium.
- The ethanoate ion is therefore a weak base.
  As with acids, strong and weak bases can exist as both **concentrated** and **dilute solutions**, provided they are sufficiently soluble in water.

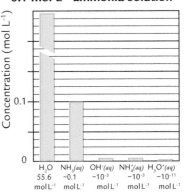

Graph 3: Species in solution for a 0.1 mol L^{-1} ammonia solution

H_2O	$NH_3(aq)$	$OH^-(aq)$	$NH_4^+(aq)$	$H_3O^+(aq)$
55.6	~0.1	~10^{-3}	~10^{-3}	~10^{-11}
mol L^{-1}	mol L^{-1}	mol L^{-1}	mol L^{-1}	mol L^{-1}

pH calculations

In any aqueous solution:
$K_w = [H_3O^+][OH^-] = 10^{-14}$
Therefore:
$-\log[H_3O^+] - \log[OH^-] = 14$
or
$pH + pOH = pK_w = 14$
'p...' stands for '$-\log$ ...'.

Meter used for finding the pH of solutions.

pH of a strong base

- If the hydroxide ion concentration is known, pOH can be calculated. Then pH = 14 – pOH.

Examples: Calculating the pH of a base that is completely ionised

Assume $K_w = 1.00 \times 10^{-14}$ and $pK_w = 14.0$.

Example 1: Calculate the pH of a 0.150 mol L^{-1} NaOH solution.
NaOH is completely ionised.
Therefore $[OH^-]$ = 0.150 mol L^{-1}
$\quad\quad$ pOH = 0.824
$\quad\quad\quad$ pH = 14.0 – 0.824 = 13.2

3.6

Example 2: Calculate the pH of a 0.00150 mol L^{-1} Ca(OH)$_2$ solution. Assume all the Ca(OH)$_2$ is dissolved. (Ca(OH)$_2$ is not sufficiently soluble to form a 0.150 mol L^{-1} solution.)

The dissolved Ca(OH)$_2$ is completely ionised.

Therefore [OH$^-$] = 2 x 0.00150 = 0.00300
pOH = 2.52
pH = 14.0 − 2.52 = 11.5

pH values for different strong bases at various concentrations are given in the following table. Use your calculator to check out the values.

Base	[OH$^-$]	pH
2.00 mol L^{-1} NaOH	2.00	14.3
0.01 (10^{-2}) mol L^{-1} NaOH	0.01	12
0.0100 mol L^{-1} Ca(OH)$_2$	0.0200	12.3
0.00352 mol L^{-1} NaOH	0.00352	11.5
0.00352 mol L^{-1} Ca(OH)$_2$	0.00704	11.8
2.3 x 10^{-5} mol L^{-1} NaOH	2.3 x 10^{-5}	9.4
1.00 x 10^{-8} mol L^{-1} NaOH	1.10 x 10^{-7}	7.04

In the last example, the contribution of OH from water is significant.
Total [OH$^-$] = 1 x 10^7 from water and 0.1 x 10^7 from the NaOH.

pH of a weak base

- In a solution of a weak base, an equilibrium exists between the dissolved base and the ions that it forms in solution.
- The equilibrium constant is the **acid constant** for the conjugate acid of the base, and is given the symbol K_a.

Extended example: Calculating the pH of a weak base

Calculate the pH of a 0.150 mol L^{-1} NH$_3$ solution.

$$K_a(NH_4^+) = 5.75 \times 10^{-10}, K_w = 1.00 \times 10^{-14}$$

NH$_3$ is a weak base forming OH$^-$(aq) ions in solution.
The ionisation reaction (with concentrations underneath) is:

$$NH_3(aq) + H_2O(l) \rightleftharpoons NH_4^+(aq) + OH^-(aq)$$
(0.15 − x) ~ 0.15 mol L^{-1} x mol L^{-1} x mol L^{-1}

where x is the number of moles of ammonia that have reacted with water.

$$x = [NH_4^+] = [OH^-] = \frac{K_w}{[H^+]} = \frac{10^{-14}}{[H^+]}$$

$$K_a(NH_4^+) = 5.75 \times 10^{-10} = \frac{[NH_3][H^+]}{[NH_4^+]} = \frac{0.150 \times [H^+]^2}{10^{-14}}$$

$$[H^+]^2 = \frac{5.75 \times 10^{-10} \times 10^{-14}}{0.150}$$

$$[H^+] = 6.19 \times 10^{-12}$$

$$pH = -\log 6.19 \times 10^{-12} = 11.2$$

ISBN: 9780170355544
PHOTOCOPYING OF THIS PAGE IS RESTRICTED UNDER LAW.

Comparing pHs of strong and weak bases

- The pH of a 0.150 mol L^{-1} solution of the strong base NaOH is 13.2, and the pH of a 0.150 mol L^{-1} solution of the weak base NH_3 is 11.2.
- The greater the concentration of hydroxide ions, the higher the pH.

KEY POINTS SUMMARY

- A **strong base** reacts completely with water to give OH^- ions in solution.
- Only a small proportion of a **weak base** reacts with water to give OH^- ions in solution.
- pOH is equal to $-\log[OH^-]$.
- In an aqueous solution, pH = 14 − pOH.

$$[H^+] \text{ or } [H_3O^+] = \frac{K_w}{[OH^-]}$$

- The conjugate acid of a weak base has a K_a value less than 10^{-7}.

ASSESSMENT ACTIVITIES

Values of K_a, K_w and pK_a given in this section are those measured at 25°C.

1 Matching terms with descriptions

a only a small portion of the substance reacts with water releasing OH^- ions	basic solution
b solution that contains more $OH^-(aq)$ ions than $H^+(aq)$ ions	strong base
c $-\log[OH^-]$	weak base
d substance that reacts completely in water releasing OH^- ions	pOH

2 Species in solution

List all the species present in the following aqueous solutions. Write their approximate concentrations; omit water.

Strong bases:

a 0.10 mol L^{-1} NaOH

b 1×10^{-4} mol L^{-1} $Ca(OH)_2$

Weak bases:

c 1 mol L^{-1} NH_3 pH ≈ 11

d 0.25 mol L^{-1} $C_2H_5NH_2$ (ethanamine*) pH ≈ 12

    * Ethanamine can be called aminoethane or ethylamine. These amines are weak bases.

ISBN: 9780170355544 PHOTOCOPYING OF THIS PAGE IS RESTRICTED UNDER LAW.

3 pH calculations

In any aqueous solution,

$pOH = -\log[OH^-]$ and $pH = 14 - pOH$

a For the following strong bases, calculate the pH of each solution.

    i 0.10 mol L^{-1} NaOH

    ii 1×10^{-4} mol L^{-1} $Ca(OH)_2$

b For the following weak bases, calculate the pH of the each solution stating any assumptions made in **i** to **iv** in your calculations.

    i 1 mol L^{-1} NH_3 ($K_a(NH_4^+) = 5.75 \times 10^{-10}$)

    ii 0.032 mol L^{-1} NH_3 ($K_a(NH_4^+) = 5.75 \times 10^{-10}$)

    iii 1 mol L^{-1} $C_2H_5NH_2$ ($K_a(C_2H_5NH_3^+) = 2.14 \times 10^{-11}$)

    iv 2.5×10^{-4} mol L^{-1} $C_2H_5NH_2$ ($K_a(C_2H_5NH_3^+) = 2.14 \times 10^{-11}$)

4 Comparing base strength

List 0.0010 mol L^{-1} solutions of the following in increasing order of pH:

$NaOH$, $Ca(OH)_2$, $C_2H_5NH_2$, NH_3

5 Calculating hydroxyl ion concentration from pH

Calculate the OH^- ion concentration in solutions with the following pH values:

a 10	**b** 13	**c** 5	**d** 9	**e** 14
f 14.5	**g** 8.8	**h** 0.25	**i** 15	**j** −0.033

6 A sparingly soluble base

As calcium hydroxide is added to water, the pH rises until it reaches 12.30. Addition of further calcium hydroxide causes no further change in pH.

a Calculate the OH^- ion concentration when the pH is 12.30.

b Suggest a reason why the pH does not rise above 12.30.

c Calculate the solubility product of calcium hydroxide.

7 Using everyday bases

Bases play important roles in everyday life.

a Low molecular weight amines are present in fish.
Explain why vinegar or slices of lemon are served with fish.

b Carbonates and bicarbonates are weak bases in indigestion mixtures. These weak bases neutralise stomach acid (hydrochloric acid).
Write equations to show the carbonate ion and the bicarbonate ion acting as bases (reacting with H^+ ions) in aqueous solution.

ISBN: 9780170355544   PHOTOCOPYING OF THIS PAGE IS RESTRICTED UNDER LAW.

Unit 16 | Titration curves

Learning Outcomes — on completing this unit you should be able to:

- recognise the shapes of titration curves for acids and bases of different strengths
- use a titration curve to find a suitable acid-base indicator for the reaction
- use a titration curve to find the value of pK_a for a weak acid
- draw titration curves of pH against volume of reagent added.

3.6

Titration curves

- A **titration curve** is a graph showing changes in the pH of a solution as an acid or base is added gradually to it.

Titration of weak acid with strong base

- A titration curve for a **weak acid** reacting with a **strong base** can be obtained by adding NaOH solution gradually to an acetic (ethanoic) acid solution and graphing the pH against the volume of NaOH added.
- For example, 0.10 mol L^{-1} NaOH is added 1 mL at a time to 20 mL 0.10 mol L^{-1} CH$_3$COOH. The pH is measured after each addition. The titration curve is obtained by graphing the pH values against the total volume of sodium hydroxide added.
- The reaction is

$$CH_3COOH(aq) + OH^-(aq) \rightleftharpoons CH_3COO^-(aq) + H_2O(l)$$

- On Graph 1 below, at point **A**, no NaOH has been added. The pH is the pH of 0.10 mol L^{-1} CH$_3$COOH, which is almost 3. Compare that with 0.10 mol L^{-1} HCl which is 1.
- From point **A**, the pH rises steeply at first as the strong base NaOH is added to a solution of the weak acid.
- In the region marked **B**, some of the CH$_3$COOH has reacted with OH$^-$ to form its conjugate base CH$_3$COO$^-$, so there are now fairly high concentrations of both the weak acid and its conjugate base. The pH changes slowly with additions of NaOH. *The solution now acts as a* **buffer solution**. If a little strong acid is added to this solution, the conjugate base CH$_3$COO$^-$ reacts with it. If a little strong base is added to it, the weak acid CH$_3$COOH reacts with it. A buffer solution's pH changes very little with the addition of small amounts of strong acid or base.

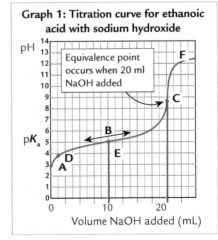

Graph 1: Titration curve for ethanoic acid with sodium hydroxide

Also, by rearranging:

$$K_a = \frac{[H^+][CH_3COO^-]}{[CH_3COOH]} \quad \text{to} \quad [H^+]_a = K_a \div \frac{[CH_3COO^-]}{[CH_3COOH]}$$

and taking –log of both sides, we get: $pH = pK_a + \log\frac{[CH_3COO^-]}{[CH_3COOH]}$

When half of the acid has reacted with the base:

$$[CH_3COOH] = [CH_3COO^-]$$
(unreacted acid)   (salt formed from the acid)

At this point, pH = pK_a (note log 1 = 0).

This is at **E**; half of the NaOH required to react with the CH_3COOH has been added and the pH = pK_a for acetic acid. Half of the acid CH_3COOH remains, half has formed its conjugate base CH_3COO^-.

- At point **C**, the amounts of CH_3COOH and NaOH are equal. This is the **equivalence point**. The solution is now a 0.05 mol L^{-1} solution of the salt CH_3COONa. Note that it has a pH between 8 and 9. At this point, the slope of the curve is at its greatest value.

- If an indicator is used for the titration, it must have a pK_a value around 8–9. Phenolphthalein is such an indicator. An acid-base indicator is a weak acid that has a different colour to its conjugate base:

$$HIn(aq) + H_2O(l) \rightleftharpoons In^-(aq) + H_3O(l)$$
colour 1   colour 2

- For phenolphthalein, colour 1 is colourless and colour 2 is bright pink. The way that an indicator works is described at the end of this unit.

- The point at which an indicator changes colour in a titration is called the **end point** of the titration. *If the correct indicator is chosen, the end point is the same as the equivalence point.*

- If an indicator with pK_a between 3 and 4 such as methyl orange is chosen, then the end point would be at **D**, which is very different from the equivalence point. Methyl orange is red in its acid (HIn) form and yellow in its ionised (In$^-$) form.

- The pH for solutions containing similar amounts of weak acid and its conjugate base can be found by using this equation that was derived earlier in this unit.

$$pH = pK_a + \log \frac{[\text{conjugate base}]}{[\text{weak acid}]}$$

- At **F**, pHs are those of excess NaOH at concentrations < 0.05 mol L^{-1}.

Titration of strong acid with strong base

- When a strong acid is titrated against a strong base, the titration curve has the shape shown in Graph 2.
- 0.10 mol L^{-1} NaOH is added to 20.0 mL 0.10 mol L^{-1} HCl.
- The reaction is: $H_3O^+(aq) + OH^-(aq) \longrightarrow 2H_2O(l)$
- At point **A** on the graph, the pH is the pH of 0.1 mol L^{-1} HCl solution, which is 1.
- At the region marked **B**, the solution is NOT a buffer solution. While there is unreacted acid to neutralise any base that is added, there is not a base that will neutralise any added acid. The pH values are those of a strong acid being gradually diluted.
- At point **C**, the amounts of acid and base are equal. This is the equivalence point of the titration. Note the choice of indicator for this titration is not important. Its pK_a may be any value between 3 and 10.

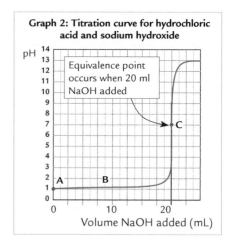

Graph 2: Titration curve for hydrochloric acid and sodium hydroxide

Equivalence point occurs when 20 ml NaOH added

ISBN: 9780170355544   PHOTOCOPYING OF THIS PAGE IS RESTRICTED UNDER LAW.

- If the conductivity of the solution is graphed against the volume of NaOH added, the graph would look like this.

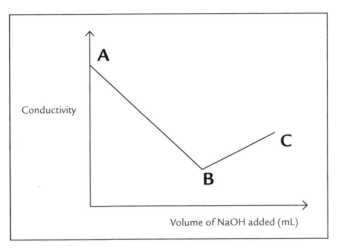

Graph shows the change in conductivity when 0.10 mol L^{-1} NaOH is added to 20 mL 0.10 mol L^{-1} HCl

- **H$^+$(aq)** and **OH$^-$(aq)** have higher conductivity than any other singly charged ions. **H$^+$(aq)** has five times the conductivity, **OH$^-$(aq)** has three times that of other ±1 ions.
- At **A**, the concentration of **H$^+$(aq)** is at its highest, so the conductivity of the solution is at its highest value.
- From **A** to **B**, the concentration of **H$^+$(aq)** is **decreasing** so the **conductivity decreases**, although the concentration of Na$^+$(aq) is increasing.
- At **B**, [Na$^+$] = [Cl$^-$] = 0.050. This is the **equivalence point** and **conductivity is at its lowest**. [H$^+$] = [OH$^-$] = 1 × 10^{-7}. Na$^+$(aq) and Cl$^-$(aq) have low conductivity compared with H$^+$(aq) and OH$^-$(aq).
- From **B** to **C**, the **conductivity increases** as the **[OH$^-$] increases**. As OH$^-$(aq) is not as conductive as H$^+$(aq), the rise is gradual.

Titration of weak base with strong acid

- The titration curve obtained when a **strong acid** such as HCl is added 1 mL at a time to a **weak base** such as NH$_3$ is shown in Graph 3.
- 0.10 mol L^{-1} HCl is added to 20 mL 0.10 mol L^{-1} NH$_3$.
- The reaction is:

$$NH_3(aq) + H_3O^+(aq) \rightleftharpoons NH_4^+(aq) + H_2O(l)$$

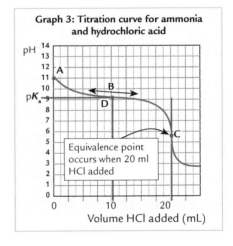

Graph 3: Titration curve for ammonia and hydrochloric acid

ammonia   'hydrogen' ion   ammonium ion   water molecule

- At point **A**, the pH is the pH of a 0.10 mol L^{-1} NH$_3$ solution, approximately 11. (A 0.10 mol L^{-1} solution of a strong base such as NaOH would be 13.)
- After point **A**, the pH falls steeply at first as the strong acid HCl is added to the solution of weak base.
- In the region marked **B**, NH$_3$ and its conjugate acid NH$_4^+$ are both present in appreciable amounts. *The solution now behaves as a **buffer solution***.

- At **C**, the amounts of NH_3 and HCl are equal and the solution contains the salt $NH_4^+(aq)$, $Cl^-(aq)$. The pH is the pH of a 0.05 mol L^{-1} NH_4Cl solution. *This is the equivalence point of the titration.* Note that a suitable indicator would have pK_a between 5 and 7.
- At **D**, half of the NH_3 has reacted:
  $[NH_3] = [NH_4^+]$
- The pH at this point is equal to pK_a of NH_4^+.

Example: Interpreting a titration curve

Titration curve for ethanoic acid with sodium hydroxide

≈ 0.10 mol L^{-1} NaOH solution is added 1 mL at a time to 20 mL 0.10 mol L^{-1} ethanoic acid.

The titration curve gives the following information.

1. At **A**, we see that the pH of 0.1 mol L^{-1} ethanoic acid is 2.9.
2. At **B**, we see that pK_a for ethanoic acid is 4.7.
3. In the region **C**, the weak acid and its conjugate base are present in sufficient quantities that the solution is a buffer solution.
4. **D** is the equivalence point, the steepest part of the curve. Here the amounts of acid and base are equal.
5. At **D**, we see that 20 mL of sodium hydroxide has reacted with 20 mL ethanoic acid. If we know the exact concentration of ethanoic acid, we can calculate the concentration of the NaOH solution using: $c_1V_1 = c_2V_2$
6. At **D** we see that the pH of sodium ethanoate (approximately 0.05 mol L^{-1}) is about 8.7. An indicator that is suitable for this titration would have pK_a around 8.7.

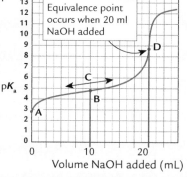

Titration curve for ethanoic acid with sodium hydroxide

Equivalence point occurs when 20 ml NaOH added

Volume NaOH added (mL)

Action of an indicator

For the indicator:

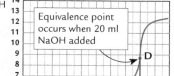

$$HIn(aq) + H_2O(l) \rightleftharpoons In^-(aq) + H_3O(l)$$

colour 1   colour 2

rearrangement of the expression for K_a gives

$$pH = pK_a + \log \frac{[In^-]}{[Hn]}$$

When: $[In^-] = [HIn]$

$$pH = pK_a \qquad as \; \frac{[In^-]}{[Hn]} = 1 \text{ and } \log 1 = 0.$$

- If **[In⁻]** is **less** than **[HIn]**, **colour 1** is seen and pH is slightly less than pK_a.
- If **[In⁻]** is **greater** than **[HIn]**, **colour 2** is seen and the pH is slightly > pK_a.
- When **[In⁻] = [HIn]**, the indicator is about to change colour and pH = pK_a.

ISBN: 9780170355544
PHOTOCOPYING OF THIS PAGE IS RESTRICTED UNDER LAW.

KEY POINTS SUMMARY

- The graph drawn to track the pH of an acid solution as base is gradually reacted with it (or of a basic solution as acid is gradually added to it) is called a **titration curve**.
- The shape of a titration curve depends on the nature of the acid and base, whether they are weak or strong.
- From a titration curve, we can obtain information such as the value of pK_a for a weak acid and pH of a salt solution From the pH of the salt solution we can decide on a suitable indicator for the titration.

ASSESSMENT ACTIVITIES

Questions 1 to 6 relate to acid-base titration curves.

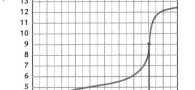

Titration curve for a weak acid with sodium hydroxide

1 **The titration curve shown represents the titration of 10 mL of a weak acid with 0.10 mol L^{-1} sodium hydroxide solution.**

Write down:

a the pH of the original acid solution

b an approximate pH value of a buffer solution made up of the weak acid and its conjugate base

c the value of pK_a for the acid and calculate K_a

d the pH of the solution of the sodium salt of the acid formed in the titration

e the concentration of the acid.

Study the data on indicators given in question **6**.

f Choose a suitable indicator for the above titration.

2 **To 20 mL of 0.10 mol L^{-1} ethanoic acid (acetic acid), 0.10 mol L^{-1} sodium hydroxide is gradually added.**

a Sketch a graph of pH against amount of sodium hydroxide added (pH of 0.10 mol L^{-1} ethanoic acid is 2.88).

b Indicate the region in which a buffer solution exists, and the approximate pH of the solution at the equivalence point.

3 **To 10 mL of 0.10 mol L^{-1} hydrochloric acid, 0.05 mol L^{-1} sodium hydroxide solution is added gradually.**

a Sketch a graph of pH against amount of sodium hydroxide added.

b Show clearly the pH of 0.10 mol L^{-1} hydrochloric acid and the pH of the solution at the equivalence point.

c Explain why the conductivity of the solution decreases as NaOH is added and is at its lowest at the equivalence point.

4 **To 20 mL of 0.10 mol L^{-1} ammonia solution, 0.10 mol L^{-1} nitric acid as gradually added.**

a Sketch a graph of pH against amount of nitric acid added (pH of 0.10 mol L^{-1} ammonia solution is 11.1).

b Indicate the region in which a buffer solution exists, and the approximate pH of the solution at the equivalence point.

5 **In three separate flasks, there are (a) 20 mL 0.10 mol HCl, (b) 20 mL 0.10 mol L^{-1} CH$_3$COOH, and (c) 20 mL water. To each flask, 0.10 mol L^{-1} NaOH solution is gradually added. For each solution, sketch graphs to show the change in pH with the volume of NaOH added.**

6 **From the list of indicators below, choose the one most suitable for:**

    a the titration of a weak acid (examples are ethanoic acid, propanoic acid) with sodium hydroxide

    b the titration of a weak base (for example ammonia) with hydrochloric acid.

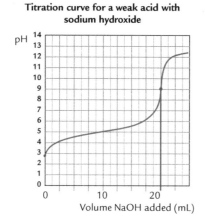

Titration curve for a weak acid with sodium hydroxide

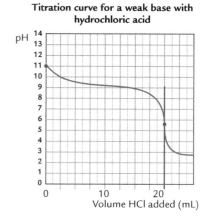

Titration curve for a weak base with hydrochloric acid

Indicator	pK$_a$
methyl orange	3.7
methyl red	5.0
bromothymol blue	7.1
phenol red	7.8
phenolphthalein	9.6

7 **Matching terms with descriptions**

a a point at which the amounts of acid and base are the same	acid-base titration curve
b a graph of pH against the amount of acid (or base) added to a solution of base (or acid)	acid-base indicator
c a point at which an acid-base indicator changes colour	equivalence point of an acid-base titration
d a compound which changes colour over a small pH range	end point of an acid-base titration

ISBN: 9780170355544 PHOTOCOPYING OF THIS PAGE IS RESTRICTED UNDER LAW.

Unit 17 | pH of buffers and salt solutions

Learning Outcomes — on completing this unit you should be able to:

- describe the function and composition of a buffer solution
- calculate the pH, and changes in pH, of buffer solutions
- relate the low pH of a salt solution to the hydrolysis of a cation that is a weak acid
- relate the high pH of a salt solution to the hydrolysis of an anion that is a weak base.

Buffer solutions are important in hydroponics.

Buffer solutions

- A **buffer solution** is made up of a weak acid and its conjugate base. It shows very little change in pH when small amounts of strong acid or base are added.
- An ammonium buffer is made up of NH_4^+ (weak acid) and NH_3 (conjugate base of NH_4^+).
- An ethanoate (acetate) buffer is made up of CH_3COOH (weak acid) and CH_3COO^- (conjugate base of CH_3COOH).
- A bicarbonate buffer is made up of H_2CO_3 (acid) and HCO_3^- (base), or HCO_3^- (acid) and CO_3^{2-} (base).

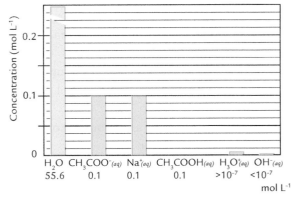

Graph 1: Species present in a 0.1 mol L^{-1} acetate buffer

	H_2O	$CH_3COO^-_{(aq)}$	$Na^+_{(aq)}$	$CH_3COOH_{(aq)}$	$H_3O^+_{(aq)}$	$OH^-_{(aq)}$
	55.6	0.1	0.1	0.1	>10^{-7}	<10^{-7}

mol L^{-1}

- A buffer solution shows only small changes in pH because:
  1. when a small amount of base (OH$^-$) is added, the weak acid in the buffer reacts with it,
  2. when a small amount of acid (H$_3$O$^+$) is added, the conjugate base in the buffer reacts with it.
- pH is affected only by the amount of hydrogen ions and hydroxyl ions in solution.
- A buffer solution can be prepared in the following two ways.
  1. Adding conjugate base to the weak acid. For example, dissolving 0.10 mol CH_3COONa in 1 L of 0.10 mol L^{-1} CH_3COOH. The solution now has 0.10 mol L^{-1} CH_3COO^- (conjugate base) and 0.10 mol L^{-1} CH_3COOH (weak acid). pH = pK_a.
  2. Neutralising half of a weak acid with a strong base. For example, add 500 mL 0.10 mol L^{-1} NaOH to 1.00 L 0.10 mol L^{-1} CH_3COOH. The solution now has 0.033 mol L^{-1} CH_3COO^- (conjugate base) and 0.033 mol L^{-1} CH_3COOH (weak acid). pH = pK_a.
     Half of the acid has been neutralised; there is 0.05 mol left in 1.50 L of solution.
     c = 0.050 mol /1.50 L = 0.033 mol L^{-1}
- The buffer prepared by the first way is a more effective buffer. There are more moles of weak acid to neutralise any added base, and more moles of conjugate base to neutralise added acid.

Buffers in everyday life

- Blood is a buffer solution containing the bicarbonate ion, HCO_3^-, and dissolved carbon dioxide, CO_2. Other natural buffers occur in living organisms; living cells are often sensitive to pH changes. Prepared buffer solutions are used in medicine, for example intravenous fluids, and in industries such as the food industry.

Example: Calculating pH changes of buffer solutions

The following example is used to:
- explain how a buffer solution works
- show how to calculate its pH
- show how its pH changes with the addition of small amounts of acid or base.

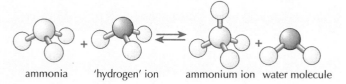

ammonia   'hydrogen' ion   ammonium ion   water molecule

Equal volumes of 0.200 mol L^{-1} ammonium chloride and 0.200 mol L^{-1} ammonia are combined to form a buffer solution. $M(NH_4Cl) = 53.5$ g mol^{-1}

1. Write a chemical equation for the reaction that occurs when a small amount of hydrochloric acid is added.
2. Write a chemical equation for the reaction that occurs when a small amount of sodium hydroxide is added.
3. Explain why the pH shows little change with the addition of small amounts of strong acid or strong base.
4. Calculate the pH of the buffer solution, given $K_a(NH_4^+) = 5.75 \times 10^{-10}$.
5. Calculate the pH if 2 mL of 1 mol L^{-1} of the strong acid HCl is added to 1 L of the buffer solution.
6. Calculate the pH of a new buffer solution made by dissolving 5 g ammonium chloride in 1 L of the original buffer.

Answering the problem:	Useful notes:
NH_4^+ is the **weak acid**. NH_3 is the **conjugate base**.	To describe the action of any buffer solution, first **identify the weak acid and its conjugate base**
1 $H_3O^+(aq) + NH_3(aq) \longrightarrow NH_4^+(aq) + H_2O(l)$	The added acid will react with the conjugate base.
2 $OH^-(aq) + NH_4^+(aq) \longrightarrow NH_3(aq) + H_2O(l)$	The added base will react with the weak acid.
3 The species that affect pH are H_3O^+ and OH^-. These are removed by the base and acid in the buffer.	pH is a measure of H_3O^+ and OH in aqueous solution. If these are removed as they are added, the pH is not changed.

ISBN: 9780170355544 PHOTOCOPYING OF THIS PAGE IS RESTRICTED UNDER LAW.

4 $[NH_3] = [NH_4^+] = 0.100$ mol L^{-1} because mixing equal volumes of solution results in a dilution factor of 2. $\dfrac{[NH_3]}{[NH_4^+]} = 1$ pH = $pK_a = -\log K_a = 9.24$	A useful equation to remember for calculating pH of buffer solutions is: pH = $pK_a + \log \dfrac{[\text{conjugate base}]}{[\text{acid}]}$
5 $n(HCl) = c \times V = 1$ mol $L^{-1} \times 0.002$ L = 0.002 mol. This amount was added to 1 L of buffer and reacts with NH_3 to form NH_4^+. New $[NH_3]$ = 0.10 − 0.002 = 0.098 New $[NH_4^+]$ = 0.10 + 0.002 = 0.102 pH = $pK_a + \log \dfrac{0.098}{0.102}$ = 9.24 − 0.017 = 9.22	If some strong acid is added, the pH will be expected to drop, but only by a very small amount.
6 $n(NH_4^+)$ added = $n(NH_4Cl)$ added $= \dfrac{5\ g}{53.5\ g\ mol^{-1}}$ = 0.093 mol This is added to 1 L of solution. $[NH_3]$ is still ~0.100 mol L^{-1} New $[NH_4^+]$ ~ 0.100 + 0.093 = 0.193 pH = $pK_a + \log \dfrac{0.100}{0.193}$ = 9.24 − 0.29 = 8.95	Adding NH_4Cl is adding the weak acid NH_4^+ to the buffer.

Salt solutions

- A solution of a **salt** that can be prepared by reacting a **strong acid** with **a strong base** will have **pH = 7**. Example: NaCl.
- A solution of a salt that can be prepared by reacting a weak acid with a **strong base** will have **pH > 7**. Example: CH_3COONa.
- A solution of a salt that can be prepared by reacting a **strong acid** with a weak base will have **pH < 7**. Example: NH_4Cl.
- The pH of a salt solution prepared by reacting a weak acid with a weak base depends on the relative strengths of the acid and base.
- All salts are completely ionised in solution. Soluble salts are strong electrolytes. Their solutions have high conductivity.
- Sodium sulfate, the salt of the strong acid H_2SO_4 and the strong base NaOH, forms sodium ions and sulfate ions.

$$Na_2SO_4(s) \longrightarrow 2Na^+(aq) + SO_4^{2-}(aq)$$

This solution has a higher conductivity than a solution of NaCl of the same concentration because it has more ions.

The pH of the solution is 7.

$$[H_3O^+] = [OH^-] = 1 \times 10^{-7}$$

3.6

- Ammonium chloride is the **salt of the strong acid HCl and the weak base NH_3.**
  In solution, $NH_4Cl(s) \longrightarrow NH_4^+(aq) + Cl^-(aq)$
  NH_4^+ ions then react with water; they are said to hydrolyse.
  $NH_4^+(aq) + H_2O(l) \rightleftharpoons NH_3(aq) + H_3O^+(aq)$ (H_3O^+ formed so pH < 7)
  The pH of 0.1 mol L^{-1} NH_4Cl is the pH of a 0.1 mol L^{-1} solution of the weak acid NH_4^+.
  The calculation of the pH of the salt solution is the calculation of pH for 0.1 mol L^{-1} NH_4^+ (a weak acid).

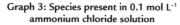

Graph 2: Species present in a 0.1 mol L^{-1} Na_2SO_4 solution

H_2O	$Na^+(aq)$	$SO_4^{2-}(aq)$	$H_3O^+(aq)$	$OH^-(aq)$
55.6	0.2	0.1	10^{-7}	10^{-7} mol L^{-1}

- Sodium ethanoate is the **salt of the weak acid CH_3COOH and the strong base NaOH.** In solution the salt is completely ionised to $CH_3COO^-(aq)$ and $Na^+(aq)$.
  CH_3COO^- ions then react with water (hydrolyse):

  $$CH_3COO^-(aq) + H_2O(l) \rightleftharpoons CH_3COOH(aq) + OH^-(aq) \text{ (OH}^- \text{ formed so pH is > 7)}$$

  The pH of 0.1 mol L^{-1} CH_3COONa is the pH of a 0.1 mol L^{-1} solution of the weak base CH_3COO^-.

  The calculation of the pH of the salt solution is the calculation of pH for 0.1 mol L^{-1} CH_3COO^- (a weak base).

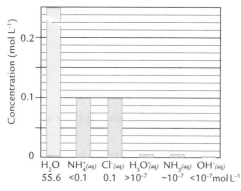

Graph 3: Species present in 0.1 mol L^{-1} ammonium chloride solution

H_2O	$NH_4^+(aq)$	$Cl^-(aq)$	$H_3O^+(aq)$	$NH_3(aq)$	$OH^-(aq)$
55.6	<0.1	0.1	>10^{-7}	~10^{-7}	<10^{-7} mol L^{-1}

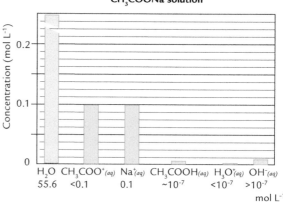

Graph 4: Species present in a 0.1 mol L^{-1} CH_3COONa solution

H_2O	$CH_3COO^-(aq)$	$Na^+(aq)$	$CH_3COOH(aq)$	$H_3O^+(aq)$	$OH^-(aq)$
55.6	<0.1	0.1	~10^{-7}	<10^{-7}	>10^{-7} mol L^{-1}

Extended examples: Calculating the pH of a salt solution

Example 1: Calculate the pH of a 0.100 mol L^{-1} ammonium chloride solution.
$K_a(NH_4^+) = 5.75 \times 10^{-10}$
The ammonium ions hydrolyse, the chloride ions do not. The reaction, with concentrations shown underneath, is:

$$NH_4^+(aq) + H_2O(l) \rightleftharpoons NH_3(aq) + H_3O^+(aq)$$
$$(0.100 - x) \sim 0.100 \text{ mol } L^{-1} \quad x \text{ mol } L^{-1} \quad x \text{ mol } L^{-1}$$

$$K_a(NH_4^+) = 5.75 \times 10^{-10} = \frac{[NH_3][H_3O^+]}{[NH_4^+]} = \frac{x^2}{0.100}$$

$\longrightarrow x^2 = 5.75 \times 10^{-11}$

$\longrightarrow x = 7.58 \times 10^{-6} = [H_3O^+] \longrightarrow$ pH = 5.12

This is the calculation of the pH of 0.100 mol L^{-1} weak acid $NH_4^+(aq)$.

ISBN: 9780170355544 PHOTOCOPYING OF THIS PAGE IS RESTRICTED UNDER LAW.

Example 2: Calculate pH of a 0.05 mol L^{-1} CH$_3$COONa solution. K_a(CH$_3$COOH) = 1.74 x 10^{-5}, K_w = 10^{-14}

$$CH_3COO^-(aq) + H_2O(l) \rightleftharpoons CH_3COOH(aq) + OH^-(aq)$$

$(0.05 - x) \simeq 0.05$ mol L^{-1} $\qquad$ x mol L^{-1} $\qquad$ x mol L^{-1} = $\dfrac{K}{[H^+]}$

$$K_a(CH_3COOH) = 1.74 \times 10^{-5} = \frac{0.05 \times [H^+]^2}{10^{-14}}$$

$\longrightarrow$ [H$^+$] = 1.865 x 10^{-9} $\longrightarrow$ pH = 8.73

This is the calculation of the pH of 0.05 mol L^{-1} weak base CH$_3$COO$^-$(aq).

Calculating the pH at a given point in a titration

For a strong acid titrated with a strong base:
1 Determine whether there is unreacted acid or unreacted base present.
2 Calculate the amount in moles of the unreacted acid or base.
3 Determine the total volume of solution that contains this unreacted acid or base.
4 Steps **2** and **3** will now give you the concentration of unreacted acid or base.
5 Calculate its pH using methods described in Units 14 and 15.

Example: 20.0 mL 0.100 mol L^{-1} HCl are titrated with 0.100 mol L^{-1} NaOH solution. Calculate the pH of the solution after 15.0 mL NaOH have been added.
1 Some HCl has not reacted with NaOH.
2 The amount is n = 0.100 x 5.00 x 10^{-3} = 5.00 x 10^{-4} mol.
3 The volume is V = (20.0 + 15.0) x 10^{-3} L.
4 The concentration is $c = \dfrac{5.00 \times 10^{-4}}{35.0 \times 10^{-3}}$ = 0.0143 mol L^{-1}.
5 pH of a 0.0143 mol L^{-1} HCl solution is 1.84.

For a weak acid titrated with a strong base:
1 Determine whether there is unreacted acid or unreacted base present.
2 Calculate the ratio of reacted acid (this will have formed the conjugate base) to the amount that has not reacted.
3 Use the relation pH = pK_a + log $\dfrac{[\text{conjugate base}]}{[\text{weak acid}]}$.

Example: 20.0 mL 0.100 mol L^{-1} CH$_3$COOH solution are titrated with 0.100 mol L^{-1} NaOH solution. Calculate the pH of the solution after 15.0 mL NaOH have been added.
K_a(CH3COOH) = 1.74 x 10^{-5}
1 Some CH$_3$COOH has not reacted with NaOH.
2 **The ratio of conjugate base : unreacted acid = 15:5.**
3 pH = –log (1.74 x 10^{-5}) + log $\dfrac{15}{5}$ = 5.24

3.6

KEY POINTS SUMMARY

- A **buffer solution** is made up of a weak acid and its **conjugate base**. The pH of a buffer solution shows little change when small amounts of a strong acid or base are added.
- A **solution of a salt** prepared by reacting a strong base with a strong acid will have a pH = 7. That formed from a weak acid and a strong base will have a pH > 7, and that formed from a strong acid and a weak base will have a pH < 7.

Intravenous solutions are often buffer solutions.

ASSESSMENT ACTIVITIES

1 Matching terms with descriptions

The following descriptions can apply to either a buffer solution or a salt solution. Copy the table below and write each of the descriptions in the appropriate column.

Descriptions:

a formed by reacting equal amounts of acid and base

b formed by mixing a weak acid and its conjugate base

c pH of solution changes little with addition of a small amount of strong base

d pH of solution increases rapidly with addition of a small amount of strong base

e can be used to change the pH of a solution

f can be used to keep the pH of a solution constant

Buffer solution	Salt solution

2 Understanding buffer solutions

A buffer solution is made up of a weak acid solution and its conjugate base. Write the formulae of those present in:

a an ethanoate (acetate) buffer

b a bicarbonate buffer

c an ammonium buffer.

3 Explaining how a buffer solution works

The pH of blood is kept approximately constant by a bicarbonate buffer. This is necessary to safeguard living cells.

a The weak acid present (H_2CO_3) is formed from dissolved carbon dioxide. Write the equation for the reaction that occurs when a small amount of a strong base (OH^-) is added.

b Write the formula of the conjugate base of H_2CO_3 and the equation for the reaction that occurs when a small amount of a strong acid (H^+) is added.

c Use the equations you have written in **a** and **b** to explain why there is very little change in pH with small additions of acid and base.

4 An ammonium buffer

A buffer solution contains equal amounts of ammonium chloride and ammonia.

a Write an equation for the reaction that occurs when a small amount of strong base is added.

b Write an equation for the reaction that occurs when a small amount of strong acid is added.

 ISBN: 9780170355544   PHOTOCOPYING OF THIS PAGE IS RESTRICTED UNDER LAW.

5 An ethanoate (acetate) buffer

a For an ethanoate buffer, write the formula of the acid and its conjugate base.

b Which species will react with added acid? Write an equation for the reaction.

c Which species will react with added base? Write an equation for the reaction.

d Explain why pH of an ethanoate buffer remains constant when small amounts of strong acid or base are added.

6 pH of an ethanoate buffer

$$pH = pK_a + \log \frac{[\text{conjugate base}]}{[\text{acid}]}$$

A buffer solution is prepared by dissolving 0.06 mole sodium ethanoate (acetate), CH_3COONa, in 1 litre of 0.05 mol L^{-1} ethanoic (acetic) acid, CH_3COOH, solution.

a Write the formula of the weak acid present in the solution.

b What is the concentration of this acid?

c Write the formula of the conjugate base of the acid.

d What is the concentration of this conjugate base?

e Use the equation:

$$pH = pK_a + \log \frac{[\text{conjugate base}]}{[\text{acid}]}$$

and the value pK_a(ethanoic acid) = 4.76

to show that the pH of this solution is 4.84.

f Why is this solution more basic than one containing equimolar quantities of sodium ethanoate and ethanoic acid?

g A buffer of the same pH Is made by adding 0.05 mol L^{-1} NaOH to 1 L 0.05 mol L^{-1} ethanoic acid. Is this a more effective buffer than the one made by adding CH_3COONa to 1 L 0.05 mol L^{-1} CH_3COOH? Explain your answer.

7 pH of an ammonium buffer

A buffer solution is prepared by dissolving 0.06 mole ammonium chloride in 1 litre of 0.05 mol L^{-1} ammonia.

a Write the formula of the weak acid present in the solution.

b What is the concentration of this acid?

c Write the formula of the conjugate base of the acid.

d What is the concentration of this conjugate base?

e Use the equation:

$$pH = pK_a + \log \frac{[\text{conjugate base}]}{[\text{acid}]}$$

and the value pK_a(ammonium ion) = 9.24

to show that the pH of this solution is 9.16.

f Why is this solution more acidic than one containing equimolar quantities of ammonium chloride and ammonia?

g A buffer of the same pH Is made by adding 0.05 mol L^{-1} HCl to 1 L 0.05 mol L^{-1} ammonia. Is this a more effective buffer than the one made by adding NH_4Cl to 1 L 0.05 mol L^{-1} ammonia? Explain your answer.

3.6

8 pH of various buffer solutions

Given that: pK_a(ethanoic acid) = 4.76 pK_a(propanoic acid) = 4.87

pK_a(ammonium ion) = 9.24 pK_a(methylammonium ion) = 9.24

calculate the pH of the following solutions.

a A solution prepared by dissolving 0.05 mole of sodium ethanoate (sodium acetate) in 1 litre of 0.05 mol L^{-1} ethanoic (acetic) acid.

b A solution prepared by dissolving 8.2 g sodium ethanoate (acetate) in 1 litre 0.05 mol L^{-1} ethanoic (acetic) acid). Is this solution more basic than the solution in part **a**? M(sodium ethanoate) = 82 g mol^{-1}

c A solution obtained by mixing 10 mL 0.10 mol L^{-1} sodium hydroxide and 20 mL 0.10 mol L^{-1} ethanoic (acetic) acid.

d A solution containing 0.1 mol L^{-1} ammonium sulfate and 0.2 mol L^{-1} ammonia.

e A solution prepared by dissolving 5.35 g ammonium chloride in 1 litre of 0.12 mol L^{-1} ammonia solution. M(ammonium chloride) = 53.5 g mol^{-1}

f A solution obtained by adding 15 mL 0.20 mol L^{-1} hydrochloric acid to 20 mL 0.25 mol L^{-1} ammonia solution.

g A solution prepared by dissolving 0.05 mole of sodium propanoate in 1 litre of 0.04 mol L^{-1} propanoic acid.

h A solution prepared by dissolving 4.8 g sodium propanoate in 1 litre 0.05 mol L^{-1} propanoic acid. Is this solution more basic than the solution in part **g**? M(sodium propanoate) = 96 g mol^{-1}

i A solution obtained by mixing 10 mL 0.12 mol L^{-1} sodium hydroxide and 20 mL 0.10 mol L^{-1} propanoic acid.

j A solution prepared by dissolving 0.06 mole of methylammonium chloride in 1 litre of 0.05 mol L^{-1} methylamine solution.

k A solution obtained by mixing 8 mL 0.10 mol L^{-1} hydrochloric acid and 20 mL 0.10 mol L^{-1} methylamine.

9 Preparing buffer solutions

Given that pK_a(ethanoic acid) = 4.76, describe how to prepare a buffer solution with pH 4.80 from:

a 0.1 mol L^{-1} ethanoic acid solution and solid sodium ethanoate

b 0.1 mol L^{-1} ethanoic acid and 0.1 mol L^{-1} sodium hydroxide.

10 Identifying species present in salt solutions

For each of the following 0.10 mol L^{-1} aqueous solutions, list the species present and give their approximate concentrations.

a calcium chloride

b sodium chloride

c ammonium chloride (pH about 5)

d sodium ethanoate (pH about 9)

11 Hydrolysis reactions

Write equations to show why the solution in question **10 c** is acidic, while the solution in **d** is basic.

12 pH of partly neutralised solutions

a 20 mL 0.100 mol L^{-1} HCl are titrated with 0.08 mol L^{-1} NaOH. Calculate the pH of the solution when 20 mL NaOH have been added.

b 15 mL 0.100 mol L^{-1} CH_3COOH are titrated with 0.100 mol L^{-1} NaOH. Calculate the pH of the solution after 10 mL NaOH have been added.

$K_a(CH_3COOH) = 1.74 \times 10^{-5}$

ISBN: 9780170355544 PHOTOCOPYING OF THIS PAGE IS RESTRICTED UNDER LAW.

Revision Four

1 Species in solution

List all the species present (omit water, H_2O) in the following solutions and their approximate concentrations. $K_w = 1.00 \times 10^{-14}$

a 0.200 mol L^{-1} CH_3COOH (pH approximately 2.73)

b 0.300 mol L^{-1} NH_3 (pH approximately 11.4)

c 1.00×10^{-4} mol L^{-1} $Ca(OH)_2$

d 0.100 mol L^{-1} NH_4Cl (pH approximately 5.12)

2 Comparing pH

Write the following lists of solutions in order of increasing pH. Justify your choice for each list.

List 1: 0.01 mol L^{-1} solutions of HCOONa, NH_3, Na_2SO_4

List 2: 0.01 mol L^{-1} solutions of NH_4Cl, CH_3COOH, NaOH

3 Comparing conductivity

a i List the following 0.10 mol L^{-1} solutions in order of decreasing conductivity. Note that the conductivity of H^+ is five times, and the conductivity of OH^- is three times, that of any other ion with a +1 or –1 charge.

$C_6H_{12}O_6$ (glucose)

CH_3COONa (sodium ethanoate)

C_2H_5COOH (propanoic acid)

CH_3NH_3Cl (methylammonium chloride)

$CaCl_2$ calcium chloride

ii Relate the conductivity to the species present in each solution.

b i List the following 0.10 mol L^{-1} solutions in order of decreasing conductivity.

HCl

CH_3COOH

NH_3

NaOH

ii Relate the conductivity to the species in each solution.

4 Solubility

The solubility constant, K_s, for silver chromate, Ag_2CrO_4, at 25°C is 1.12×10^{-12}.

a Write an equation for the equilibrium that exists in a saturated solution of Ag_2CrO_4.

b Write an expression for $K_s(Ag_2CrO_4)$.

c Calculate the solubility of Ag_2CrO_4 in water.

d Calculate the solubility of Ag_2CrO_4 in a solution of 0.100 mol L^{-1} K_2CrO_4.

e Explain the difference in the two values obtained in **c** and **d**.

f Ag_2CrO_4 is a red solid. A saturated solution of it contains such a small amount that it is colourless. Describe what you would observe when a solution of K_2CrO_4 is added to a saturated solution of Ag_2CrO_4.

g A saturated solution of Ag_2CrO_4 is in equilibrium with solid Ag_2CrO_4 in a container. Ammonia solution is added.

What change, if any, would occur to each of the following? Explain your answer.

i The amount of solid Ag_2CrO_4 in the container.

ii The value of $K_s(Ag_2CrO_4)$.

ISBN: 9780170355544 PHOTOCOPYING OF THIS PAGE IS RESTRICTED UNDER LAW.

3.6

5 pH calculations

Calculate the pH of

a   2.05 mol L^{-1} HNO_3

b   0.0512 mol L^{-1} KOH (K_w = 10^{-14})

c   0.115 mol L^{-1} CH_3COOH ($K_a(CH_3COOH)$ = 1.74 x 10^{-5})

d   0.300 mol L^{-1} NH_3 ($K_a(NH_4^+)$ = 5.75 x 10^{-10}, K_w = 10^{-14})

e   0.500 mol L^{-1} $(NH_4)_2SO_4$ ($K_a(NH_4^+)$ = 5.75 x 10^{-10})

6 Buffer solutions

An ethanoate buffer is made by dissolving 4.00 g solid sodium ethanoate in 100.0 mL 0.500 mol L^{-1} CH_3COOH.

$pK_a(CH_3COOH)$ = 4.76 $M(CH_3COONa)$ = 82.0 g mol^{-1}

a   What is the purpose of a buffer solution?

b   Write equations for the reaction that takes place in the buffer when

    i   a small amount of acid is added

    ii   a small amount of base is added.

c   Calculate the pH of the ethanoate buffer described above.

d   Another ethanoate buffer is prepared by adding 0.0500 mol L^{-1} NaOH to 100.0 mL 0.0500 mol L^{-1} CH_3COOH until the pH is the same as the one prepared above. Which is the more effective buffer? Explain your answer.

7 Titration curves

To 20.0 mL 0.100 mol L^{-1} HCOOH, 0.100 mol L^{-1} NaOH is added.

$pK_a(HCOOH)$ = 3.76, K_w = 1 × 10^{-14}

a   Write the formula of the salt formed at the equivalence point of the titration and give its concentration in mol L^{-1}.

b   Calculate the pH of the solution at the equivalence point.

c   Calculate the pH of the solution after 10.0 mL NaOH has been added.

d   Calculate the pH of the solution after 12 mL NaOH has been added.

e   If the pH of 0.100 mol L^{-1} HCOOH is approximately 2.40, copy the axes at right and draw a graph of pH against volume of NaOH added if a total of 25 ml NaOH are added.

f   Circle the buffer region on your graph.

g   Given pK_a of the following indicators, which one is/ones are suitable for detecting the end point of this titration?

Methyl orange pK_a = 3.7

Phenolphthalein pK_a = 9.6

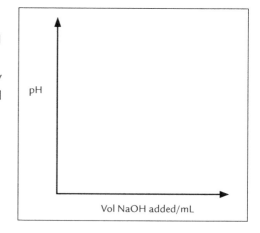

Question II

1 Species in solution

List all the species present (omit water) in the following solutions and their approximate concentrations. K_w = 1 x 10^{-14}

a   0.25 mol L^{-1} HCl

b   2.5 mol L^{-1} NaOH

c   0.10 mol L^{-1} CH_3COONa (pH approximately 8.9)

d   0.10 mol L^{-1} $(CH_3NH_3)_2SO_4$ (pH approximately 5.7)

e   0.0050 mol L^{-1} K_2SO_4

ISBN: 9780170355544   PHOTOCOPYING OF THIS PAGE IS RESTRICTED UNDER LAW.

2 Conductivity

a For each list, write the solutions in decreasing order of conductivity.

List 1: 0.10 mol L⁻¹ solutions of C_2H_5OH (ethanol), NH_4Cl, CH_3NH_2

List 2: 0.10 mol L⁻¹ solutions of KNO_3, HCl, CH_3COONa, $HCOOH$

b For each list, relate the conductivity to the species present.

3 Solubility

The solubility constant, K_s, for lead hydroxide $Pb(OH)_2$ at 25°C is 1.43×10^{-20}.

a Write an equation for the equilibrium that exists in a saturated solution of $Pb(OH)_2$.

b Write an expression for $K_s(Pb(OH)_2)$.

c Calculate the solubility of $Pb(OH)_2$ in water.

d Calculate the pH of a saturated solution $Pb(OH)_2$.

e Calculate the solubility of $Pb(OH)_2$ in a solution of pH 9.

f Explain the difference in the values obtained in **c** and **e**.

g As the pH increases above 9, the solubility of $Pb(OH)_2$ is seen to increase. Explain this. Write an equation for any reaction that occurs.

4 pH calculations

Calculate the pH of

a 1.0 mol L⁻¹ HCl

b 0.010 mol L⁻¹ NaOH ($K_w = 10^{-14}$)

For questions **c** to **f**, state any assumptions that you make in your calculation.

c 0.200 mol L⁻¹ CH_3COOH ($K_a(CH_3COOH) = 1.74 \times 10^{-5}$).

d 0.100 mol L⁻¹ CH_3COONa ($K_a(CH_3COOH) = 1.74 \times 10^{-5}$, $K_w = 10^{-14}$)

e 0.0100 mol L⁻¹ NH_3 ($K_a(NH_4^+) = 5.75 \times 10^{-10}$, $K_w = 10^{-14}$)

f 1.5 mol L⁻¹ NH_4Cl ($K_a(NH_4^+) = 5.75 \times 10^{-10}$)

5 Buffer solutions

a What is the purpose of a buffer solution?

b Write equations for the reactions that takes place in a $CH_3NH_3^+/CH_3NH_2$ buffer when

    i a small amount of acid (H^+) is added

    ii a small amount of base (OH^-) is added.

c A buffer solution is made by dissolving 0.25 mol solid CH_3NH_3Cl in 1 L 0.25 mol L⁻¹ CH_3NH_2. Calculate the pH of this buffer. $pK_a(CH_3NH_3^+) = 10.6$

d Calculate the pH of a buffer made by adding 10.0 mL 0.10 mol L⁻¹ HCl to 20 mL 0.10 mol L⁻¹ CH_3NH_2. $pK_a(CH_3NH_3^+) = 10.6$

e Calculate the pH of the buffer made by dissolving 42.0 g CH_3NH_3Cl in 1 L 0.500 mol L⁻¹ CH_3NH_2. ($M(CH_3NH_3Cl) = 67.5$ g mol⁻¹, $pK_a(CH_3NH_3^+) = 10.6$)

6 Titration curves

To 30.0 mL 0.10 mol L⁻¹ NH_3, 0.10 mol L⁻¹ HCl is added. $pK_a(NH_4^+) = 9.25$

a Calculate the pH of the solution after 15 mL HCl has been added.

b Calculate the pH of the solution after 20 mL HCl has been added.

c If the pH of 0.10 mol L⁻¹ NH_3 is approximately 11, copy the axes at right and draw a graph of pH against volume of HCl added until a total of 40 ml HCl are added. pH at the equivalence point is approximately 5.

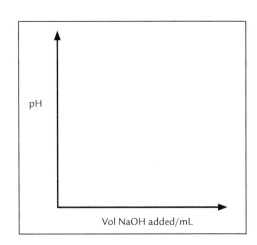

d Circle the buffer region on your graph.

e Write the formula of the salt formed at the equivalence point of the titration and give its concentration in mol L^{-1}.

f Given pK_a of the following indicators, choose the one that is the most suitable for detecting the end point of this titration.

Methyl red pK_a = 5.0

Phenol red pK_a = 7.8

Phenolphthalein pK_a = 9.6

ISBN: 9780170355544 PHOTOCOPYING OF THIS PAGE IS RESTRICTED UNDER LAW.

3.1

Carry out an investigation in Chemistry involving quantitative analysis

Internal assessment   **4 credits**

Unit	Content	Aim
18	**Investigations using volumetric analyses**	To understand the main points in carrying out analyses that depend on chemical reactions, and to interpret results obtained by the accurate measurement of volumes
19	**Gravimetric and colorimetric analyses**	To interpret results obtained by the accurate weighing of solids, and to interpret results obtained by the accurate measurement of light absorbance by coloured compounds

Unit 18 | Investigations using volumetric analyses

You are required to use chemistry to investigate a trend in the amount of a substance in a product.

As a guide for carrying out an investigation you should take the following steps:
- decide what you will analyse
- plan the method you will use for analysis
- try out the method to obtain the best conditions to carry out the investigation
- carry out the investigation
- look at sources of error
- make a conclusion from your results
- write a report of your investigation.

First steps

- Begin a log book. It is useful to write the date every time you start to write in it.
- Over the course of the investigation, you should write what you decided to analyse, your method, results, calculations and conclusions.
- Write any thoughts or comments you may have on such things as the procedure or the product as you go.
- Write down all references that you have used. Include internet references.

Deciding what to analyse

- Your investigation must be able to be carried out in a school laboratory. Using volumetric analysis you can analyse:
  - Acids. These are in foods — ethanoic acid in vinegars of different varieties, fatty acids in oils and fats, lactic acid in milk or yoghurt, citric acid or tartaric acid in fruit such as lemons or grapes, tartaric acid in wine. They are also in medicines — aspirin in a tablet.
  - Bases. These are in cleaners — sodium hydroxide in drain cleaner and oven cleaner, sodium carbonate in washing soda, ammonia in commercial ammonia solutions. They are also in indigestion remedies — calcium carbonate in a tablet.
  - Reducing agents (antioxidants). These are in food and drink — vitamin C in a drink or juice from a fruit, ethanol in wine. They are also in medicines — iron (II) in tablets for anaemia.
  - Oxidising agents. These are in cleaners and antibacterial solutions such as chlorine bleaches and iodine antiseptics.
- You are required to measure a trend or pattern of the amount of compound in the product that you are analysing.
  The amount of a compound in a product may vary with
  - exposure to light or to heat
  - number of months before and after 'Best before' or 'Use by' date
  - exposure to the atmosphere.

  ISBN: 9780170355544 PHOTOCOPYING OF THIS PAGE IS RESTRICTED UNDER LAW.

Planning the investigation

You will be analysing a compound by titration. To plan a method, you need to know the approximate amount or concentration of the compound. The container of the product may tell you what the manufacturer states as the approximate concentration. If it is a natural product, look online to find an average value. Then carry out the following.

*   Write equations for the reaction(s) to be used.
*   Decide on a primary standard (see page 144). What concentration and how much is needed?
*   Make a list of chemicals and equipment needed, including any indicators.
*   What are the safety precautions needed? Are material safety data sheets available for the chemicals used?
*   Decide on any storage requirements that may be needed during the course of the experiment.
*   Write out your method so that the volumes you use in each titration will be between 5 and 25 mL. There may be other compounds in your product that will interfere with your result (such as preservatives in food). Try to allow for this.
*   Try out the method by doing a 'rough' titration. Use measuring cylinders and variable pipettes. Adjust your method if necessary.

Carrying out the investigation

*   To get accurate data, clean equipment and a tidy and orderly method are important. Results will depend on an accurately prepared primary standard.
*   Report all of your results, even ones that seem out of line. You could comment why you think a result is out of line.

Sources of error

*   Look critically at your equipment. How accurately does the balance weigh? How accurately can you read the burette? Are there values of uncertainty written on the pipettes and volumetric flasks?
*   Is there an error in the procedure that can be minimised but not prevented?

Calculating results

*   Results are calculated from a balanced chemical equation.

Example: Processing quantitative data for an acid-base titration

0.0750 g of impure sodium bicarbonate is dissolved in water and titrated with 0.032 mol L^{-1} sulfuric acid using methyl orange indicator. 12.8 mL of sulfuric acid was required.
Calculate the percentage of sodium bicarbonate in the solid sample.
$M(Na) = 23.0$ g mol^{-1}, $M(H) = 1.0$ g mol^{-1}, $M(C) = 12.0$ g mol^{-1}, $M(O) = 16.0$ g mol^{-1}

Step 1	**Write a balanced equation showing the compounds involved.**	$2Na^+(aq) + 2HCO_3^-(aq) + 2H^+(aq) + SO_3^{2-}(aq) \longrightarrow$ $2Na^+(aq) + SO_4^{2-}(aq) + 2CO_2(g) + 2H_2O(l)$ **or** $2NaHCO_3 + H_2SO_4 \longrightarrow Na_2SO_4 + 2CO_2 + 2H_2O$

Step 2 Underline two compounds: $\underline{2NaHCO_3} + \underline{H_2SO_4} \longrightarrow Na_2SO_4 + 2CO_2 + 2H_2O$

a) the compound of interest

b) the compound for which measurements have been made.

Write expressions for the amounts n of these two substances using the symbols m for mass (g), c for concentration (mol L^{-1}), and V for volume (L).

For NaHCO$_3$ (in a solid), $n = \dfrac{m}{M}$

For H$_2$SO$_4$ (in a solution), $n = cV$

Step 3 Relate the amount (moles) of the underlined compounds using the chemical equation.

The equation has 2NaHCO$_3$ and 1H$_2$SO$_4$. So the amount of NaHCO$_3$ ÷ 2 equals the amount of H$_2$SO$_4$ ÷ 1.

The numbers 2 and 1 are stoichiometric numbers of NaHCO$_3$ and H$_2$SO$_4$, respectively.

$$\dfrac{n(NaHCO_3)}{2} = \dfrac{n(H_2SO_4)}{1}$$

∴ $n(NaHCO_3) = 2 \times n(H_2SO_4)$

∴ $\dfrac{m(NaHCO_3)}{M(NaHCO_3)} = 2 \times c(H_2SO_4) \times V(H_2SO_4)$

Substitute the known values into the equation.

For NaHCO$_3$, m is required. (Note that 0.0750 g is not the mass of NaHCO$_3$; it is the mass of NaHCO$_3$ + impurities.)

$M = 23.0 + 1.0 + 12.0 + 48.0 = 84.0$ g mol^{-1}

For H$_2$SO$_4$, c and V are given:

$c = 0.032$ mol L^{-1}, $V = 0.0128$ L

Complete the calculation.

We now have:

$$\dfrac{m(NaHCO_3)}{84.0 \text{ g mol}^{-1}} = 2 \times 0.032 \text{ mol L}^{-1} \times 0.0128 \text{ L}$$

$m = (2 \times 0.032 \times 0.0128)$ mol $\times$ 84.0 g mol^{-1}

$m = 0.0688$ g

This is the mass of pure NaHCO$_3$ in 0.075 g impure solid.

% NaHCO$_3$ $= \dfrac{0.0688}{0.0750} \times 100 = 91.7\%$

For any quantitative analysis, the experiment is never carried out just once. It should be repeated until three concordant (range* of titres $\leq$ 0.3 mL) results are obtained.

* This range may vary depending on the compound analysed and the product that contains it.

 ISBN: 9780170355544 PHOTOCOPYING OF THIS PAGE IS RESTRICTED UNDER LAW.

Example: Processing quantitative data for redox titration

The concentration of a ferrous sulfate solution is found by titrating it with a 0.05 mol L⁻¹ solution of potassium permanganate. 20 mL of the ferrous sulfate solution is measured with a pipette into a titration flask, and the solution acidified by adding 15 mL of 2 mol L⁻¹ sulfuric acid. Potassium permanganate solution is added from a burette until a faint pink colour persists for at least 10 minutes. This experiment is repeated to obtain three concordant (similar) results. Volumes of potassium permanganate solution required were 24.5 mL, 23.8 mL, 23.6 mL, and 23.8 mL.

Calculate the concentration of ferrous sulfate in the solution in:

$$\text{mol L}^{-1} \quad \text{and} \quad \text{g L}^{-1} \quad (M(\text{FeSO}_4) = 151.9 \text{ g mol}^{-1})$$

Steps:

1 Write a balanced equation for the reaction.
$$8\text{H}^+(aq) + \text{MnO}_4^-(aq) + 5\text{e} \longrightarrow \text{Mn}^{2+}(aq) + 4\text{H}_2\text{O}(l)$$
$$5 \times [\text{Fe}^{2+}(aq) \longrightarrow \text{Fe}^{3+}(aq) + \text{e}]$$

$$8\text{H}^+(aq) + \text{MnO}_4^-(aq) + 5\text{Fe}^{2+}(aq) \longrightarrow \text{Mn}^{2+}(aq) + 5\text{Fe}^{3+}(aq) + 4\text{H}_2\text{O}(l)$$

2 Calculate known amounts of substances.

Both the concentration and volume of potassium permanganate are known. So calculate the amount (number of moles) of potassium permanganate reacted.

The average volume V is:

$(23.8 + 23.6 + 23.8) \div 3$ mL = 23.7 mL = 0.0237 L

The first value is too different from the others and is not used.

$n(\text{mol}) = c(\text{mol L}^{-1}) \times V(\text{L}) = 0.05 \times 0.0237 = 1.185 \times 10^{-3}$ mol

3 Use the balanced equation to relate known and unknown amounts.

Relate the amount of permanganate (which is the same as the amount of potassium permanganate, KMnO₄) to the unknown amount of ferrous ions (which is the same as the amount of ferrous sulfate, FeSO₄).

$$\frac{n(\text{Fe}^{2+})}{5} = \frac{n(\text{MnO}_4^-)}{1}$$

These numbers are from the balanced chemical equation.

4 Substitute the known $n(\text{MnO}_4^-)$ and calculate the unknown value.

$n(\text{Fe}^{2+}) = 5 \times n(\text{MnO}_4^-) = 5 \times 1.185 \times 10^{-3} = 5.93 \times 10^{-3}$ mol

5 Use this formula to calculate the concentration of Fe²⁺ in the solution:

$n(\text{mol}) = c(\text{mol L}^{-1}) \times V(\text{L})$

20 mL = 0.020 L of ferrous sulfate solution was reacted.

5.93×10^{-3} mol $= c \times 0.020$ L $\longrightarrow c = \dfrac{5.93 \times 10^{-3}}{0.020} = 0.30$ mol L⁻¹

6 Use the following formula to calculate mass m of ferrous sulfate in 1 L solution:

$n(\text{mol}) = m(\text{g}) \div M(\text{g mol}^{-1})$

In 1 L, there is 0.30 mol.

0.30 mol $= m \div 151.9$ g mol⁻¹ $\longrightarrow m = 45.6$ g

The concentration of ferrous sulfate is 45.6 g L⁻¹.

ISBN: 9780170355544 PHOTOCOPYING OF THIS PAGE IS RESTRICTED UNDER LAW.

3.1

Example: Processing quantitative data for a titration involving two reactions

To determine the concentration of chlorine in a bleach solution, 10 mL were measured with a pipette into a titration flask. 1.5 g of solid potassium iodide was added. In this mass there is more than sufficient iodide ions to reduce chlorine to chloride ions. The iodine formed was then titrated with sodium thiosulfate. A 0.50 mol L^{-1} sodium thiosulfate solution was added from a burette until the brown colour of iodine had been reduced to a yellow colour. At this stage, starch solution was added as an indicator and the solution became blue-black. Sodium thiosulfate was then added gradually until the blue-black solution became colourless. This experiment was repeated to obtain three concordant results. These results were 20.6 mL, 20.4 mL, and 20.3 mL (average 20.4 mL).

The equations for the reactions are:

$$Cl_2(aq) + 2I^-(aq) \longrightarrow 2Cl^-(aq) + I_2(aq)$$
$$I_2(aq) + 2S_2O_3^{2-}(aq) \longrightarrow S_4O_6^{2-}(aq) + 2I^-(aq)$$

Calculate the concentration of chlorine in the original solution in mol L^{-1} and g L^{-1}. ($M(Cl_2)$ = 71.0 g mol^{-1})

Steps:

1 Add the two equations so that all I_2 formed in the first reaction is used up by the second.

$$Cl_2(aq) + \cancel{2I^-(aq)} + \cancel{I_2(aq)} + 2S_2O_3^{2-}(aq) \longrightarrow 2Cl^-(aq) + \cancel{I_2(aq)} + S_4O_6^{2-}(aq) + \cancel{2I^-(aq)}$$

$$Cl_2(aq) + 2S_2O_3^{2-}(aq) \longrightarrow 2Cl^-(aq) + S_4O_6^{2-}(aq)$$

2 Calculate known amounts of substances. The volume and concentration of sodium thiosulfate are known.

$$n(\text{mol}) = c(\text{mol } L^{-1}) \times V(\text{L}) = 0.50 \text{ mol } L^{-1} \times 0.0204 \text{ L} = 0.0102 \text{ mol}$$

3 Use the balanced chemical equation to relate the known amount of reactant to the reactant that is being analysed.

$$\frac{n(Cl_2)}{1} = \frac{n(S_2O_3^{2-})}{2} = \frac{0.0102 \text{ mol}}{2} = 0.0051 \text{ mol}$$

4 Use the amount present in 10 mL to calculate the amount of Cl_2 in 1 L.

0.0051 mol in 10 mL $\longrightarrow$ 0.0051 $\times \dfrac{1000}{10}$ mol L^{-1} = 0.51 mol L^{-1}

Since $M(Cl_2)$ = 71.0 g mol^{-1},

the concentration = 0.51 mol L^{-1} × 71.0 g mol^{-1} = 36.2 g L^{-1}

The concentration of ferrous sulfate is 45.6 g L^{-1}.

ISBN: 9780170355544 PHOTOCOPYING OF THIS PAGE IS RESTRICTED UNDER LAW.

Example: Processing quantitative data from a back titration

3.52 g of slaked lime are dissolved in 50.0 mL of 2.00 mol L^{-1} hydrochloric acid. Unreacted hydrochloric acid was neutralised with 19.5 mL of 1.00 mol L^{-1} sodium hydroxide solution using phenolphthalein indicator.

Calculate the percentage of calcium hydroxide in the sample of slaked lime.
(M(Ca(OH)$_2$) = 74.1 g mol^{-1})

Steps:

1 Write balanced equations for the reactions.

(In quantitative chemistry, it is often more useful to write equations that relate the amounts of the compounds reacting rather than equations that relate the amounts of ions reacting. This is because we need quantitative information about compounds and not ions.)

$$Ca(OH)_2(s) + 2HCl(aq) \longrightarrow CaCl_2(aq) + 2H_2O(l)$$
$$NaOH(aq) + HCl(aq) \longrightarrow NaCl(aq) + H_2O(l)$$

2 Calculate known amounts of substances.

The total amount of hydrochloric acid used (50.0 mL of 2.0 mol L^{-1}) is given by:

$$n = 2.0 \times \frac{50.0}{1000} = 0.100 \text{ mol}$$

The amount of sodium hydroxide needed to titrate excess HCl is:

$$n = 1.0 \times \frac{19.5}{1000} = 0.0195 \text{ mol}$$

3 Use the balanced equations to relate known amounts to unknown amounts.

The known amount of sodium hydroxide used in the titration equals the amount of remaining hydrochloric acid = 0.0195 mol (second equation).

The amount of hydrochloric acid reacted with calcium hydroxide is therefore:

0.100 mol (total amount) – 0.0195 mol (amount remaining) = 0.0805 mol

This amount of HCl is related to the amount of Ca(OH)$_2$ by the first equation.

$$\frac{n(Ca(OH)_2)}{1} = \frac{n(HCl)}{2} = \frac{0.0805}{2} \text{ mol} = 0.0403 \text{ mol}$$

4 Use $n = m \div M$ to calculate the mass, and hence the percentage, of Ca(OH)$_2$ in the sample.

$$0.0403 \text{ mol} = \frac{m}{74.1 \text{ g mol}^{-1}} \quad \therefore \quad m = 0.0403 \times 74.1 = 2.99 \text{ g}$$

$$\text{Percentage Ca(OH)}_2 = \frac{2.99 \text{ g}}{3.52 \text{ g}} \times 100 = 84.9\%$$

Making a conclusion

- From your results, state the amount of compound in your product.
- Compare it with known results. Discuss reasons for the similarity or difference. Natural products may vary widely in their composition.

Writing a report

The report should include the following.

- the title of your investigation
- a description of the compound and the product
- the analytical method including
    - safety precautions
    - results
    - sources of error
    - calculations
- your conclusion

Primary standards

- Substances whose exact amounts can be calculated from their weights are used to make solutions with known concentrations. These substances are used as **primary standards**.
- Compounds which can be obtained in a very pure form are needed. If an accurate weight of it is known, then the amount (moles) can be found accurately. The compound can be reacted quantitatively with another compound to determine that compound's concentration or purity.
- Compounds suitable for primary standards are:
  - stable during storage and handling
  - able to be weighed accurately (compounds with small molar masses are less suitable)
  - able to react immediately and completely in the titration.
- Compounds suitable for primary standards include:
  - potassium hydrogen phthalate $KHC_8H_4O_4$ — for standardising solutions of alkalis
  - oxalic acid — for standardising alkalis and oxidising agents
  - sodium carbonate (anhydrous) — for standardising acids
  - potassium bromate and potassium iodate — for standardising oxidising and reducing agents (in particular, titrations involving iodine).
- Compounds not suitable for primary standards include:
  - HCl — volatile, readily lost to the surroundings
  - HNO_3 — volatile, readily lost to the surroundings
  - H_2SO_4 — absorbs water from the air
  - NaOH — absorbs water and CO_2 from the air
  - $Na_2S_2O_3$ — difficult to obtain in a pure form
  - $KMnO_4$ — decomposes on storage, loses oxygen
  - $Na_2CO_3.10H_2O$ — sodium carbonate crystals lose water on storage
  - Fe^{2+} compounds — oxidise to Fe^{3+} with O_2 in the air.

KEY POINTS SUMMARY

- A balanced chemical equation tells us the amounts of substances that are reacting and being formed in a reaction.
- Amounts n of substance are given by $n = m/M$ for substances that are weighed and by $n = cV$ for substances dissolved in solution.
- A balanced chemical equation allows us to relate the quantities of reactants and products.
- Quantitative analyses should always start with a primary standard.

ASSESSMENT ACTIVITIES

Interpreting acid-base reactions

1 Analysing the amount of acetic acid in vinegar

Although vinegar is regarded as acetic acid, vinegar contains only about 3% to 5% acetic acid. To determine the concentration of acid in a commercial vinegar, 10.0 mL was measured with a pipette into a titration flask. 0.10 mL phenolphthalein indicator was added, and the solution was titrated with 0.500 mol L^{-1} sodium hydroxide solution. The titration was repeated until three results within a range of 0.50 mL was obtained. The three results were 14.2 mL, 14.4 mL, 14.4 mL.
M(H)=1.01 g mol^{-1}, M(C) =12.00 g mol^{-1}, M(O) = 16 g mol^{-1}

ISBN: 9780170355544 PHOTOCOPYING OF THIS PAGE IS RESTRICTED UNDER LAW.

a Calculate the concentration of acetic acid in vinegar in mol L^{-1}.

   b Calculate the concentration of acetic acid in vinegar in g L^{-1}.

   c Calculate the percentage of acetic acid in the vinegar.

2 Analysing the amount of acid in wine

Wine contains a mixture of acids. To determine the concentration of acid in a sample of white wine, 25.0 mL are pipetted into a 100 mL flask and 0.2 mL phenolphthalein indicator added. 0.100 mol L^{-1} sodium hydroxide is added from a burette. 8.30 mL sodium hydroxide was required to just change the indicator from colourless to pink.

   a Calculate the concentration of acid (general formula HA) in mol L^{-1}.

   b The specification requires the concentration of acid calculated in this way to be 0.025 ± 0.005 mol L^{-1}. Is the wine within the specification?

   c If the average molar mass of the acids in wine is 74.9 g mol^{-1}, calculate the percentage of acid (g in 100 mL) in wine.

3 Analysing the amount of ammonia in a cleaner

Ammonia cleaners are used to clean greasy surfaces. The manufacturer of one cleaner claims that his product contains 5% ammonia. To check this, 5.00 mL of the cleaner are pipetted into a 100.0 mL volumetric flask and the solution made up to the mark with water. 20.0 mL of this diluted solution was titrated against 0.210 mol L^{-1} hydrochloric acid solution using 0.1 mL methyl orange as indicator. This was repeated to obtain three results within a range of 0.5 mL. An average of 14.1 mL acid was added to just change the colour of the indicator from yellow to red.

   a Calculate the concentration of ammonia in the diluted solution.

   b Does the concentration of ammonia in the cleaner comply with the manufacturer's claim?

     M(N) = 14.01 g mol^{-1}, M(H) = 1.00 g mol^{-1}

4 Analysing salt in sausage

The amount of salt in a sausage was estimated as follows. 100 g of sausage that was cooked on a barbecue was cut into small pieces and soaked overnight in 50 mL water. The water was decanted from the sausage. The sausage was washed with a further 30 mL of water. The combined filtrates were made up to 100.0 mL in a volumetric flask. 20.0 mL portions of the solution ware titrated against 0.100 mol L^{-1} silver nitrate solution with a few drops of 5% sodium chromate solution added as an indicator. Chloride ions in the salt were precipitated as silver chloride. When all the chloride ions had been precipitated, a red colour due to the formation of silver chromate appeared. Further 20.0 mL portions were titrated until three concordant results were obtained. These results were 9.80 mL, 9.60 mL, and 9.80 mL.

M(NaCl) = 58.5 g mol^{-1}

   a Calculate the average titre of silver nitrate.

   b Calculate the amount (in moles) of silver nitrate in this titre.

   c Write an equation for the reaction of sodium chloride with silver nitrate.

   d What is the amount of chloride in 20 mL of filtrate? Assume this equals the amount of sodium chloride.

   e Calculate the total amount and mass of sodium chloride in 100 g of sausage.

   f The butcher who supplied the sausage claims that there is less than 0.5% salt in his sausages. Does the sample analysed comply with this claim?

5 Analysing the amount of sodium carbonate in washing soda

To determine the sodium carbonate content in washing soda, 2.00 g of washing soda was dissolved in water and made up to 250 mL in a volumetric flask. 25.0 mL of the solution was titrated with 0.100 mol L^{-1} hydrochloric acid using methyl orange as indicator. The titration was repeated until three close results were obtained. Average titre of three close results was 20.4 mL.

M(Na$_2$CO$_3$) = 106.0 g mol^{-1}, M(H$_2$O) = 18.0 g mol^{-1}

   a Calculate the concentration of sodium carbonate in the solution (in mol L^{-1}).

3.1

b Calculate the mass of sodium carbonate present in the 2.00 g sample (that is, in the 250 mL solution).

c Calculate the percentage of sodium carbonate in the sample.

d Freshly prepared washing soda crystals have the formula $Na_2CO_3.10H_2O$. On storage they lose water of crystallisation. Was the sample analysed a freshly prepared sample? Give reasons for your answer.

6 Analysing the amount of citric acid

Citric acid was purchased from a supermarket for making homemade lemonade. Its purity was found by dissolving 0.500 g of it in 50.0 mL water and then titrating the solution with 1.00 mol L^{-1} sodium hydroxide solution. Phenolphthalein was used as the indicator. Phenolphthalein is colourless in water. The volume of sodium hydroxide required to change the colour of a solution of phenolphthalein in 50.0 mL water is called the blank titre. This volume is subtracted from the titre obtained for the citric acid solution to determine the volume of sodium hydroxide reacted with citric acid. The blank titre was found to be 0.50 mL and the sample titre 8.30 mL.

a Calculate the volume of 1 mol L^{-1} sodium hydroxide that reacted with citric acid.

b Calculate the amount of sodium hydroxide in this volume.

c One mole of citric acid reacts with three moles of NaOH.
What is the amount of citric acid that reacts with this amount of sodium hydroxide?

d The formula of citric acid is $C_6H_8O_7$.
$M(H) = 1.01$, $M(C) = 12.01$, $M(O) = 16.00$ g mol^{-1}
Calculate the weight of citric acid reacted.

e Calculate the percentage of citric acid in the 0.500 g.

Interpreting redox titrations

7 Analysing the amount of iron (II) in tablets

Anaemia can be treated by taking iron tablets containing iron as Fe^{2+}. The amount of iron in a tablet was found as follows. A primary standard solution of oxalic acid was prepared to standardise a potassium permanganate solution. The standardised potassium permanganate solution was used to titrate the iron from a tablet.

1.318 g oxalic acid crystals $(COOH)_2.2H_2O$ were dissolved in distilled water in 100 mL volumetric flask. Water was then added to the 100 mL mark. To 20.0 mL of this solution measured with a pipette, approximately 5 mL of dilute sulfuric acid was added, and the solution heated.

Potassium permanganate solution (made by dissolving about 1.2 g of potassium permanganate and making it up to 200 mL solution) was added from a burette to the hot solution until a faint pink colour remained. The average of three concordant results was 24.60 mL.

A tablet was then powdered and mixed with 20 mL of water. 5 mL of dilute sulfuric acid was added. Potassium permanganate solution was added from the burette until a faint pink colour remained. For three different tablets, the results were: 12.60 mL, 12.65 mL, and 12.65 mL.

$$M(C) = 12.01, M(O) = 16.00, M(H) = 1.008, M(Fe) = 55.85 \text{ g mol}^{-1}$$

a Write balanced redox equations for the oxidation of oxalic acid to carbon dioxide in acidified potassium permanganate solution, and for the oxidation of ferrous (Fe^{2+}) ions in acidified potassium permanganate solution.

ISBN: 9780170355544
PHOTOCOPYING OF THIS PAGE IS RESTRICTED UNDER LAW.

b From the mass of oxalic acid crystals and the volume of solution, calculate the concentration of the oxalic acid solution in mol L^{-1}.

c Use the titration results for potassium permanganate with oxalic acid to calculate the concentration of potassium permanganate in mol L^{-1}.

d Calculate the mass of iron in the tablet.

e The supplier claims there are 120 mg of iron in each tablet. Do the experimental results support the claim?

8 Analysing the amount of vitamin C in a tablet

Vitamin C, ascorbic acid, helps to protect the body against bacterial diseases and aids in the synthesis of collagen in bone cartilage. It is a reducing agent, one of the 'antioxidant' vitamins believed to slow aging. A tablet claims to contain 250 mg of vitamin C. It was dissolved in about 60 mL of water in a 100.0 mL volumetric flask. When the tablet was completely dissolved, water was added to make the solution up to 100.0 mL. 25.0 mL of this solution was measured out with a pipette into a titration flask. 1 mL of starch indicator was added and the solution was titrated with 0.0250 mol L^{-1} iodine solution until a violet-blue colour was obtained. Three concordant results for the iodine solution obtained in this experiment were 14.20 mL, 14.10 mL, and 14.10 mL.

$$M(\text{vitamin C}) = 176 \text{ g mol}^{-1}$$

a Calculate the average titre of iodine.

b Calculate the amount of iodine in this volume of 0.0250 mol L^{-1} solution.

c If 1 mol vitamin C reacts with 1 mol iodine, what is the amount of vitamin C reacted?

d Calculate the mass of vitamin C in 25.0 mL of solution.

e What is the mass of vitamin C in the tablet? Is this amount consistent with the manufacturer's claim?

Interpreting titrations involving two reactions

9 Analysing the amount of copper in a copper salt

0.132 g of copper salt was dissolved in 50 mL of water. 1 g of potassium iodide was added and the solution mixed. The iodine formed in the reaction was titrated against 0.0200 mol L^{-1} sodium thiosulfate solution. 0.5 mL of starch indicator was added when the colour of iodine paled to a yellow colour. 26.4 mL of sodium thiosulfate was required to decolourise the blue-black starch-iodine colour.

$$M(\text{Cu}) = 63.5 \text{ g mol}^{-1}, \; M(\text{CuSO}_4.5\text{H}_2\text{O}) = 249.69 \text{ g mol}^{-1}$$

a Write a balanced redox equation for the reaction of copper ions with iodide ions. $Cu^{2+}(aq)$ is reduced to $CuI(s)$, and $I^-(aq)$ is oxidised to $I_2(aq)$ or $I_3^-(aq)$.

b The iodine produced in the reaction is titrated with thiosulfate ions. Write the balanced redox equation.

c Using the fact that the iodine formed in the first reaction is completely used up in the second reaction, write an equation that relates copper (II) ions to the amount of thiosulfate reacted.

d Calculate the amount of thiosulfate that reacted with iodine.

e What is the amount of copper reacted? Calculate the mass of copper in this amount.

f Calculate the percentage copper in the salt.

g Is it possible that the copper salt is $CuSO_4.5H_2O$?

h Why is it unwise to draw conclusions from this experiment?

10 Analysing the amount of chorine in a bleach

5.00 mL of a solution of chlorine containing bleach was diluted to 200 mL with distilled water. 20.0 mL of this diluted solution was measured with a pipette into a titration flask and 1 g of potassium iodide added. The iodine formed in the reaction was titrated with 0.500 mol L^{-1} sodium thiosulfate solution. 5 drops of starch solution were added when the iodine colour had faded to a pale yellow. Sodium thiosulfate was then added slowly until the blue-black colour disappeared. This titration was repeated with 20.0 mL portions of the diluted bleach solution to obtain three concordant results. The average of the three results was 9.75 mL.

$$M(Cl_2) = 71.0 \text{ g mol}^{-1}$$

a   Why is distilled water used to dilute the solution?

b   Write a balanced redox equation for the reaction of chlorine with iodide ions.

c   The iodine produced is titrated with thiosulfate ions. Write a balanced redox equation for the reaction.

d   Using the fact that the iodine formed in the first reaction is completely used up in the second reaction, write a balanced equation that relates the amount of chlorine to the amount of thiosulfate reacted.

e   Calculate the amount of thiosulfate reacted.

f   What is the amount of chlorine in the original undiluted bleach solution?

g   The manufacturer claims that the bleach contained no less than 35% chlorine. Do the results support this claim?

Interpreting a back titration

11 Analysing the amount of aspirin in a tablet

30.0 mL of 0.500 mol L^{-1} sodium hydroxide solution is added to an amount of powdered aspirin equivalent to three aspirin tablets. The mixture was warmed gently for 10 minutes. Each mole of aspirin hydrolyses to give one mole of ethanoic (acetic) acid and one mole of salicylic acid. Both acids react with the sodium hydroxide. The unreacted sodium hydroxide is then titrated against 0.50 mol L^{-1} hydrochloric acid solution using phenol red indicator. This experiment is carried out until three close results are obtained. The volumes of HCl used are 19.0 mL, 19.2 mL, and 19.1 mL.

a   Calculate the amount of sodium hydroxide in 30.0 mL of 0.500 mol L^{-1} solution.

b   Calculate the amount of sodium hydroxide reacted with hydrochloric acid in the titration.

c   What is the amount of sodium hydroxide that has reacted with aspirin?

d   How many moles of acid from aspirin reacted with this amount of sodium hydroxide?

e   Each mole of aspirin gives 2 moles of acid. How many moles of aspirin was in the original sample?

f   If M(aspirin) = 180.2 g mol^{-1}, calculate the mass of aspirin reacted.

g   This mass of aspirin was contained in three tablets. The manufacturer claims each tablet contains 150 mg aspirin. Do the results agree with the manufacturer's specifications?

ISBN: 9780170355544   PHOTOCOPYING OF THIS PAGE IS RESTRICTED UNDER LAW.

Unit 19 | Gravimetric and colorimetric analyses

You are required to use chemistry to investigate a trend in the amount of a substance in a product. This unit looks at two procedures:

- Colorimetric analysis. The intensity of colour in a solution can be measured accurately. The colour is related to the concentration of a coloured ion or compound.
- Gravimetric analysis. Many dissolved ions can form insoluble compounds. The insoluble compound can be collected by filtration and weighed accurately.

Colorimetric analysis

- A spectrophotometer such as a colorimeter is used to accurately measure colour intensity of solutions. It measures the amount of colour absorbed by the solution. A blue solution of copper absorbs wavelengths in the red-orange-yellow part of the spectrum. The wavelengths that are not absorbed are the ones that you see.
- The concentration of a coloured compound is related to the intensity of the colour of the solution. For example, the more copper ions there are in solution, the greater the amount of red-orange-yellow light absorbed, and the more blue the solution will appear.
- The absorbances of a range of known concentrations of a compound are measured.
- A graph of concentration/absorbance is drawn.
- The absorbance of the unknown solution is measured. Its concentration is read from the graph.

SPECIAL SKILL: DETERMINING THE CONCENTRATION OF IONS USING COLORIMETRIC ANALYSIS

Steps:
1 Measure or obtain data for the absorbances of a range of standard solutions of the chemical being anaysed.
2 Plot a graph of absorbance versus concentration for the standard solutions.
3 Measure the absorbance of the solution of unknown concentration.
4 On the graph, find the concentration of the chemical which matches the measured absorbance.
5 Determine the concentration of the ion in the solution.

Example: Absorbances of different concentrations of the copper ammonium complex ion in aqueous solution are given in the table. The absorbance of a solution with an unknown concentration of the ion is 1.05 absorbance units (au). Find the concentration of Cu^{2+} ions in this solution in mol L^{-1} and g L^{-1}.

Concentration (mol L⁻¹)	0	0.10	0.020	0.030
Absorbance (au)	0	0.452	0.916	1.340

Note: $M(Cu)$ = 63.5 g mol⁻¹.

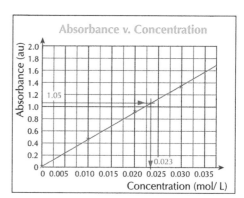

Applying the steps:

1 The data is supplied.
2 Plotting the data for the standard solutions on a graph of absorbance versus concentration.
3 The absorbance of the unknown solution has been measured: 1.05 absorbance units (au).
4 Finding the concentration which matches the absorbance: 0.023 mol L⁻¹.
5 Determining the concentration of the ion in the solution:
    in every mole of $Cu(NH_3)_4^{2+}$, there is one mole of Cu^{2+}
    so $n(Cu^{2+})$ = 0.023 mol L⁻¹ or 0.023 mol L⁻¹ × 63.5 g mol⁻¹ = 1.46 g L⁻¹.

Using colorimetric analysis

- This procedure can be used to:
  - Determine the amount of copper. The copper could be in a copper salt. Copper nitrate exposed to air absorbs water and this is noticeable over a few days. The copper could be in metal objects such as coins. The object needs to be dissolved in concentrated nitric acid in a fume hood and made up into a solution. Note that if other colours mask the blue colour, the experiment is still possible if the absorbance is measured in the correct wavelength range. Ammonia is often added so that the absorbance of the complex ion $Cu(NH_3)_4^{2+}$ is measured.
  - Determine the amount of iron (III). Iron in metal objects dissolved in nitric acid will be in the +3 oxidation state. KSCN is added to give a dark-red colour. The absorbance of the dark-red solutions are measured.

Gravimetric analysis

- This procedure is based on accurate measurements of weights.
- The compound to be analysed forms an insoluble compound that can be collected and weighed. For example, sodium chloride will form a precipitate of silver chloride when it reacts with silver nitrate solution. All of the precipitate is then collected and weighed.
- The precipitate formed must have very low solubility and it must not decompose while it is dried, often in an oven at 100°C.

Using gravimetric analysis

- This procedure can be used to:
  - Determine the amount of chloride. Sodium chloride is in sea water and nearly all prepared foods.
  - Determine the amount of sulphate. Sulfate is found in fertilisers and compounds such as Epsom salts.

Sodium chloride concentration in the Dead Sea is high.

ISBN: 9780170355544

PHOTOCOPYING OF THIS PAGE IS RESTRICTED UNDER LAW.

Example: A gravimetric analysis

In a gravimetric analysis for magnesium sulfate in Epsom salts, 2.509 g of Epsom salts (which contain magnesium sulfate) were dissolved in 100 mL water. Barium chloride solution was added. After 25 mL of 0.1 mol L^{-1} barium chloride solution was added, the precipitate was allowed to settle. It was found that the solution produced no more precipitation with further addition of barium chloride. The precipitate was obtained after filtering through a weighed filter paper. To remove excess barium chloride, the filter paper was washed several times with distilled water. The filter paper was dried at 105°C overnight, cooled in a desiccator and weighed.

Weight of filter paper = 1.324 g
Weight of filter paper + precipitate = 3.697 g

Calculate the percentage of magnesium sulfate present in the sample of Epsom salts.

$$M(O) = 16.0 \text{ g mol}^{-1}, M(S) = 32.1 \text{ g mol}^{-1}, M(Ba) = 137.4 \text{ g mol}^{-1}, M(Mg) = 24.3 \text{ g mol}^{-1}$$

1 The equation which relates reactants and products is:

$$Ba^{2+}(aq) + 2Cl^-(aq) + Mg^{2+}(aq) + SO_4^{2-}(aq) \longrightarrow BaSO_4(s) + Mg^{2+}(aq) + 2Cl^-(aq)$$

or

$$BaCl_2 + MgSO_4 \longrightarrow BaSO_4 + MgCl_2$$

2 $BaSO_4$ is the precipitate in the dried filter paper. Its weight is known so $BaSO_4$ is underlined. The amount of $MgSO_4$ is of interest, so it is also underlined.

$$BaCl_2 + \underline{Mg\,SO_4} \longrightarrow \underline{BaSO_4}(s) + MgCl_2$$

We require the mass of $MgSO_4$, $m(MgSO_4)$, so use $n = \dfrac{m}{M}$

BaSO4 is a solid, so again use $n = \dfrac{m}{M}$

3 From the equation for the reaction, $\dfrac{n(MgSO_4)}{1} = \dfrac{n(BaSO_4)}{1}$

$$\frac{m(MgSO_4)}{M(MgSO_4)} = \frac{m(BaSO_4)}{M(BaSO_4)}$$

$$\frac{m(MgSO_4)}{24.3 + 32.1 + 4 \times 16.0 \text{ g mol}^{-1}} = \frac{(3.697 - 1.324)}{137.4 + 32.1 + 4 \times 16.0 \text{ g mol}^{-1}}$$

$$m(MgSO_4) = 1.224 \text{ g}$$

$$\% MgSO_4 \text{ in Epsom salts} = \frac{1.224 \times 100}{2.509} = 48.78$$

The intensity of colour is proportional to the concentration.

KEY POINTS SUMMARY

- **Colorimetric analysis** relies on the fact that the intensity of colour depends on the concentration of the coloured compound.
- **Gravimetric analysis** depends upon the low solubility of a compound. It must be sufficiently stable to be dried and weighed without decomposing.

ASSESSMENT ACTIVITIES

Colorimetric analysis

1 **Absorbances of different concentrations of copper in the blue copper ammonium complex ion in aqueous solution are given in the following table. Absobances are measure in absorbance units (au).**

Cu^{2+}concentration (mol L^{-1})	0	0.005	0.020	0.040
Absorbance (au)	0	0.40	1.64	3.198

To determine the concentration of a solution of copper sulfate, ammonia was added to form the complex ion.

To 10.0 mL of solution measure with a pipette, 5.0 mL of ammonia solution was added, also using a pipette. The absorbance of the solution was found to be 2.50 au.

A spectrophotometer which is used for colorimetry.

    a Draw a graph to display the data in the table. Plot absorbance units on the vertical axis.

    b From the graph, determine the concentration of copper in the copper sulfate plus ammonia solution.

    c Calculate the concentration of copper in the original solution in mol L^{-1} and g L^{-1}.

$$M(Cu) = 63.5 \text{ g mol}^{-1}$$

2 **Formaldehyde is a substance which forms during the setting of resins used in fibreboards and glues. It is readily absorbed into water.**

A solution of formaldehyde can be analysed by reacting it with a compound called acetylacetone to give a yellow solution. The intensity of the yellow colour is directly related to the concentration of formaldehyde.

Absorbances of the yellow complex prepared from different concentrations of standard soltuions of formaldehyde are given in the following table.

Formaldehyde concentration (mg L^{-1})	0	0.20	0.60	1.00
Absorbance (au)	0	0.146	0.450	0.725

Formaldehyde emitted by a 3.0g block of fibreboard at 40°C (this temperature is above the boiling point of formaldehyde) was absorbed into 50 mL of distilled water. When treated in the same way as the standard solutions, the solution had an absorbance of 0.35 au.

    a Draw a graph to display the data in the table. Plot absorbance units on the vertical axis.

    b From the graph, determine the concentration of formaldehyde in the solution in mg L^{-1}.

    c Calculate the mass (mg) of formaldehyde emitted by the block in mg g^{-1}.

    d Solutions analysed by this method should have absorbance values in the range of the values in the table.

       Suggest what could be done with a solution of formaldehyde that was found by this method to have an absorbance of 1.58 au.

ISBN: 9780170355544   PHOTOCOPYING OF THIS PAGE IS RESTRICTED UNDER LAW.

Gravimetric analysis

3 Analysing the amount of ammonium sulfate in a fertiliser

Ammonium sulfate is a water-soluble fertiliser for plants that prefer acid conditions. 1.24 g industrial-grade ammonium sulfate was dissolved in 20 mL water. A barium chloride solution (containing 10 g in 100 mL) was added until no further precipitation occurred. The precipitate was collected in a weighed filter paper and dried at 105°C overnight.

Weight of filter paper = 1.073 g

Weight of filter paper + barium sulfate = 3.133 g

$M(H) = 1.01$, $M(N) = 14.01$, $M(S) = 32.06$, $M(O) = 16.00$, $M(Ba) = 137.3$

a Calculate the weight of ammonium sulfate in the 1.24 g sample.

b Calculate the percentage of ammonium sulfate in the sample.

4 Analysing the salt in potato chips

To determine the percentage of sodium chloride in potato chips, 100 g of potato chips were mashed with 50 mL water and filtered. The residue in the filter paper was washed with three further 50 mL portions of water. Silver nitrate solution is added to the clear filtrate until a drop removed from the filtrate no longer reacted with silver nitrate. The precipitate formed was collected in a weighed filter paper and washed with 200 mL water. After drying at 105°C, the precipitate in the filter paper was weighed.

Weight of filter paper = 0.867 g

Weight of filter paper + silver chloride precipitate = 3.542 g

$M(Ag) = 107.9$ g mol^{-1}, $M(Cl) = 35.5$ g mol^{-1}, $M(Na) = 23.0$ g mol^{-1}

(Assume that sodium chloride is the only compound in the chips that forms a precipitate with silver ions.)

a Calculate the weight of silver chloride precipitated.

b Calculate the weight of sodium chloride (g in 100 g) in the chips.

c Calculate the percentage of sodium chloride in the chips.

d How could you check that chloride has been effectively extracted from the chips?

3.1

3.3

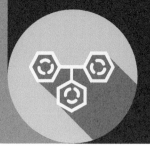

Demonstrate an understanding of chemical processes in the world around us

Internal assessment   **3 credits**

Unit	Content	Aim
20	**Chemical processes in the world around us**	To provide examples of natural and industrial chemical processes for you to investigate and discuss their effect on people and the environment

Unit 20 | Chemical processes in the world around us

Learning Outcomes — on completing this unit you should be able to:

- process and interpret information for one, or several related, chemical process(es)
- explain the consequences of the chemical processes to the environment or to people
- use the correct chemical vocabulary, symbols, conventions and equations to describe the processes and their consequences.

Chemical processes

Chemical processes occur all the time in the natural world. Many other processes are developed in response to man's needs. The processes described below are listed under the number of the NCEA Achievement Standard (AS) that is best related to it. The chemistry and related vocabulary, symbols and conventions will be described in the section covering that Achievement Standard.

Processes related to Chemistry AS3.4 — Units 1 to 7

Treated timber analysed by AA spectroscopy

Timber that survives wet conditions without rotting are usually treated with copper, chromium and arsenic compounds. The amount of these elements in the wood can be analysed by atomic absorption spectroscopy (AAS). AAS relies on the fact that electrons occupy certain energy levels in which they have discrete amounts of energy. Measurement of the energy that electrons absorb when they move to higher levels allows determination of the amount of an element that is present. Different amounts of chemicals in the timber gives it different hazard ratings. Elements present in water supplies and industrial waste are also analysed this way.

Treated timber can survive harsh wet conditions.

A consequence of treating timber in this way is that the chemicals used are harmful. The treated timber must not be burned so that chemicals are released into the air. The chemicals must not leach out into the soil, drains and the sea. If they did leach out, the timber would no longer be treated timber and it would rot. Timber treatment sites ensure that spills are minimal. Testing kits are available for determining whether timber has been treated.

Dissolved ions analysed by ion chromatography

The process of ion chromatography enables ions to be separated according to their size and charge. The following is a chromatogram from the separation of anions. From it we can deduce which ions are present and their concentrations.

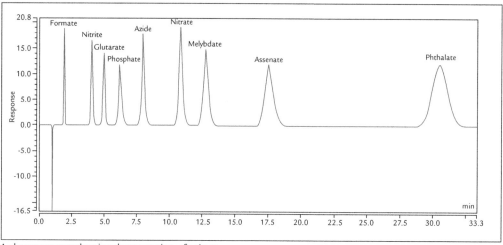

A chromatogram showing the separation of anions

As a consequence, dissolved substances in streams, farm run-offs and stormwater drains can be analysed and the necessary remedies put in place.

Pharmaceuticals analysed by gas and liquid chromatography

Molecules with different polarities have different solubilities in a solvent. The processes of gas and liquid chromatography enable molecules to be separated according to their sizes and polarities. HPLC (high-performance liquid chromatography) is a common tool in the pharmaceutical industry to analyse the purity of compounds.

Our medicines can be constantly checked that they meet their specifications on manufacturing and their 'best before' dates can be determined for them.

Processes related to Chemistry AS3.5 — Units 21 to 26

Plastic conductors

Carbon compounds with conjugated double bonds (alternate single and double carbon-carbon bonds) conduct electricity. Graphite is an example.

The chemical process involves the polymerisation of compounds with C≡C triple bonds such as acetylene to give a polymer with conjugated double bonds.

$$H-C \equiv C-H + H-C \equiv C-H + H-C \equiv C-H \longrightarrow$$
$$-CH=CH-CH=CH-CH=CH-$$

The polymer itself then needs to be 'doped' to become a good conductor.

ISBN: 9780170355544 PHOTOCOPYING OF THIS PAGE IS RESTRICTED UNDER LAW.

Three scientists, including New Zealander Professor Alan MacDiarmid, were awarded the Nobel Prize in 2000 for their role in developing conducting polymers from organic compounds.

Alan J. Heeger

Alan G. MacDiarmid

Hideki Shirakawa

Scientists who developed conducting polymers have changed our lives. The consequences have given a huge increase in the products we have in our daily lives. We have flat-screen and curved-screen televisions, electroluminescent displays on mobile phones, and new ways to inhibit corrosion to mention a few recent developments. Rapid development of more desirable products has resulted in disposal problems for the gadgets we are tired of.

Other polymers

Some polymers are made by natural processes. All large molecules in living things are natural polymers. These include proteins (polyamides) and cellulose (polysaccharides).

Man-made polymers are constantly being developed to meet our needs. These include bulletproof vests, materials that will withstand the temperatures of a spacecraft re-entering the earth's atmosphere as well as the exposure to UV and sunlight in space, or just cheaper plastic bags.

Natural polymers are also modified to meet man's needs. Vulcanised rubber is an example.

Natural polymers are biodegradable, but some man-made polymers remain in landfills for long periods of time.

Enantiomers

Enantiomers are compounds with the same structural formula but the atoms have a different spatial arrangement. One enantiomer is the non-superimposable mirror image of the other. They are described on page 175.

Three common drugs that can exist as enantiomers are ibuprofen, naproxen and thalidomide.

The chemical processes involved can be described by giving the starting material, the reagents and the final product. Note the asymmetric carbon atom in the final product.

The consequences of patients taking the enantiomers varies from a small effect (ibuprofen) to serious (naproxen) and very serious (thalidomide). Enantiomers may revert from one form to the other in the human body.

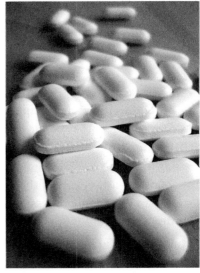
Tablets containing ibuprofen, a common anti-inflammatory drug.

Processes related to Chemistry AS3.6 — Units 12 to 17

Calcium carbonate equilibria

Chemical processes involving equilibrium reactions of calcium carbonate occur:

- in the sea, for example in shells and pearls
- in the earth's crust, for example in the soil and in caves
- in land animals, for example in snail shells and eggshells
- in man-made structures such as concrete buildings, paths and swimming pools.

Factors that affect the equilibrium include:

- the amount of CO_2. Carbon dioxide dissolved in rain and in the sea can vary. The variation may be natural or man-made.
- the pH. The pH of rain water and sea water depends on other dissolved substances.
- the temperature. This is naturally different in different parts of the world but it could be changing with time. Natural calcium carbonate deposits are found in cold climates and in hot springs.

Acid rain affects structures that contain calcium carbonate.

Consequences:

- decreasing amounts of calcium carbonate affects the strengths of shells and concrete structures
- increasing amounts in calcium carbonate can result in unwanted brittleness.

Processes related to Chemistry AS3.7 — Units 8 to 11

Applications of fuel cells

The chemical process involves the reaction of a fuel with oxygen to produce energy. One example is the breathalyser device, which measures the amount of ethanol (the fuel) in a person's breath. Different fuel cells have been developed for different purposes. Cars, toys and spacecraft use fuel cells today!

A consequence of fuel cells is that we can obtain electrical energy from traditional fuels. Carbon containing fuels will still result in emission of carbon dioxide although fuel cells using waste that normally go to landfills have advantages.

To demonstrate understanding of a chemical process

- State the chemical principles and/or the reactions involved.
- Describe how the chemical behaviour of substances apply to the process.
- Explain how removal of reactants and/or the release of products affect people and the environment.
- Consider any effects of the process itself.
- For an industrial process, explain why the advantages of the process must outweigh the disadvantages.
- For a natural process, explain whether the process is affected by changes in climate and other conditions such as acid rain. The changes may occur naturally or they may be a consequence of the actions of humans.

 ISBN: 9780170355544 PHOTOCOPYING OF THIS PAGE IS RESTRICTED UNDER LAW.

3.5

Demonstrate an understanding of the properties of organic compounds

External assessment   **5 credits**

Unit	Content	Aim
21	**Alkanes, alkenes and addition polymers of alkenes**	To demonstrate understanding of these compounds by being able to give their **systematic names**, write their **structures**, describe their **physical properties** and relate their **chemistry** to **functional** groups
22	**Haloalkanes**	
23	**Isomers**	To understand the difference between **constitutional isomers** and **stereoisomers** To understand differences in two types of stereoisomers, **cis-trans** isomers and **enantiomers**
24	**Alcohols, aldehydes and ketones**	To demonstrate understanding of these compounds by giving **systematic names**, drawing **structures**, describing their **physical properties** and relating their **chemistry** to **functional** groups To distinguish between these compounds by their chemistry
25	**Carboxylic acids and amines**	To demonstrate understanding of these compounds by giving **systematic names**, drawing **structures**, and relating their **physical** and **chemical properties** to **functional** groups, and to observe similarities with **inorganic acids** and **bases**
26	**Polymers**	To understand the difference between **addition polymerisation** and **condensation polymerisation** To study reactions that form **polyamides** and **polyesters** To understand the formation of **peptides** and **proteins**

Unit 21 | Alkanes, alkenes and addition polymers of alkenes

Learning Outcomes — on completing this unit you should be able to:

- write systematic names of alkanes
- describe substitution reactions of alkanes
- write systematic names of alkenes
- describe addition reactions of alkenes
- understand Markovnikov addition to unsymmetric alkenes.

Petrol contains liquid alkanes.

Alkanes

- Alkanes are **hydrocarbons** (compounds of **carbon and hydrogen**) in which all the **carbon-carbon bonds are single bonds**.
- Each carbon atom has four bonds. The arrangement of bonds around each carbon atom is tetrahedral and the bond angles are approximately 109°.
- An example is butane, C_4H_{10}, which is the fuel in disposable lighters.

$$H-\overset{\displaystyle H}{\underset{\displaystyle H}{\overset{|}{\underset{|}{C}}}}-\overset{\displaystyle H}{\underset{\displaystyle H}{\overset{|}{\underset{|}{C}}}}-\overset{\displaystyle H}{\underset{\displaystyle H}{\overset{|}{\underset{|}{C}}}}-\overset{\displaystyle H}{\underset{\displaystyle H}{\overset{|}{\underset{|}{C}}}}-H$$

- The **structural formula** of butane is written (as above)

but its shape is more like those shown by the molecular models. The 3D structures look very different to the linear chain drawn above.

- Methane, in natural gas, is CH_4. The **3D structural formula** of methane is

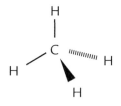

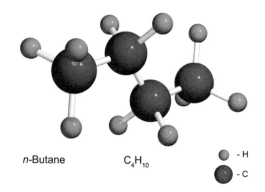

n-Butane C_4H_{10}

• - H
• - C

- Alkanes form a **homologous series**. Two consecutive compounds differ by CH_2.
- The general formula of an alkane is C_nH_{2n+2}.
- A **structural formula** can be abbreviated. For example, butane can be written $CH_3CH_2CH_2CH_3$.
- Butane is an example of a straight chain alkane. Each C atom is bonded to no more than 2 C atoms.
- If a C atom is bonded to 3 or 4 C atoms, the compound is a branched chain alkane.

This can be written $(CH_3)_3CH$.

ISBN: 9780170355544 PHOTOCOPYING OF THIS PAGE IS RESTRICTED UNDER LAW.

Name	Molecular Formula	Straight Chain Structural Formula	Molar Mass (g mol^{-1})	Boiling Point (° C)	Uses
methane	CH_4	H–C–H (with H above and below)	16	–162	major part of natural gas
ethane	C_2H_6	H–C–C–H	30	–88.5	minor part of natural gas
propane	C_3H_8	H–C–C–C–H	44	–42	component of LPG
butane	C_4H_{10}	H–C–C–C–C–H	58	0	component of LPG
pentane	C_5H_{12}	H–C–C–C–C–C–H	72	36	component of petrol
hexane	C_6H_{14}	H–C–C–C–C–C–C–H	86	69	component of petrol
heptane	C_7H_{16}	H–C–C–C–C–C–C–C–H	100	98	component of petrol
octane	C_8H_{18}	H–C–C–C–C–C–C–C–C–H	114	126	component of petrol

- To name branched chain alkanes, name the longest continuous carbon chain first. Side chains are named methyl (1 C atom), ethyl (2 C atoms), propyl (3 C atoms), and so on.
  If side chains are the same, the prefix 'di' is used if there are two, 'tri' for three, 'tetra' for four, and so on.

$$CH_3 - C - CH_2 - CH - CH_3 \quad \text{is 2,2,4-trimethylpentane}$$

(with CH_3 groups branching from the C atoms)

The octane rating of petrol depends on the presence of this compound. This compound has 8 carbon atoms and is a constitutional isomer of octane. See page 173 for constitutional isomers.

Properties of alkanes

- All are non polar and therefore insoluble in water.
- In the chart above, the first four, methane to butane, are gases at room temperature. Pentane to octane are liquids. The boiling point of straight chain alkanes increases with molar mass. A straight chain alkane has a higher boiling point than its branched chain isomer. This is because molecules can be in contact over a greater surface area. This is not necessarily the case with melting points, which depend on how the alkanes can pack together.
- Alkanes burn in air. If they burn completely, CO_2 and H_2O are formed.
- Alkanes undergo substitution reactions.
- For example, methane undergoes a series of substitution reactions with chlorine gas (Cl_2) in the presence of ultraviolet (UV) light.
  Note: the systematic name of each product is written under it with its common name in brackets. A systematic name describes the formula of a compound; it is based on a system that is accepted internationally.

methane → chloromethane (methyl chloride) → dichloromethane (methylene chloride) → trichloromethane (chloroform) → tetrachloromethane (carbon tetrachloride)

ISBN: 9780170355544   PHOTOCOPYING OF THIS PAGE IS RESTRICTED UNDER LAW.

Alkenes

- Alkenes are **hydrocarbons** in which at least one **carbon-carbon bond is a double bond**.
- The arrangement of atoms about the double bond is planar. The bond angles are 120°.
- The homologous series of alkenes have the general formula C_nH_{2n}.
- The first member of the series of alkenes is ethene, C_2H_4. Its structural formula can be written $H_2C=CH_2$. Its structure is given below.

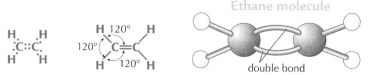

Ethane molecule

double bond

- Formulae of the first three alkenes are shown below.

Name	Molecular Formula	Structural Formula	Molar Mass (g mol⁻¹)	Boiling Point (° C)

ethene	C_2H_4	H H
		  | |
		C=C
		  | |
		H H
			28	–103

propene	C_3H_6	H H | | C=C–C–H | | | H H H	42	–47
butene	C_4H_8	H H H | | | C=C–C–C–H | | | | H H H H	56	–6.5

- Alkenes are named giving the double bond the lowest possible number. For example, $CH_3CH(CH_3)CH_2CH=CHCH_3$ is 5-methylhex-2-ene.
- The double bond C=C gives the alkenes its characteristic reactions. The C=C group is its **functional group.**

Properties of alkenes

- All are **non polar** and therefore **insoluble in water**.
- The seven alkenes that will be discussed in this book (ethene to octene) are all gases at room temperature.
- Alkenes **burn in air**. If they burn completely, CO_2 and H_2O are formed.
- Alkenes undergo **addition reactions**. Molecules add to the double bond.

Bromination

- When ethene is shaken with orange bromine water (made by shaking a little liquid Br_2 with water), the bromine is decolorised.
- Bromine has added to ethene across the double bond.

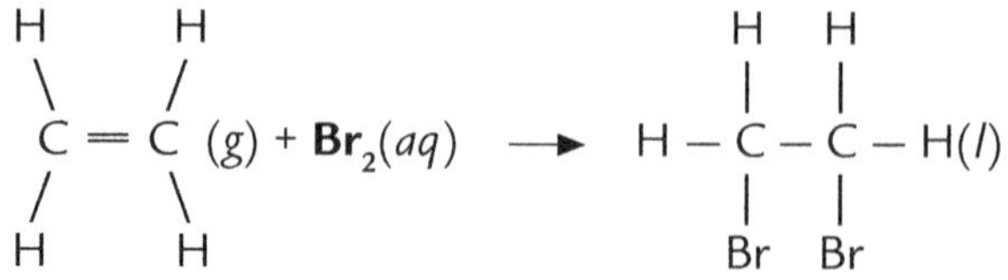

$$
\begin{array}{c}
H \quad\quad H \\
\backslash \quad\; / \\
C = C \;(g) + \mathbf{Br_2}(aq) \longrightarrow H - C - C - H(l) \\
/ \quad\quad \backslash \\
H \quad\quad H
\end{array}
$$

ethene   bromine   1,2-dibromoethane

ISBN: 9780170355544 PHOTOCOPYING OF THIS PAGE IS RESTRICTED UNDER LAW.

Hydrogenation

- When ethene and hydrogen are passed over a heated catalyst such as platinum (Pt) or nickel (Ni), hydrogen (H_2) is added to ethene.

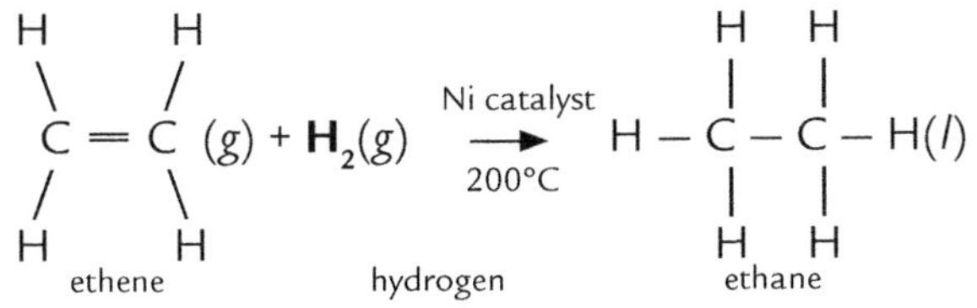

$$\underset{\text{ethene}}{\overset{H}{\underset{H}{>}}C = C\overset{H}{\underset{H}{<}}} (g) + \mathbf{H_2}(g) \xrightarrow[\text{200°C}]{\text{Ni catalyst}} \underset{\text{ethane}}{H - \overset{\overset{H}{|}}{\underset{\underset{H}{|}}{C}} - \overset{\overset{H}{|}}{\underset{\underset{H}{|}}{C}} - H(l)}$$

Addition of water (hydrolysis)

- When ethene is reacted with aqueous sulfuric acid and the mixture is then warmed, water adds to ethene. Sulfuric acid is a catalyst for the reaction.

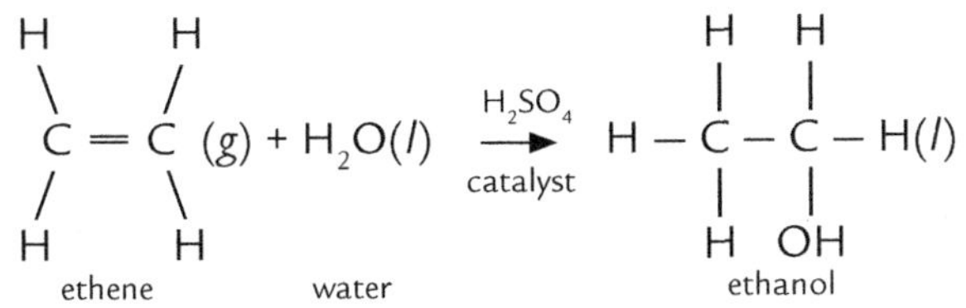

$$\underset{\text{ethene}}{\overset{H}{\underset{H}{>}}C = C\overset{H}{\underset{H}{<}}} (g) + H_2O(l) \xrightarrow[\text{catalyst}]{H_2SO_4} \underset{\text{ethanol}}{H - \overset{\overset{H}{|}}{\underset{\underset{H}{|}}{C}} - \overset{\overset{H}{|}}{\underset{\underset{OH}{|}}{C}} - H(l)}$$

- An important addition reaction is **addition polymerisation** described in Unit 26.
- Addition to unsymmetric alkenes:

  When an **unsymmetric alkene** such as propene reacts with molecules such as HCl, HBr, or BrCl, the major product is formed when the more electronegative atom bonds to the C atom with the least number of H's. This is called **Markovnikov addition**.

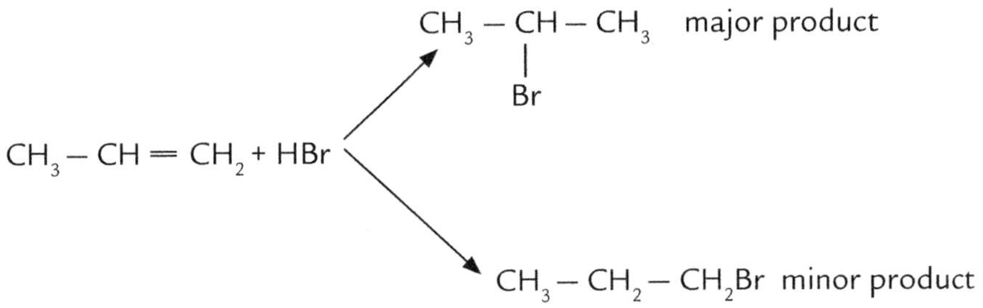

$$CH_3 - CH = CH_2 + HBr \nearrow \underset{\overset{|}{Br}}{CH_3 - CH - CH_3} \quad \text{major product}$$
$$\searrow CH_3 - CH_2 - CH_2Br \quad \text{minor product}$$

Note that H has bonded to the C with more H's; 'the rich get richer'.

When water adds to an unsymmetric alkene, –OH bonds to the C atom with the least number of H's. Again, H bonds to the C with more H's.

Example: Markovnikov addition

Problem:

Write the formula of the major product formed when HBr adds to the alkene molecule limonene drawn below.

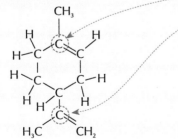

Steps:

1 Circle the carbon atoms in the C=C double bonds with the least H atoms.
2 In the diatomic molecule, circle the more electronegative atom.

   In HBr, the more electronegative atom is Br.

   H–(Br)

ISBN: 9780170355544   PHOTOCOPYING OF THIS PAGE IS RESTRICTED UNDER LAW.

3.5

3 The circled atoms go together.
The less electronegative atom of the diatomic molecule goes to the other carbon atom of the double bond.

In normal addition reactions to unsymmetric alkenes, about 96–97% of the major product forms. If the minor product is required, then the pathway of the reaction is changed to block the formation of the major product.

Addition reactions are typical of alkenes:

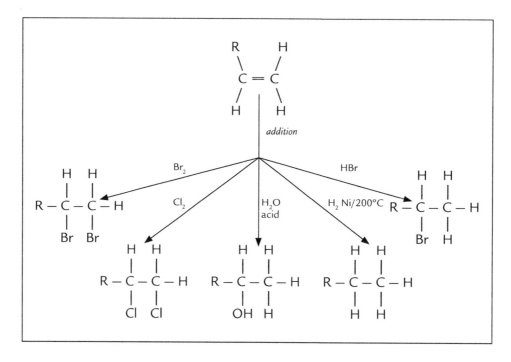

KEY POINTS SUMMARY

- Alkanes are hydrocarbons with only C–C single bonds.
- Alkenes are hydrocarbons with at least one C=C double bond.
- Alkanes undergo substitution reactions.
- Alkenes undergo addition reactions.
- In Markovnikov addition to unsymmetric alkenes, H of the added molecule will bond to the C that has more H (the rich get richer) or the negative part of the added molecule bonds to the C with fewer H atoms.

ISBN: 9780170355544 PHOTOCOPYING OF THIS PAGE IS RESTRICTED UNDER LAW.

1 Naming alkanes

Write the systematic names of the following alkanes.

a

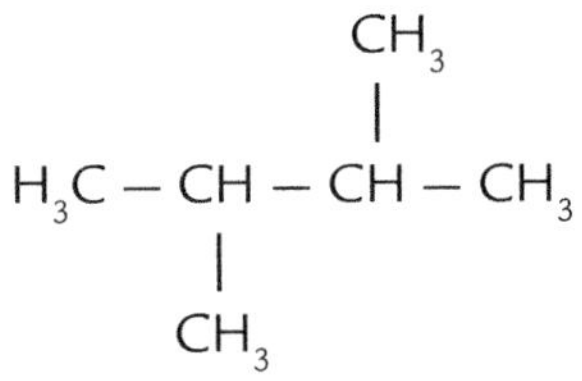

$$CH_3$$
$$|$$
$$H_3C - CH - CH - CH_3$$
$$|$$
$$CH_3$$

b

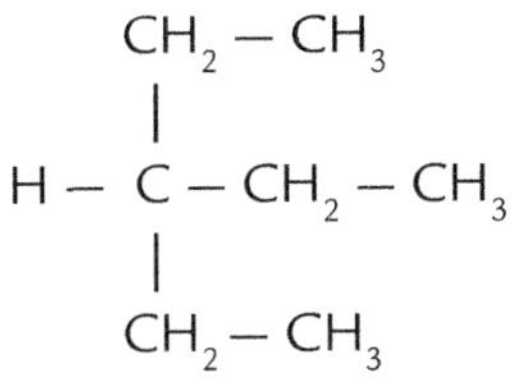

$$CH_2 - CH_3$$
$$|$$
$$H - C - CH_2 - CH_3$$
$$|$$
$$CH_2 - CH_3$$

c

$$CH_3 - CH_2 - CH_2 - CH - CH_3$$
$$|$$
$$CH_3 - CH_2 - CH - CH_3$$

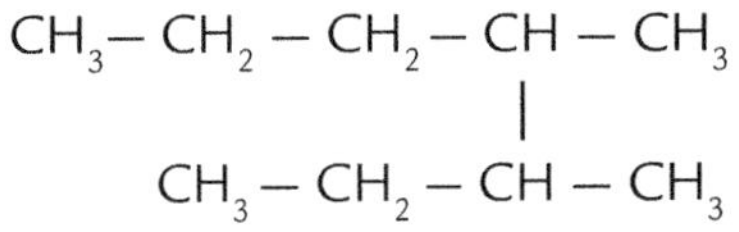

2 Naming alkenes

Write the systematic names of the following alkenes.

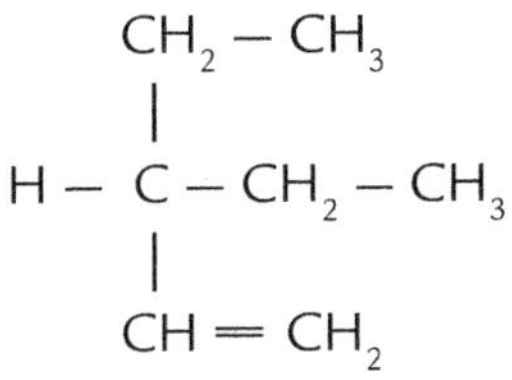

a

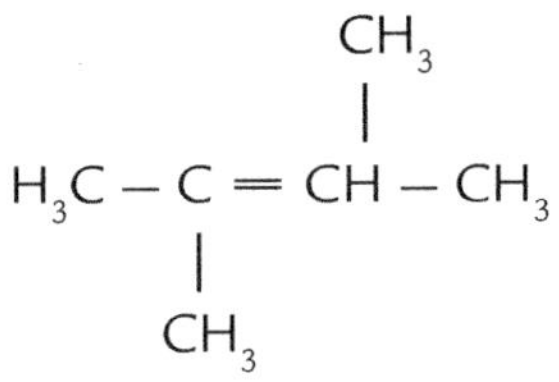

$$CH_3$$
$$|$$
$$H_3C - C = CH - CH_3$$
$$|$$
$$CH_3$$

b

$$CH_2 - CH_3$$
$$|$$
$$H - C - CH_2 - CH_3$$
$$|$$
$$CH = CH_2$$

c

$$CH_3 - CH = CH - CH - CH_3$$
$$|$$
$$CH_3 - CH_2 - CH - CH_3$$

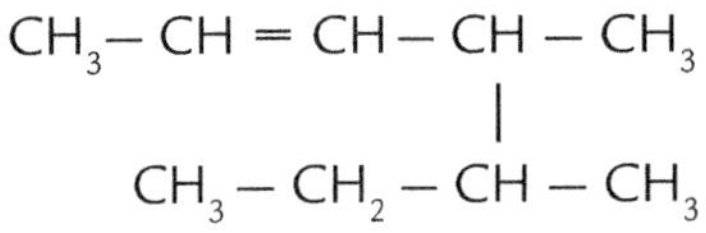

3 Naming alkanes and alkenes

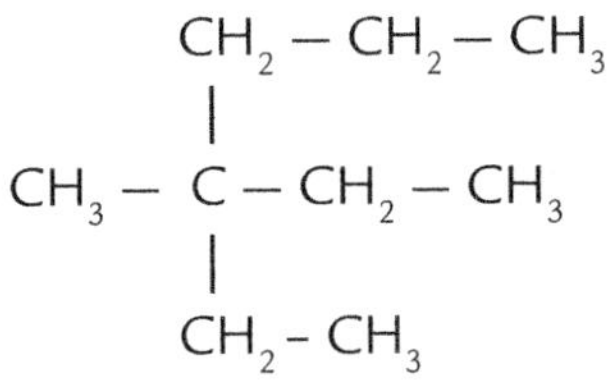

Write the systematic names of the following.

a

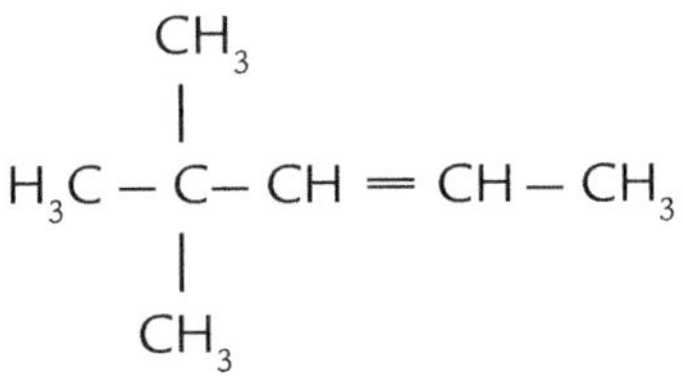

$$CH_3$$
$$|$$
$$H_3C - C - CH = CH - CH_3$$
$$|$$
$$CH_3$$

b

$$CH_2 - CH_2 - CH_3$$
$$|$$
$$CH_3 - C - CH_2 - CH_3$$
$$|$$
$$CH_2 - CH_3$$

c

$$CH_3 - CH_2 - CH_2 - CH - CH_3$$
$$|$$
$$CH_3 - CH_2 - CH - CH_3$$
$$|$$
$$CH_3$$

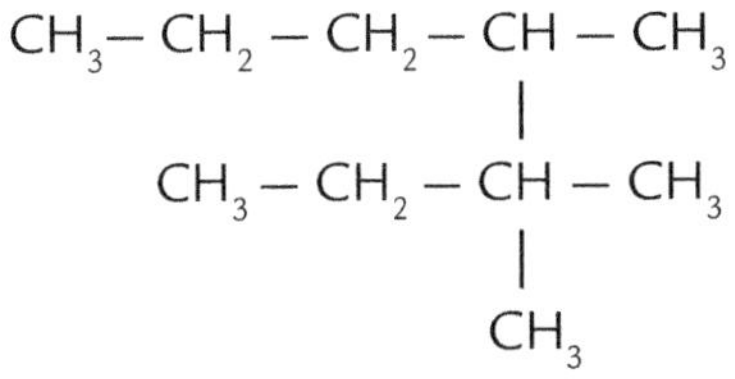

4 Reactions of alkanes and alkenes

For the following reactions, identify the reagent required and state the type of reaction occurring (whether it is 'substitution' or 'addition').

a $CH_3CH_2CH_3 \longrightarrow CH_3CHClCH_2Cl$

b $CH_3CH=CH_2 \longrightarrow CH_3CHClCH_2Cl$

c $H_3CCH=CHCH_3 \longrightarrow CH_3CH_2CHOHCH_3$

5 Understanding Markovnikov addition

a Draw the structural formula of the major product when butene undergoes addition reactions with the following.

  i HBr

  ii H_2O

  iii BrCl

  iv BrI

b Draw the structural formula of the major product when methyl propene undergoes addition reactions with the following.

  i HBr

  ii H_2O

  iii BrCl

  iv BrI

3.5

c   Draw the structural formula of the major product when vitamin D3 undergoes addition
    reactions with the following. The formula of vitamin D3 is:

    i   HBr
    ii H_2O
    iii BrCl
    iv BrI

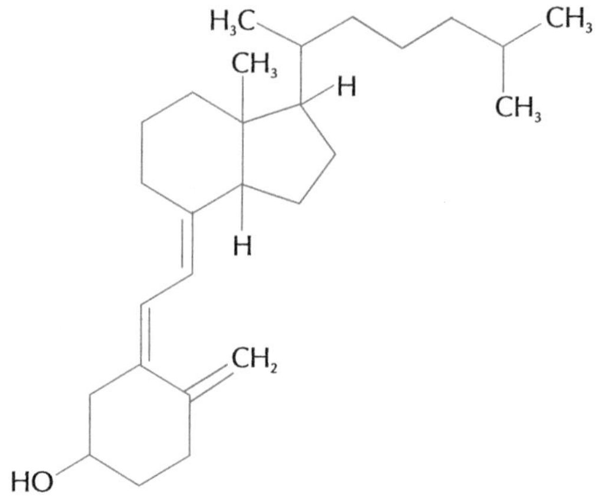

ISBN: 9780170355544   PHOTOCOPYING OF THIS PAGE IS RESTRICTED UNDER LAW.

Unit 22 | Haloalkanes

Learning Outcomes — on completing this unit you should be able to:

- write systematic names of haloalkanes
- classify haloalkanes as primary, secondary or tertiary
- describe substitution reactions of haloalkanes
- describe elimination reactions of haloalkanes
- explain why haloalkanes are useful starting compounds in chemical syntheses
- recognise the environmental impact of some haloalkanes.

Haloalkanes

- A haloalkane is an alkane in which one or more H atoms has been replaced by a halogen atom, F, Cl, Br or I.

  The general formula for a haloalkane is **R–X**, where **R** is an alkyl group and **X** is a halogen. **–X** is the functional group. The molecule shown opposite is 2-chloro-propane, $CH_3CHClCH_3$.

2-chloro-propane

- Haloalkanes are named as shown in the table below (the common names are in brackets).

Formula	Name	Uses
CH_3Cl	chloromethane (methyl chloride)	local anaesthetic — rapid evaporation soothes when sprayed onto injury
CH_2Cl_2	dichloromethane (methylene chloride)	common laboratory solvent, paint stripper
$CHCl_3$	trichloromethane (chloroform)	laboratory solvent, used to be used as anaesthetic but it is too toxic
CCl_4	tetrachloromethane (carbon tetrachloride)	was widely used as drycleaning fluid but it is too toxic, fire extinguishers
CH_3CH_2Cl	chloroethane	used for same purpose as chloromethane
$CH_3CH_2CH_2Cl$	chloropropane	No common uses, causes chronic health problems

- Haloalkanes are classified as primary, secondary or tertiary by looking at the **C** bonded to **X**.
- In a primary haloalkane, the **C** bonded to **X** is bonded to one carbon atom only. For example, 1-chlorobutane, C_4H_9Cl:

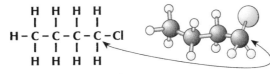

As the **C** bonded to Cl is bonded to only one carbon atom, 1-chlorobutane is a primary haloalkane.

3.5

- In a secondary haloalkane, the **C** bonded to **X** is bonded to two carbon atoms. For example, 2-chlorobutane, $C_2H_5CHClCH_3$:

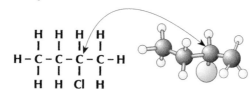

  As the **C** bonded to Cl is bonded to two carbon atoms, 2-chlorobutane is a secondary haloalkane.
- In a tertiary haloalkane, the **C** bonded to **X** is bonded to three carbon atoms. For example, 2-chloro-2-methylpropane, $CH_3CCl(CH_3)_2$:

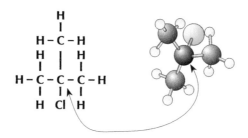

  As the **C** bonded to Cl is bonded to three carbon atoms, 2-chloro-2-methylpropane is a tertiary haloalkane.
- Haloalkanes are not produced by living organisms. **They are not soluble in water and are important non-polar solvents.** They are important because of their many uses. **Because the halogen atom can be readily substituted, they are used to synthesise new compounds.**
- Freon-12, CF_2Cl_2, is dichlorodifluoromethane (a chlorofluorocarbon, or CFC), a gas that is non-toxic, non-corrosive and unreactive. It has been used as a refrigerant in refrigerators and air-conditioners, and as a propellant in aerosol sprays. However, it is implicated in the destruction of parts of the ozone layer in the upper atmosphere.
- Ozone absorbs some of the harmful UV rays from the sun before they reach Earth's surface. It has been internationally agreed the Freon-12 will no longer be used.
- CF_3CH_2F, 1,1,1,2-tetrafluoroethane (a hydrofluorocarbon, or HFC), has replaced it as a refrigerant, and hydrocarbons are used as aerosol propellants.

Preparation of haloalkanes

- Haloalkanes can be prepared by the following methods.
  ### 1 Reaction of an alkane with a halogen
  An H of the alkane is **substituted** with the halogen. For example, the reaction of chlorine gas with methane gas to give chloromethane:

$$H-\overset{\overset{\displaystyle H}{|}}{\underset{\underset{\displaystyle H}{|}}{C}}-H + Cl_2 \xrightarrow{\text{UV}} H-\overset{\overset{\displaystyle H}{|}}{\underset{\underset{\displaystyle H}{|}}{C}}-Cl + HCl \qquad \text{Chloromethane}$$

  If sufficient chlorine is available, substitution can continue to produce dichloromethane, trichloromethane and tetrachloromethane as shown earlier in Unit 21.

ISBN: 9780170355544 PHOTOCOPYING OF THIS PAGE IS RESTRICTED UNDER LAW.

2 Reaction of an alkene with a halogen or hydrogen halide

Addition across the double bond occurs. For example, in the reaction between ethene and bromine 1,2-dibromoethane is produced:

$$CH_2=CH_2 \quad + \quad Br_2 \quad \longrightarrow \quad CH_2Br-CH_2Br$$

colourless gas   orange solution   colourless liquid

Another example is the reaction between propene and hydrogen bromide to yield mostly 2-bromopropane:

$$CH_3CH=CH_2 + HBr \longrightarrow CH_3CHBrCH_3$$

2-bromopropane, $CH_3CHBrCH_3$, is the major product of this Markovnikov addition reaction (see the example box on page 163), and 1-bromopropane, $CH_3CH_2CH_2Br$, is the minor product.

Reactions of haloalkanes

- The main reactions that haloalkanes undergo are either substitution or elimination reactions.

Substitution reactions

- The halogen $-X$ is readily substituted when haloalkanes (usually chloro- or bromo-alkanes) are reacted with various reagents.
- With an aqueous solution of a strong base, an alcohol is formed:

$$CH_3CH_2Br \xrightarrow{NaOH(aq)} CH_3CH_2OH$$

ethanol

- With concentrated ammonia, amines are formed:

$$CH_3CH_2Br \xrightarrow{NH_3(aq)} CH_3CH_2NH_2 \text{ or } CH_3CH_2NHCH_2CH_3 \text{ or } (CH_3CH_2)_3N$$

ethylamine   diethylamine   triethylamine

The actual amine product depends on the relative amounts of alkyl halide and ammonia.

Elimination reactions

- Only **secondary** and **tertiary** haloalkanes undergo elimination reactions.
- On reaction with a solution of KOH in ethanol, HX is eliminated and an alkene is formed. For example, in the reaction between 2-bromopropane and potassium hydroxide, propene is formed:

$$CH_3CHBrCH_3(l) + KOH(ethanol) + CH_3CH=CH_2(g) + KBr(ethanol) + H_2O(ethanol)$$

Some 2-propanol, $CH_3CHOHCH_3$, is also formed.
- If two possible products can form, the major product is the one formed when the C atom with the fewer hydrogens loses one more.

2-butene major product

1-butene minor product

- In this reaction, HBr Is eliminated. H can come from C1 or C3. C1 has three H atoms bonded to it, C3 has only two. So C3 tends to lose the H.

3.5

- Reactions typical of haloalkanes are shown below.

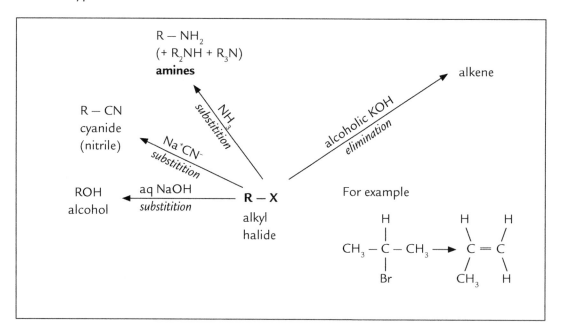

KEY POINTS SUMMARY

- **Haloalkanes** have the general formula **R — X.**
- They are useful as solvents and in the synthesis of other compounds. However, many are toxic or detrimental to the environment.
- They are prepared by **substitution reactions** of alkanes or by **addition** to alkenes.
- All haloalkanes undergo **substitution reactions**. The halogen is substituted for another function group.
- Secondary and tertiary haloalkanes also undergo **elimination reactions**. HX is eliminated in the presence of KOH.

ASSESSMENT ACTIVITIES

1 Matching terms with descriptions

a	used as a refrigerant	haloalkane
b	used to be used as an anaesthetic	primary haloalkane
c	compound made up of carbon, hydrogen and halogen atoms	addition reaction
d	reaction in which two molecules form one molecule	substitution reaction
e	was used as a drycleaner until its toxic properties became apparent	elimination reaction
f	reaction in which an atom or group of atoms is removed from a molecule	freon
g	reaction in which an atom or group of atoms in a molecule is replaced by another	chloroform
h	molecule in which a carbon atom bonded to a halogen is bonded to one carbon atom only	tetrachloromethane

ISBN: 9780170355544   PHOTOCOPYING OF THIS PAGE IS RESTRICTED UNDER LAW.

2 Naming haloalkanes

Write the systematic names of the following haloalkanes.

a

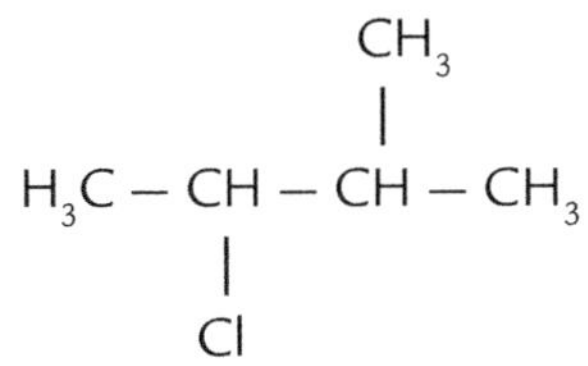

$$CH_3$$
$$|$$
$$H_3C - CH - CH - CH_3$$
$$|$$
$$Cl$$

b

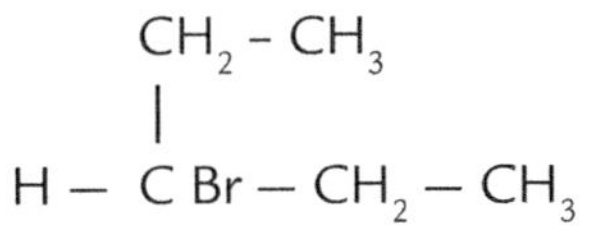

$$CH_2 - CH_3$$
$$|$$
$$H - C\,Br - CH_2 - CH_3$$

c

$$CH_3 - CH_2 - CH_2 - CH - CH_3$$
$$|$$
$$CH_2Cl - CH_2 - CH - CH_3$$

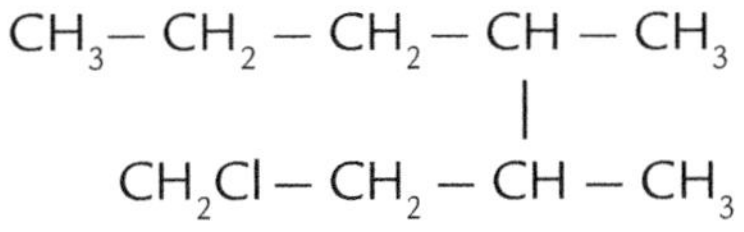

3 Primary, secondary and tertiary alkanes

Classify the following haloalkanes as primary, secondary or tertiary.

a bromopentane

b 2-bromopentane

c 3-bromopentane

d 2-bromo-2-methylpentane

e bromocyclohexane

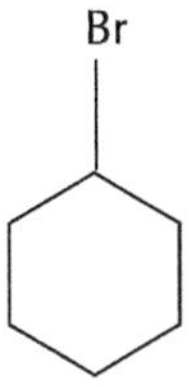

f 1-methylbromocyclohexane

g 1-bromo-2-cyclohexyl-propane

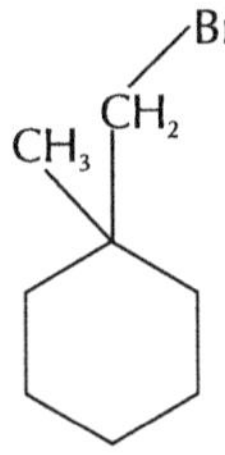

4 Substitution reactions of haloalkanes

Balance the following equations for these gas phase reactions.

a

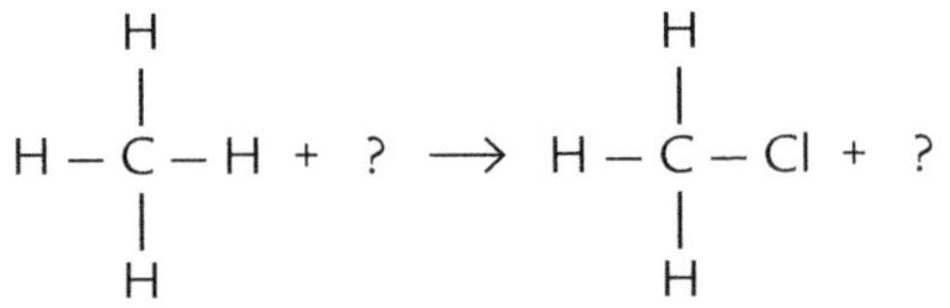

$$H$$
$$|$$
$$H - C - H + ? \longrightarrow H - C - Cl + ?$$
$$|\qquad\qquad\qquad\qquad |$$
$$H\qquad\qquad\qquad\qquad H$$

b

$$H\qquad\qquad\qquad\qquad H$$
$$|\qquad\qquad\qquad\qquad |$$
$$H - C - H + ? \longrightarrow Cl - C - Cl + ?$$
$$|\qquad\qquad\qquad\qquad |$$
$$H\qquad\qquad\qquad\qquad Cl$$

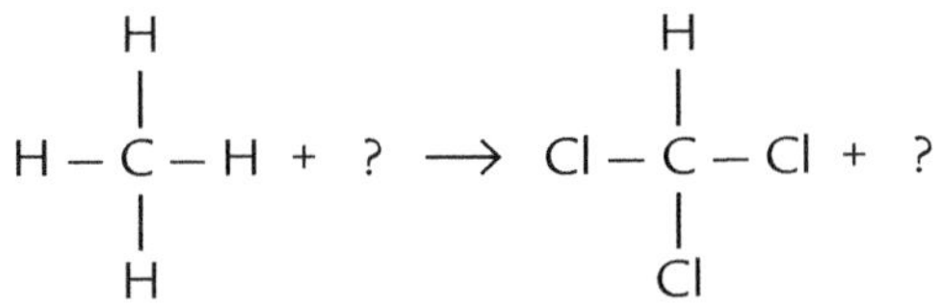

5 Substitution reactions of haloalkanes

Draw the major product of the following reactions.

a

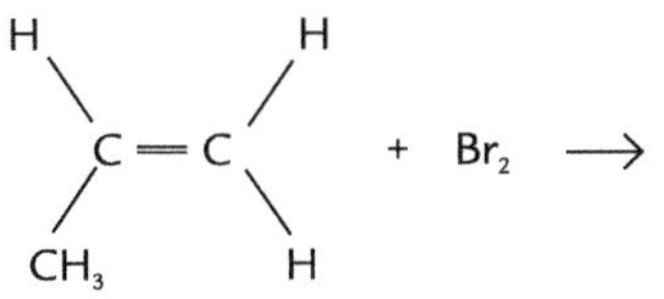

$$H\qquad\qquad H$$
$$\diagdown\qquad\diagup$$
$$C = C \qquad + \ Br_2 \ \longrightarrow$$
$$\diagup\qquad\diagdown$$
$$CH_3\qquad\quad H$$

b

$$H\qquad\qquad H$$
$$\diagdown\qquad\diagup$$
$$C = C \qquad + \ HBr \ \longrightarrow$$
$$\diagup\qquad\diagdown$$
$$CH_3\qquad\quad H$$

c

$$+ \ HBr \ \longrightarrow$$

3.5

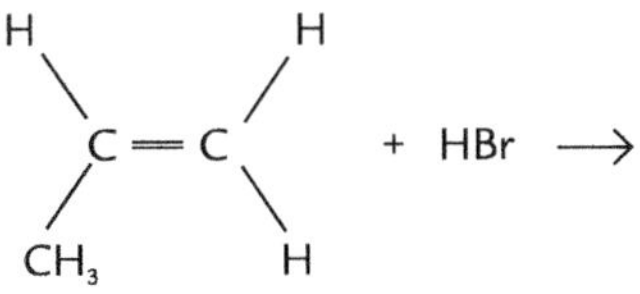

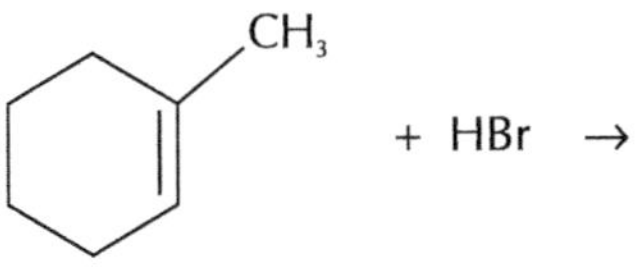

6 Minor products of addition reactions
Draw the minor products for the each of the reactions in question **4** above.

7 Haloalkanes and alcohols
Primary haloalkanes may form alcohols on reaction with NaOH or KOH. What physical property can differentiate between a haloalkane and an alcohol?

8 Reactions of haloalkanes
Name the products of the following reactions. If there are major and minor products, name both and state which is the major product.

a $CH_3CH_2CH_2CH_2Br + OH^-(aq) \longrightarrow$

b $CH_3CH_2CHBrCH_3 + KOH(ethanol) \longrightarrow$

c 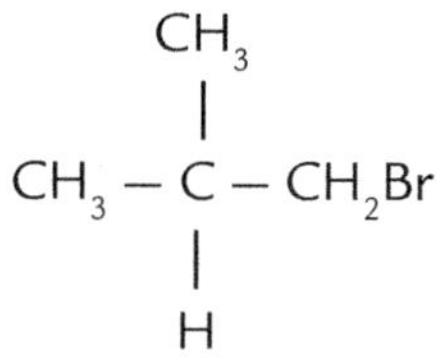
$$CH_3 - \underset{\underset{H}{|}}{\overset{\overset{CH_3}{|}}{C}} - CH_2Br \quad + \quad OH^-(aq) \longrightarrow$$

d 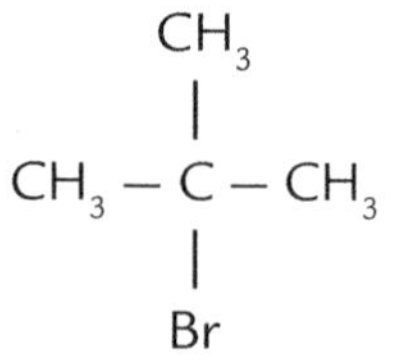
$$CH_3 - \underset{\underset{Br}{|}}{\overset{\overset{CH_3}{|}}{C}} - CH_3 \quad + \quad KOH(ethanol) \longrightarrow$$

e
$$CH_3 - CH - CHBr - CH_3 \quad + \quad KOH(ethanol) \longrightarrow$$
with CH_3 above the CH

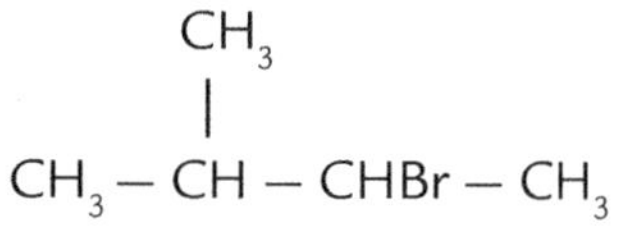

f 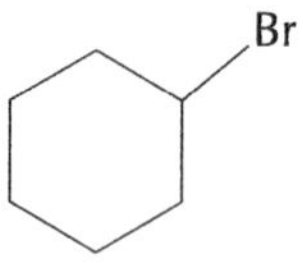
cyclohexane with Br $\quad + \quad KOH(ethanol) \longrightarrow$

g 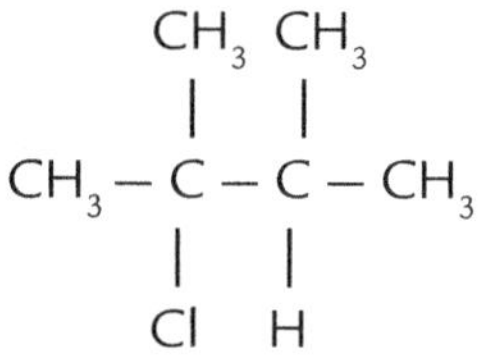
$$CH_3 - \underset{\underset{Cl}{|}}{\overset{\overset{CH_3}{|}}{C}} - \underset{\underset{H}{|}}{\overset{\overset{CH_3}{|}}{C}} - CH_3 \quad + \quad KOH(ethanol) \longrightarrow$$

9 Thinking critically
Read the following text:

CFCs, or chlorofluorocarbons, are fairly inert and unreactive molecules. In the ozone layer, high-energy UV light causes these molecules to split. One of the products is a fluorine atom that is very reactive and destroys ozone. This results in a hole in the ozone layer.

However, due to high-energy sparks in car engines that cause oxygen to form ozone, all major cities in the world have an excess of ozone (which is toxic).

Photocopiers also cause oxygen to form ozone.

Why does the ozone formed by car engines and photocopiers not react with CFCs? If it did, would it reduce the ozone pollution and remove CFCs?

Ozone research being conducted on the ground.

ISBN: 9780170355544   PHOTOCOPYING OF THIS PAGE IS RESTRICTED UNDER LAW.

Unit 23 | Isomers

Learning Outcomes — on completing this unit you should be able to:

- distinguish between constitutional isomers and stereoisomers
- recognise compounds that have stereoisomers which are cis and trans isomers
- recognise compounds that have stereoisomers which are enantiomers
- draw and name cis/trans isomers and enantiomers
- state how isomers differ in physical, chemical and biological properties.

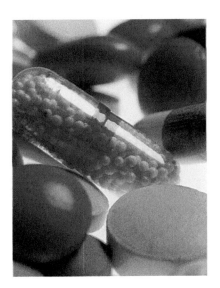

Constitutional isomers

- Isomers of a compound have the same molecular formula, for example a number of compounds have the formula C_4H_9Cl.
- In **constitutional isomers**, the atoms are bonded in a different sequence. There are four **constitutional** isomers with the molecular formula C_4H_9Cl:

$$
\begin{array}{cccc}
& H & H & H & H \\
& | & | & | & | \\
H- & C- & C- & C- & C-H \\
& | & | & | & | \\
& H & H & H & Cl
\end{array}
$$

$(CH_3)CH_2CH_2CH_2Cl$

1-chlorobutane

$$
\begin{array}{cccc}
& H & H & H & H \\
& | & | & | & | \\
H- & C- & C- & C- & C-H \\
& | & | & | & | \\
& H & H & Cl & H
\end{array}
$$

$(CH_3)CH_2CHClCH_3$

2-chlorobutane

$$
\begin{array}{ccc}
& H & H & Cl \\
& | & | & | \\
H- & C- & C- & C-H \\
& | & | & | \\
& H & CH_3 & H
\end{array}
$$

$CH_3CH(CH_3)CH_2Cl$

2-methyl-1-chloropropane

$$
\begin{array}{ccc}
& H & Cl & H \\
& | & | & | \\
H- & C- & C- & C-H \\
& | & | & | \\
& H & CH_3 & H
\end{array}
$$

$CH_3CCl(CH_3)CH_3$

2-methyl-2-chloropropane

- **Constitutional** isomers have **different structural formulae**.

Stereoisomers

- In stereoisomers, the atoms are bonded in the same sequence but are arranged differently in space. **Stereoisomers have the same structural formulae.**
- Stereoisomers occur in the following situations.
  1. When alkenes have bulky groups bonded to both carbon atoms of the double bond. For example, in 2-butene $CH_3CH = CHCH_3$, there is a CH_3- bonded to each carbon of the $C=C$ bond. Also, each carbon atom of the double bond has different groups bonded to it, a CH_3- and an $-H$. The result is the occurrence of geometric isomers. Geometric isomers of the same compound are termed **cis** and **trans** isomers.
  2. When a carbon atom has four different groups bonded to it. For example, in 2-butanol $CH_3CH_2CHOHCH_3$, the **C** atom has a CH_3CH_2-, an $-H$, an $-OH$ and a CH_3- bonded to it. This results in enantiomers, which are also called optical isomers.

ISBN: 9780170355544
PHOTOCOPYING OF THIS PAGE IS RESTRICTED UNDER LAW.

3.5

Cis-trans (cis/trans) isomers or geometric isomers

- These stereoisomers occur in compounds that have C=C double bonds, and **each carbon atom** of the double bond has **two different groups** bonded to it.
- They arise because the double bond is rigid. Rotation about a double bond is not possible. Such isomers of 2-butene are called **cis** 2-butene and **trans** 2-butene.

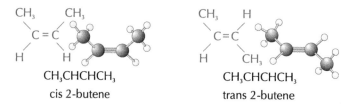

$CH_3CHCHCH_3$
cis 2-butene

$CH_3CHCHCH_3$
trans 2-butene

- In the cis compound, bulky groups are on the same side. In the trans compound, they are on opposite sides.
- The bond angles around the carbons of the double bond are approximately 120° and the arrangement is planar.
- Rotation about the double bond is **not possible.** So the cis isomer cannot change into the trans isomer.
- **Cis and trans 2-butene have different physical properties.**

Property	cis 2-butene	trans 2-butene
Melting point	−139°C	−106°C
Boiling point at 101.3 kPa	3.7°C	0.9°C
Density at 20°C	0.62 gmL⁻¹	0.60 gmL⁻¹

- In the diagrams below, **R** and **R'** represent bulky groups.

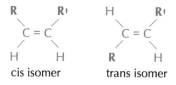

cis isomer   trans isomer

The two isomers may also differ in their chemical properties. For example, if the **R** and **R'** groups are large hydrocarbons, addition to the double bond may be hindered in the trans isomer.

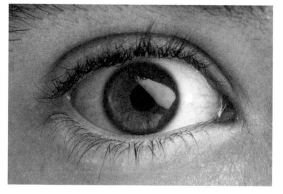

- Cis and trans isomers are found in nature where they serve a variety of purposes. Visual pigment in the retina of the eye contains a derivative of vitamin A called retinal. Retinal exists as cis and trans isomers. Light causes a reaction in which the cis isomer forms the trans isomer and the pigment fades. An enzyme catalyses a reaction which forms the trans isomer again when light is not present.
- Many molecules made by living organisms are cis isomers. Enzymes are surface catalysts and a cis isomer would adhere better to a surface.

ISBN: 9780170355544 PHOTOCOPYING OF THIS PAGE IS RESTRICTED UNDER LAW.

SPECIAL SKILL 1: SPOTTING CIS-TRANS ISOMERS

A compound will have cis-trans isomers if:

a there is a C=C double bond

b each carbon atom in the double bond has two different groups bonded to it.

If the compound does have cis-trans isomers, the **cis** isomer has bulky groups on the **same** side of the double bond.

Examples:

State whether these compounds will have cis-trans isomers.

1 C_4H_{10}

This compound is a saturated hydrocarbon; it has only single bonds. It does not have cis-trans isomers.

2 Propene

One carbon atom of the double bond is bonded to two H's. The compound does not have cis-trans isomers.

3 $C_3H_7CH=CHC_2H_5$

This carbon has a double bond. Each carbon atom of the C=C bond has different groups bonded to it. One has a C_3H_7 and an H, the other has a C_2H_5 and an H. This compound has cis-trans isomers.

$$\begin{array}{cc} C_3H_7 \quad C_2H_5 & C_3H_7 \quad H \\ \diagdown \diagup & \diagdown \diagup \\ C=C & C=C \\ \diagup \diagdown & \diagup \diagdown \\ H \quad H & H \quad C_2H_5 \end{array}$$

cis 3-heptene   trans 3-heptene

4 $(C_2H_5)_2C=CH(CH_3)$

One carbon has two C_2H_5 groups bonded to it. This compound does not have cis-trans isomers.

Enantiomers, or optical isomers

- These stereoisomers occur in compounds that have a carbon atom with **four different groups** bonded to it.
- The four groups are in a tetrahedral arrangement around the carbon atom, and one isomer is the mirror image of the other. 2-butanol has **enantiomers**.

+2-butanol
$C_2H_5CHOHCH_3$

−2-butanol
$C_2H_5CHOHCH_3$

| denotes a bond in the plane of the page.

↖ denotes a bond coming out of the page plane.

↗ denotes a bond going behind the page plane.

- The carbon atom **C** with four different groups bonded to it is said to be **asymmetric**. The molecule is said to be **chiral**. The two forms of the molecule are **enantiomers.** The enantiomers are **non-superimposable mirror images of each other**. This means that if the molecule on the right is rotated so that the OH groups coincide, then the CH_3– groups go into the page on one molecule and out of the page on the other. **The non-superimposable images are enantiomers.**
- Your right hand is a mirror image of your left hand but they are not superimposable. Your right hand does not fit perfectly (not at all, in fact!) on top of your left hand.

- The two isomers **differ in the way that they rotate the plane of polarised light**. Light waves travel in all planes. After passing through a polariser, waves travelling in all but one plane are filtered out. If this polarised light is passed through a solution containing one of the enantiomers, the plane of the light is rotated.

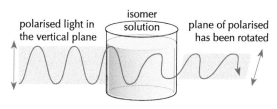

The compound is said to be optically active.

- If the plane is rotated to the right (clockwise), the enantiomer is given a positive (+) sign and said to be dextrorotatory, **D**.
- If the plane is rotated to the left (anticlockwise), the enantiomer is given a negative (−) sign and is said to be laevorotatory, **L**.
- Chiral molecules rotate the plane of polarised light. +2-butanol is a chiral molecule as is -2-butanol. **Two enantiomers will rotate plane polarised light equal amounts but in opposite directions.**

SPECIAL SKILL 2: SPOTTING ENANTIOMERS

A compound will have enantiomers if four different groups are bonded to one carbon atom. Determine whether the following compounds will have enantiomers.

1 $(C_3H_7)CHOH(CH_3)$

   In the compound $(C_3H_7)\textbf{C}HOH(CH_3)$, the **C** has four different groups (C_3H_7, H, OH and CH_3) bonded to it. This compound has enantiomers.

2
$$Cl-\underset{\underset{H}{|}}{\overset{\overset{Cl}{|}}{C}}-CH_3$$
   Neither of the two carbon atoms has four different groups bonded to it. One has two Cl, an H and a CH_3, the other has three H and a $CHCl_2$. This compound does not have enantiomers.

3

[structure of limonene]   which can be shown as   [skeletal structure of limonene]

The **C** marked in bold has four different groups bonded to it:

$$-\underset{\underset{CH_3}{|}}{C}=CH_2 \quad -CH_2CH=CCH_3 \quad -CH_2CH_2\overset{\overset{||}{}}{C}CH_3 \quad \text{and} \quad -H$$

This compound is limonene, an oil found in lemon and orange peel and also in caraway, dill and bergamot. The + isomer is extracted from mandarin peel. This compound does have enantiomers.

ISBN: 9780170355544
PHOTOCOPYING OF THIS PAGE IS RESTRICTED UNDER LAW.

- An equimolar mixture of the two enantiomers therefore has no effect on plane polarised light. Such a mixture is called a **racemic mixture**.
- Molecules made by living organisms tend to consist of one enantiomer only. Enzymes are stereospecific. They hold molecules in a certain way for them to react. Synthetic substances tend to be racemic mixtures.
- In cholesterol there are 8 asymmetric carbon atoms so there are $2^8 = 256$ possible enantiomers. Living organisms make only one of them. With other compounds, the enantiomers may differ from one organism to another, and sometimes racemic mixtures are found although this is very rare.
- The most striking effects of different enantiomers in recent times are effects of the drug thalidomide. In the 1960s, a racemic mixture of this drug was used in a sleeping pill, a tranquiliser and a cure for morning sickness. The + isomer has no ill effects. The – isomer causes severe birth defects when taken by pregnant women. Even if the + isomer only is taken, the body converts it to a racemic mixture.
- However, the drug has now been found to relieve skin inflammations and sores in leprosy and AIDS patients. So it has made a comeback, but it is not administered to pregnant women.
- Many common drugs have enantiomers. Although one isomer is usually more active than the other, most are sold as racemic mixtures. The chemical and physical properties of enantiomers are so similar, the cost of separating them is high. Therefore, unless one isomer has an undesired effect, they are not separated.

KEY POINTS SUMMARY

- **Constitutional isomers** are molecules with the same molecular formula but different structural formulae.
- **Stereoisomers** are molecules with the same structural formula, but their atoms are arranged differently in space. They can be **cis/trans** (geometric) isomers or **enantiomers** (optical isomers).
- **Cis/trans isomers** result from the rigidity of a double bond. They occur when each carbon atom of the double bond has two different groups bonded to it.
- Cis/trans isomers differ in their physical properties such as boiling point and sometimes in their chemical properties.
- **Enantiomers**, or **optical isomers**, occur when four different groups are bonded to one carbon atom.
- Enantiomers differ in the way that they rotate the plane of **polarised light**.
- Terms associated with enantiomers are:
  - **asymmetric carbon atom** — a carbon atom bonded to four different groups
  - **chiral** — molecule is able to exist as non-superimposable mirror images
  - **+ and – isomers** — isomers which rotate the plane of polarised light equal amounts but in opposite directions
  - **racemic mixture** — an equimolar mixture of two enantiomers.

3.5

ASSESSMENT ACTIVITIES

1 Matching terms with descriptions

a result from each carbon atom of a double bond being bonded to two different groups	constitutional isomers
b a carbon atom that is bonded to four different groups	structural isomers
c light waves travelling in one plane only	cis and trans isomers
d an equimolar mixture of two compounds, each the optical isomer of the other	enantiomers
e compounds with the same molecular formula but different structural formulae	asymmetric carbon
f able to exist as two non-superimposable mirror images	polarised light
g have the same structural formula, but different spatial arrangements of atoms	racemic mixture

2 Cis-trans isomers of 2-pentene
a Draw the cis-trans isomers of 2-pentene.
b Label them as cis and trans 2-pentene.

3 Other cis-trans isomers
Some of the alkenes in the following list have cis and trans isomers. Draw formulae for the cis and trans isomers where they exist.
a 1-butene
b propene
c 1,2-dichloroethene
d 2-methyl-2-butene
e 2-pentene
f fumaric acid, $CH(COOH)=CH(COOH)$, an important acid in the metabolism of carbohydrates
g oleic acid, $CH_3(CH_2)_7CH=CH(CH_2)_7COOH$, an unsaturated fatty acid in vegetable oils

4 Enantiomers of ibuprofen and naproxen
Ibuprofen is both an analgesic (pain reliever) and an anti-inflammatory (reduces swelling in joints) medicine that is readily available. Naproxen has the same effect as ibuprofen and is sold in different formulations.
a Copy the group of atoms that contains the carbon atom responsible for the optical activity of these compounds. Mark the asymmetric carbon atom with an asterisk.
b The isomer –ibuprofen is the active form while the + form is not. The − form acts in 12 minutes, while the racemic mixture takes three times as long. However, ibuprofen is sold as a racemic mixture. Offer an explanation for this.

Ibuprofen

Naproxen

ISBN: 9780170355544 PHOTOCOPYING OF THIS PAGE IS RESTRICTED UNDER LAW.

c While one enantiomer of naproxen is an effective drug, the other causes liver damage. How can a sample of naproxen be examined to show that only one enantiomer is present?

5 Identifying the asymmetric carbon atoms

Draw the part of the molecule containing the asymmetric carbon atom and mark the carbon atom with an asterisk for each of the following.

a Cysteine, an amino acid found in collagen and hair protein.

Cysteine

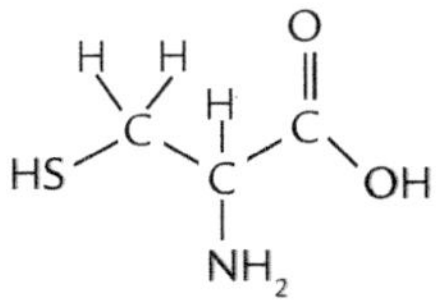

b Thalidomide, a tranquiliser. One enantiomer causes severe defects in developing foetuses.

Thalidomide

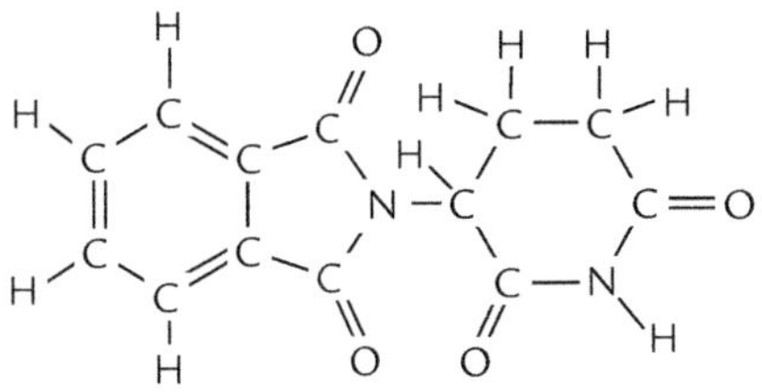

6 Drawing structural formulae and identifying asymmetric carbon atoms

For each of the following molecules, draw the structural formula and mark the asymmetric carbon atom with an asterisk.

a 3-methylhexane

b 2-pentanol

c 1-chloroethanol

3.5

Unit 24 | Alcohols, aldehydes and ketones

Learning Outcomes — on completing this unit you should be able to:

- demonstrate an understanding of properties of alcohols, aldehydes and ketones
- recognise these compounds by their functional groups
- name the compounds and write their formulae
- state the reactions that relate, or distinguish between, them.

Alcohols

The general formula for an alcohol is **R−OH**, where **R** is an alkyl group. The −OH is the functional group of an alcohol. The diagram shows a molecule of methanol, CH_3OH; the alkyl group is CH_3- (meth-). The suffix for naming alcohols is -anol.

Methanol

Primary, secondary and tertiary alcohols

- The functional group in alcohols is O–H bonded to a carbon atom.
- In a primary alcohol, the -OH is bonded to a C atom that is bonded to only one carbon atom. For example, in 1-butanol C_4H_9OH:

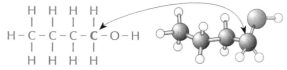

  the **C** bonded to the OH group **is bonded to only one carbon atom**.

- In a secondary alcohol, the −OH is bonded to a C atom that is bonded to two carbon atoms. For example, in 2-butanol C_4H_9OH:

  the **C** bonded to the OH group **is bonded to two carbon atoms**.

- In a tertiary alcohol, the –OH is bonded to a C atom that is bonded to three carbon atoms. For example, in 2-methyl propan-2-ol, C_4H_9OH:

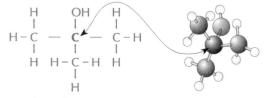

  the **C** bonded to the OH group **is bonded to three carbon atoms**.

- Alcohols are termed primary, secondary and tertiary in the same way that haloalkanes are.

ISBN: 9780170355544 PHOTOCOPYING OF THIS PAGE IS RESTRICTED UNDER LAW.

Diols and triols

- In the above examples, the alcohols have only one –OH group. Those with two –OH groups are called diols, and those with three, triols.
- An example of a diol is 1,2 dihydroxyethane (common name ethylene glycol or just glycol):

- An example of a triol is 1,2,3 trihydroxypropane (common name glycerol or glycerine):

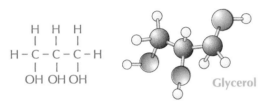

- Both glycol and glycerol are sticky liquids with a sweet taste. Glycol has been inadvertently added to wine to sweeten it. This is not legal as glycol is toxic. Glycol is used as antifreeze in car radiators. Glycerol is not toxic and is used as a sweetener in foods and medicines. Naturally occurring esters of glycerine and fatty acids are fats and oils. These are described in Unit 26.

Preparation of alcohols

1 Addition of water to an alkene

The reaction of an alkene with dilute sulfuric acid results in **water adding across the double bond to form an alcohol**.

For example, when sulfuric acid is added to ethene:

$$CH_2 = CH_2 + H_2O \xrightarrow{\text{sulfuric acid}} CH_3CH_2OH$$

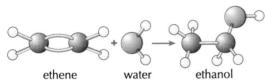

ethene   water   ethanol

- 2-propanol is formed when sulfuric acid is added to propene:

$$CH_3CH = CH_2 + H_2O \xrightarrow{\text{sulfuric acid}} CH_3CHOHCH_3$$
propene   2-propanol

A small amount of 1-propanol is also formed. The yield is approximately 97% 2-propanol and 3% 1-propanol.

- Propene is an unsymmetric alkene. Water is slightly ionised in solution into H^+ and OH. The negative –OH adds to the carbon with the least number of attached hydrogen atoms, for example the **C** in $CH_3CH=CH_2$.

2 Fermentation of sugars with enzymes

- Some alcohols can be produced when carbohydrates are fermented using specific enzymes.
- For example, glucose can be converted into ethanol. The overall reaction is:

$$C_6H_{12}O_6(aq) \xrightarrow{\text{enzymes}} 2C_2H_5OH(aq) + 2CO_2(g)$$

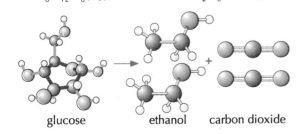

glucose   ethanol   carbon dioxide

3.5

3 Reduction of aldehydes or ketones with hydrogen

- H_2 and a catalyst (for example, Ni or Pt) is used. Other reducing reagents are zinc and ethanoic acid (Zn/CH_3COOH) or lithium aluminium hydride ($LiAlH_4$).
- Aldehydes and ketones have a $-C=O$ group, called a carbonyl group. When hydrogen adds to the $C=O$ double bond, an alcohol is formed.
- **Aldehydes** have the general formula:

$$RCHO \quad or \quad \begin{matrix} RC=O \\ | \\ H \end{matrix}$$

where **R** is an alkyl group. They are named in a similar way to alcohols but with the suffix - al. For example, CH_3CHO is ethanal.

- **Ketones** have the general formula:

$$RCHO \quad or \quad \begin{matrix} RCR' \\ \| \\ O \end{matrix}$$

Ethanal

where **R** and **R'** are both alkyl groups. They are named in a similar way to alcohols but with the suffix **- one**. For example, CH_3COCH_3 is propan**one** (common name acetone).

- **The reduction of an aldehyde gives a primary alcohol.**
  For example, reacting propanal with hydrogen gives 1- propanol:

$$CH_3CH_2C=O + H_2 \xrightarrow{Ni} CH_3CH_2CH_2OH$$
$$| \atop H \quad or \quad \xrightarrow{+2H}$$

Propanone

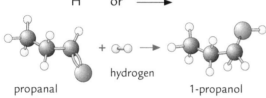

propanal   + hydrogen →   1-propanol

- **Reduction of a ketone gives a secondary alcohol.** For example, reacting propanal with hydrogen gives 1-propanol:

$$CH_3CCH_3 + H_2 \xrightarrow{Ni} CH_3CHOHCH_3$$
$$\| \atop O \quad or \quad \xrightarrow{+2H}$$

propanone   + hydrogen →   2- propanol

Reactions of alcohols

1 Combustion in oxygen and air

- Alcohols burn readily in oxygen and air.
- Complete combustion produces carbon dioxide and water. Incomplete combustion may produce a mixture of C, CO, CO_2 and H_2O.

2 Oxidation to other organic compounds

- Acidified potassium dichromate or potassium permanganate solutions are suitable oxidising agents.
- With the orange dichromate solution, a green solution of chromic ions forms. With purple potassium permanganate solution, a colourless (or very pale pink) solution of manganous ions forms.

- **Primary alcohols are oxidised to aldehydes and then to acids**. For example, propanol is oxidised to propanal and then to propanoic acid.

$$CH_3CH_2CH_2OH \xrightarrow{-2H} CH_3CH_2\underset{\underset{H}{|}}{C}=O \xrightarrow{+O} CH_3CH_2COOH$$

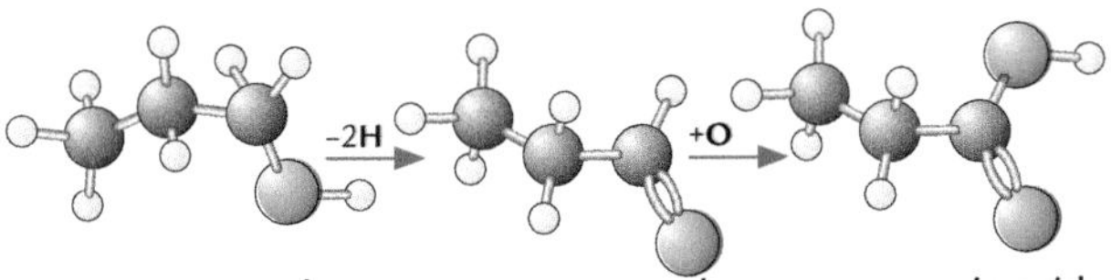

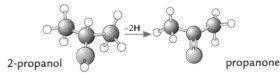

propanol   propanal   propanoic acid

In the first step, two H atoms are removed: one from the – OH group and one from the C bonded to the – OH.

In the second step, an O atom is inserted between the H and C in the functional group of the aldehyde.

- **Secondary alcohols are oxidised to ketones.** For example, 2-propanol is oxidised to propanone.

$$CH_3CHOHCH_3 \xrightarrow{-2H} CH_3COCH_3$$

2-propanol   propanone

In the above reaction, two H atoms are removed: one from the – OH group ($-C=O$) and one from the C bonded to – OH.

An O atom cannot be inserted between an H and a C of the carbonyl group as the C has no H bonded to it.

- **Tertiary alcohols are resistant to oxidation under these conditions.** For example, 2-methylpropan-2-ol.

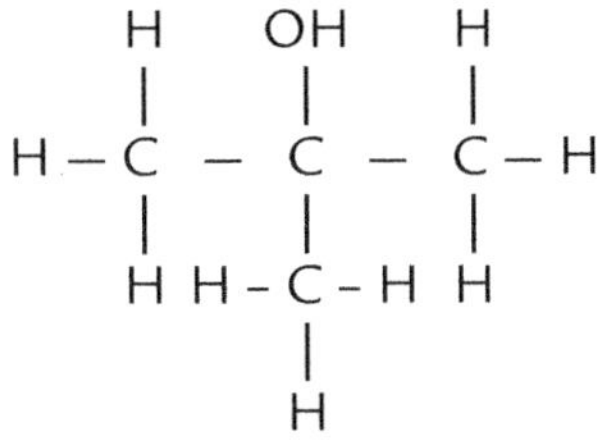

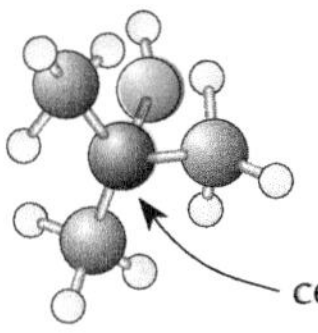

central C atom

Two H atoms (one from the – OH group and one from the C bonded to the – OH group) cannot be removed from central C atom, as there is no H on the central C atom.

3 Reaction with $ZnCl_2$ and HCl acid (Lucas reagent)

- In the reaction with zinc chloride and concentrated hydrochloric acid, the OH of an alcohol is replaced with Cl. **The soluble alcohol becomes an insoluble haloalkane.** The solution separates out into two layers.

$$RCH(OH)R' \xrightarrow{ZnCl_2/HCl} RCH(Cl)R'$$

For example, 2-propanol (a secondary alcohol) is converted into 2-chloropropane in the reaction:

2-propanol   2-chloropropane

3.5

- The reaction is rapid (immediate) with tertiary alcohols, moderately fast (takes a few minutes) with secondary alcohols, and so slow with primary alcohols that no reaction is seen to occur. **Primary, secondary and tertiary alcohols can be distinguished using the Lucas reaction. However, alcohols with more than four carbon atoms are insoluble. Two layers form whether there is or is not a reaction. This test is therefore very limited!**

4 Ester formation

- Alcohols react with carboxylic acids to form a series of compounds called esters.
- Esters have the general formula: **RCOOR'**

  where **R and R'** are alkyl groups. Esters names have two words; the second word ends in the suffix - oate. For example, $CH_3CH_2CH_2OCOCH_3$ is named propyl ethanoate. It is the ester of propanol and ethanoic acid.

- Any inorganic acid (for example, H_2SO_4, HNO_3 or HCl) will catalyse the reaction of an alcohol with a carboxylic acid. A water molecule is eliminated and the two molecules condense together.

  For example, propanol reacts with ethanoic acid to give propyl ethanoate in the reaction:

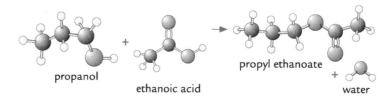

propanol + ethanoic acid → propyl ethanoate + water

Recognising chemical differences between primary, secondary and tertiary alcohols

- Primary and secondary alcohols react with acidified potassium dichromate causing a colour change from orange to green. With acidified permanganate, the colour change is from purple to colourless. Tertiary alcohols do not react.
- Primary alcohols do not form an insoluble layer with **zinc chloride and concentrated hydrochloric acid** (Lucas reagent). Secondary alcohols react after a few minutes; tertiary alcohols react immediately. This test is valid only with alcohols that have four or fewer C atoms.
- Reactions typical of primary alcohols are shown below.

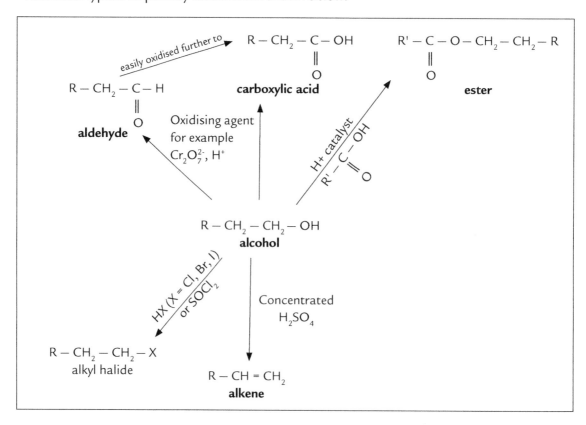

ISBN: 9780170355544 PHOTOCOPYING OF THIS PAGE IS RESTRICTED UNDER LAW.

Aldehydes and ketones

- Aldehydes and ketones are known as carbonyl compounds. They contain the carbonyl group $\diagup\!\!\!\!\diagdown C = O$ with the remainder of the molecule consisting of C and H atoms only.

- If **R** represents an alkyl group, the general formula for an aldehyde is **RCHO** or $R-\overset{}{\underset{O}{C}}-H$
  The functional group –CHO is circled. The suffix for naming aldehydes is -anal.

- The first three aldehydes are described in the following table.

Aldehyde	Formula	Structural formula	M (g mol⁻¹)
methanal (formaldehyde)	HCHO	$H - \underset{\parallel}{\underset{O}{C}} - H$	30
ethanal (acetaldehyde)	CH₃CHO	$H - \underset{\underset{H}{\vert}}{\overset{\overset{H}{\vert}}{C}} - \underset{\parallel}{\underset{O}{C}} - H$	44
propanal	C₂H₅CHO	$H - \underset{\underset{H}{\vert}}{\overset{\overset{H}{\vert}}{C}} - \underset{\underset{H}{\vert}}{\overset{\overset{H}{\vert}}{C}} - \underset{\parallel}{\underset{O}{C}} - H$	58

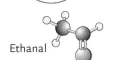

Ethanal

The perfume Chanel No. 5 contains a man-made aldehyde. Ernest Beaux, a Russian-French chemist, compounded the perfume for Coco Chanel. The perfume was launched on 5 May 1921. Today, more than one bottle is sold every 30 seconds.

- If **R** and **R'** represent alkyl groups, the general formula for a **ketone** is **RCOR'** or $R-\overset{}{\underset{O}{C}}-R'$
  The functional group –CO– is circled. The suffix for naming ketones is -anone.

Ketone	Formula	Structural formula	M (g mol⁻¹)
propanone (acetone)	CH₃COCH₃	$H - \underset{\underset{H}{\vert}}{\overset{\overset{H}{\vert}}{C}} - \underset{\parallel}{\underset{O}{C}} - \underset{\underset{H}{\vert}}{\overset{\overset{H}{\vert}}{C}} - H$	58
butanone	C₂H₅COCH₃	$H - \underset{\underset{H}{\vert}}{\overset{\overset{H}{\vert}}{C}} - \underset{\underset{H}{\vert}}{\overset{\overset{H}{\vert}}{C}} - \underset{\parallel}{\underset{O}{C}} - \underset{\underset{H}{\vert}}{\overset{\overset{H}{\vert}}{C}} - H$	72
2-pentanone	C₃H₇COCH₃	$H - \underset{\underset{H}{\vert}}{\overset{\overset{H}{\vert}}{C}} - \underset{\underset{H}{\vert}}{\overset{\overset{H}{\vert}}{C}} - \underset{\underset{H}{\vert}}{\overset{\overset{H}{\vert}}{C}} - \underset{\parallel}{\underset{O}{C}} - \underset{\underset{H}{\vert}}{\overset{\overset{H}{\vert}}{C}} - H$	86
3-pentanone	C₂H₅COC₂H₅	$H - \underset{\underset{H}{\vert}}{\overset{\overset{H}{\vert}}{C}} - \underset{\underset{H}{\vert}}{\overset{\overset{H}{\vert}}{C}} - \underset{\parallel}{\underset{O}{C}} - \underset{\underset{H}{\vert}}{\overset{\overset{H}{\vert}}{C}} - \underset{\underset{H}{\vert}}{\overset{\overset{H}{\vert}}{C}} - H$	86

Propanone

3.5

ISBN: 9780170355544 PHOTOCOPYING OF THIS PAGE IS RESTRICTED UNDER LAW.

- The first four ketones are described in the table on page 185.
- Each aldehyde with three or more carbon atoms has a ketone as a constitutional isomer. For example, the constitutional isomer of propanal (C_2H_5CHO) is propanone (CH_3COCH_3).

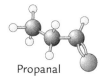

Propanal

Propanone

Preparation of aldehydes and ketones

Oxidation of alcohols

- An aldehyde is formed by the oxidation of a primary alcohol.
- A ketone is formed by the oxidation of a secondary alcohol.

Reactions of aldehydes and ketones

1 Oxidation
- **Aldehydes can be oxidised to carboxylic acids. Ketones are resistant to oxidation.** While aldehydes react with even mild oxidising agents, **ketones give no reaction**.
- Mild oxidising agents include:
  - **Tollens' reagent**
    This is a solution of **ammoniacal silver nitrate**. It is prepared by gradually adding ammonia solution to silver nitrate, so that the precipitate that forms just redissolves. On warming with an aldehyde, silver ions are reduced to silver metal. This forms a mirror on the inside of the test tube. The test is called the 'silver mirror test'. The aldehyde is oxidised to a carboxylic acid.

$$Ag(NH_3)_2^+(aq) + e \longrightarrow Ag(s) + 2NH_3(aq)$$
$$RCHO(aq) + 2OH^-(aq) \longrightarrow RCOOH(aq) + H_2O(l) + 2e$$

  - **Benedict's and Fehling's solutions**
    Both solutions contain aqueous copper (II) ions, which are reduced by aldehydes to the red insoluble compound copper (I) (cuprous) oxide, Cu_2O.

$$2Cu^{2+}(aq) + 2OH^-(aq) + 2e \longrightarrow Cu_2O(s) + H_2O(l)$$

- Aldehydes also react with the stronger oxidising agents, **acidified potassium dichromate** and **acidified potassium permanganate** solutions. Ketones do not. Aldehydes produce a colour change from orange to green with dichromate, and from purple to colourless with permanganate. Ketones give no colour change.

2 Reduction
- **Aldehydes are reduced to primary alcohols. Ketones are reduced to secondary alcohols.**
- Sodium borohydride ($NaBH_4$, also known as sodium tetrahydroborate) or zinc and acetic acid (Zn/CH_3COOH) are two common laboratory reducing agents used for this purpose. 2 H from the reagent reduce –C=O to –CH–OH.

Aldehydes and ketone chemical differences

Aldehydes:
- **form a silver mirror (Ag) with Tollens' reagent**, which is ammoniacal silver nitrate ($Ag(NH_3)_2^+$)
- **give a red precipitate (Cu_2O) with Benedict's solution**, which contains $Cu^{2+}(aq)$ ions
- **give a red precipitate (Cu_2O) with Fehling's solution**, which contains $Cu^{2+}(aq)$ ions also
- **decolourise** purple acidified **potassium permanganate**
- turn **orange** acidified **dichromate green**.

Ketones do none of the above.

Aldehydes are reduced by $NaBH_4$ to primary alcohols and ketones to secondary alcohols.

ISBN: 9780170355544 PHOTOCOPYING OF THIS PAGE IS RESTRICTED UNDER LAW.

KEY POINTS SUMMARY

- The general formula for an alcohol is **ROH**.
- In a **primary alcohol** the OH group is bonded to a C, which in turn is bonded to one C only. In a **secondary alcohol** the OH group is bonded to a C that is bonded to two C's. In a **tertiary alcohol** the OH group is bonded to a C that is bonded to three C's.
- Alcohols with two OH groups are called **diols**, those with three, **triols**.
- Alcohols can be formed by the **addition of water** to an alkene, the **fermentation** of sugar, or by the **reduction** of an aldehyde or ketone with hydrogen.
- Primary alcohols are **oxidised** to **aldehydes**, then to **carboxylic acids**. Secondary alcohols are oxidised to **ketones**. Tertiary alcohols are **not** oxidised.
- The reaction with **Lucas reagent** or with **acidified potassium dichromate** can be used to distinguish between primary, secondary and tertiary alcohols. Reaction with Lucas reagent is useful only for alcohols with four or fewer C atoms.
- Alcohols react with carboxylic acids to form **esters**, which have the general formula **RCOOR'**.
- Aldehydes and ketones are **carbonyl compounds**. The general formula for an aldehyde is **RCHO**, and that for a ketone is **RCOR'**.
- **Aldehydes can be oxidised** to carboxylic acids; **ketones are resistant** to oxidation.
- **Aldehydes** are **reduced** on reaction with H to **primary alcohols**; **ketones** are reduced to **secondary alcohols**.

ASSESSMENT ACTIVITIES

1 Matching terms with descriptions

a	compound with two OH groups on one molecule	alcohol
b	reagent made up of $ZnCl_2$ and concentrated HCl	primary alcohol
c	reagent containing ammoniacal silver nitrate	secondary alcohol
d	reagent containing $Cu^{2+}(aq)$	tertiary alcohol
e	test for aldehydes	diol
f	compound with the general formula R – CO – R'	triol
g	compound with the functional group – OH	Lucas reagent
h	C atom bonded to OH is bonded to three other carbon atoms	ester
i	C atom bonded to OH is bonded to only one other carbon atom	aldehyde
j	compound with the general formula R – CO – OR'	ketone
k	compound with the functional group – CHO	Tollens' reagent
l	compound with three OH groups on one molecule	Fehling's solution
m	C atom bonded to OH is bonded to only two other carbon atoms	silver mirror test

3.5

Understanding alcohols

2 **Write the systematic names of the following alcohols.**

a

$$CH_3$$
$$|$$
$$H_3C - C - CH - CH_2OH$$
$$|$$
$$CH_3$$

(with CH$_3$ also attached above)

b

$$CH_2 - CH_3$$
$$|$$
$$H - C - CH_2 - CH_3$$
$$|$$
$$OH$$

c

$$CH_3 - CH_2 - CH_2 - CH - CH_3$$
$$|$$
$$CH_3 - CH - CH - CH_3$$
$$|$$
$$OH$$

3 **Classify the following alcohols as primary, secondary or tertiary.**
  a octanol
  b 2-methylbutanol
  c dimethylpropanol
  d 2-pentanol
  e 3-pentanol
  f methylpropan-2-ol

4 **Describe what you would observe if each of the compounds in question 3 is reacted with acidified potassium permanganate solution.**

5 **An alcohol is prepared by reacting 2-methylpropene, $CH_3 - C = CH_2$ with dilute sulfuric acid.**
$$|$$
$$CH_3$$

  a Write the structural formula and name of the major product.
  b Describe the reaction (if any) of this product with acidified potassium permanganate. If there is a reaction, write the formula of the substance formed.
  c Describe the reaction (if any) of this product with Lucas reagent (zinc chloride and hydrochloric acid). If there is a reaction, write the formula of the substance formed.
  d Write the structural formula and name of the minor product of the original alkene and sulfuric acid.
  e Describe the reaction (if any) of the minor product with acidified potassium permanganate. If there is a reaction, write the formula of the substance formed.
  f Describe the reaction (if any) of the minor product with Lucas reagent (zinc chloride and hydrochloric acid). If there is a reaction, write the formula of the substance formed.

Understanding aldehydes and ketones

6 **Write the systematic names of the following aldehydes.**

a

$$CH_3$$
$$|$$
$$H - C - CH - CH_2 - CH_3$$
$$\|$$
$$O$$

b

$$CH_3$$
$$|$$
$$H - C - CH_3$$
$$|$$
$$O = C - H$$

c

$$O$$
$$\|$$
$$H - C - CH_2 - CH_2 - CH - CH_3$$
$$|$$
$$CH_3 - CH_2 - CH - CH_3$$

ISBN: 9780170355544   PHOTOCOPYING OF THIS PAGE IS RESTRICTED UNDER LAW.

7 **Write the systematic names of the following ketones.**

a

$H_3C - C - CH_2 - CH - CH_3$
with CH_3 branch above the CH and O double bond below the second carbon

b

$H_3C - C - CH_2 - CH_3$
with O double bond below the second carbon

c

$H_3C - CH_2 - CH_2 - CH_2$
and $CH_3 - CH_2 - C$ with O double bond below

8 **Sort this list of compounds into aldehydes and ketones. Name each compound.**

a CH_3COCH_3

b CH_3CH_2CHO

c $H - C - (CH_2)_2CH_3$
with O double bond below the C

d $CH_3(CH_2)_2CO(CH_2)_3CH_3$

e $CH_3(CH_2)_2CO(CH_2)_2CH_3$

9 **Various tests are used to distinguish between an aldehyde and a ketone. Describe the observed result when each compound in question 5 is reacted with:**

a Tollens' reagent

b Fehling's solution.

Relating alcohols, aldehydes and ketones

10 **A colourless liquid A reacted with acidified potassium dichromate to form a compound B with a vinegar-like odour. The same liquid A, when warmed with ethanoic acid in the presence of a small amount of sulfuric acid, formed a pleasant-smelling compound C.**

a Identify compounds A, B and C and write their structural formulae.

b Write balanced equations for the reactions.

c Describe any colour changes that you would observe.

11 **Compound X has the molecular formula C_4H_8O. It forms a silver mirror when warmed with Tollens' reagent. An unpleasant-smelling compound Y resulted from the reaction. Reduction of X with sodium borohydride gave a compound Z. Reaction of Y and Z in the presence of dilute sulfuric acid gave a pleasant-smelling compound, W.**

a To which four classes of compounds do X, Y, Z and W belong?

b Write the structural formulae of X, Y, Z and W.

c Name the compounds X, Y, Z and W.

12 **Compounds M and N both have the molecular formula $C_4H_{10}O$. Compound M decolourised acidified potassium permanganate solution forming compound Q. Compound Q gave no reaction with Fehling's solution. Compound N did not decolourise acidified potassium permanganate, but both M and N reacted with zinc chloride in hydrochloric acid forming an insoluble oily-looking layer.**

Write the structural formulae and names of compounds M, N and Q.

13 **Name the following compounds and describe tests to distinguish between them. Name the reagents required and describe the expected results.**

a $CH_3CH_2CH_2OH$ and $CH_3CHOHCH_3$

b $(CH_3)_3COH$ and CH_3CHO

c CH_3CH_2CHO and CH_3COCH_3

d $CH_3CH_2CH_2OH$ and $(CH_3)_3COH$

Unit 25 | Carboxylic acids and amines

Learning Outcomes — on completing this unit you should be able to:

- demonstrate an understanding of the properties of carboxylic acids (organic acids)
- demonstrate an understanding of the properties of the acid derivatives: acyl chlorides, esters and amides
- demonstrate an understanding of the properties of amines (organic bases)
- recognise these compounds by their functional groups
- name the compounds and write their formulae
- write equations for typical reactions of these compounds.

Butanoic acid gives some cheeses their characteristic smell.

Organic acids: carboxylic acids

- The general formula for a carboxylic acid is R—C—OH
  $\overset{\|}{O}$

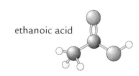
ethanoic acid

  where **R** is an alkyl group. The circled **–COOH** is the functional group of the **carboxylic acid**. The carboxylic acid shown above is ethanoic acid, CH_3COOH. The suffix for a carboxylic acid is -anoic acid.

- Carboxylic acids are also called organic acids.

Preparation of carboxylic acids

- Carboxylic acids can be prepared by a variety of methods.

  1 Oxidation of a primary alcohol or an aldehyde with a strong oxidising agent
  The oxidation of a primary alcohol or an aldehyde with a strong oxidising agent such as acidified potassium dichromate or acidified potassium permanganate solutions will yield the related carboxylic acid.
  For example:

$$CH_3CH_2CH_2OH \xrightarrow{-2H} CH_3CH_2C{=}O \xrightarrow{+O} CH_3CH_2COOH$$
$$\overset{|}{H}$$

propanol $\xrightarrow{-2H}$ propanal $\xrightarrow{+O}$ propanoic acid

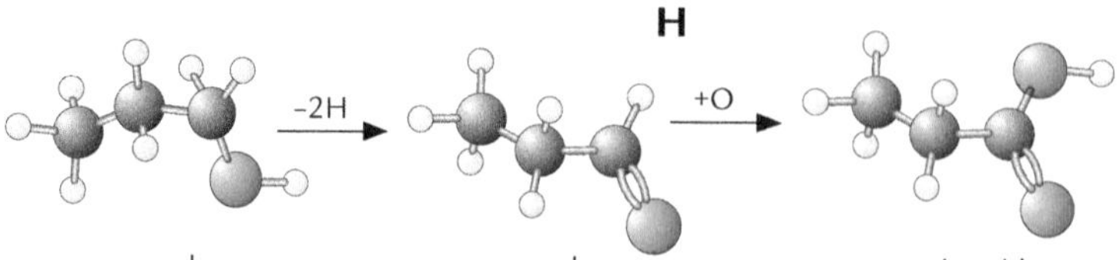

  2 Oxidation of an aldehyde with a mild oxidising agent
  - The oxidation of an aldehyde with a mild oxidising agent such as Tollens' reagent, Fehling's solution or Benedict's solution will yield the related carboxylic acid.

ISBN: 9780170355544 PHOTOCOPYING OF THIS PAGE IS RESTRICTED UNDER LAW.

Properties of carboxylic acids

- **Organic acids that are soluble in water are weak acids.** Those with one to three carbon atoms (methanoic to propanoic acid) are totally soluble in water, but the solubility drops off and octanoic acid is almost insoluble.
- Methanoic and ethanoic (acetic) acids have sour but not unpleasant odours. Acids from propanoic to decanoic acids have very unpleasant odours. Butanoic is in rancid butter, parmesan cheese and stale perspiration; octanoic acid gives goats their characteristic odour.

Fatty acids

- Esters of acids with between 3 and 20 carbon atoms are found in fats. Historically, these acids have therefore been termed fatty acids. All are 'straight chain' acids.
- Those with only **C–C** single bonds are termed saturated fatty acids. Those with one or more **C=C** double bonds in the chain are termed unsaturated and polyunsaturated fatty acids, respectively.
- A diet rich in unsaturated and polyunsaturated fatty acids is thought to prevent heart attacks. Canola oil and olive oil are rich in these.
- Fatty acids, which are liquid at room temperature, either have small molecules or are unsaturated. A cis arrangement around a double bond hinders the regular arrangement required for a solid.
- Cis-oleic acid in olive oil is

$$CH_3(CH_2)_7 \quad (CH_2)_7COOH$$
$$\backslash \qquad\qquad /$$
$$C = C$$
$$/ \qquad\qquad \backslash$$
$$H \qquad\qquad H$$

Acid derivatives

- If the −**OH** group of −**COOH** is replaced, various acid derivatives are formed.
- Replacement by −**Cl** gives a series of compounds called **acyl** (or acid) **chlorides**. These are highly reactive. They react rapidly with water, alcohols and amines.
- Replacement by −**OR** gives esters, which are less reactive than acids.
- Replacement by −**NH₂** gives a series of compounds called amides. These are the least reactive of the acid derivatives.

Acyl chlorides

- The functional group of an acyl chloride is −**COCl** or
$$\begin{array}{c} -C-Cl \\ \| \\ O \end{array}$$

ethanoyl chloride

The acyl chloride shown is ethanoyl chloride, CH_3COCl.
- The suffix for naming **acyl chorides** is **-anoyl chloride**.
- Acyl chlorides are prepared by reacting the acid with compounds that are high in chlorine content such as $SOCl_2$.

For example, when propanoic acid reacts with $SOCl_2$, propanoyl chloride is formed.

$$CH_3CH_2COOH \xrightarrow{\ SOCl_2\ } CH_3CH_2CO\,Cl$$

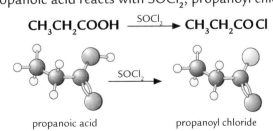

propanoic acid   propanoyl chloride

- The above reaction is an example of a substitution reaction. The −**OH** group has been substituted by −**Cl**.

3.5

- Acyl chlorides are **highly reactive**.
  1 They react rapidly with water forming acidic solutions. Moisture in the air causes them to fume.
  For example, propanoyl chloride reacts with water to form propanoic acid in the reaction.

$$CH_3CH_2COCl(l) + H_2O(l) \longrightarrow CH_3CH_2COOH(aq) + H^+(aq) + Cl^-(aq)$$

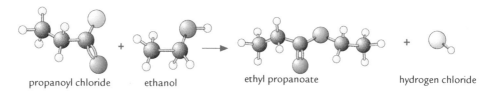

propanoyl chloride   propanoic acid

  2 They react rapidly with alcohols to form esters. No catalyst is needed.
  For example, propanoyl chloride reacts with ethanol to form ethyl propanoate in the reaction:

$$CH_3CH_2COCl(l) + CH_3CH_2OH(l) \longrightarrow CH_3CH_2COOCH_2CH_3(l) + HCl(g)$$

propanoyl chloride   ethanol   ethyl propanoate   hydrogen chloride

This is an example of a **condensation** reaction. **Two molecules condense together (join) with the elimination of one small molecule**, HCl in this case.

  3 They react rapidly with ammonia (in ethanol solution) and amines to form amides. The functional group of amides is − **CONH**$_2$ or −**C** − **NH**$_2$

$$\underset{O}{\overset{\|}{}}$$

The amide shown is propanamide, $CH_3CH_2CONH_2$.
For example, propanoyl chloride reacts with ammonia to form propanamide in the reaction:

$$CH_3CH_2COCl(l) + 2NH_3(aq) \longrightarrow CH_3CH_2CONH_2(aq) + NH_4^+(aq) + Cl^-(aq)$$

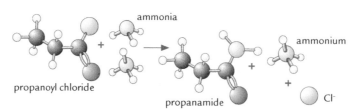

ammonia

ammonium

propanoyl chloride

propanamide

Cl$^-$

Propanoyl chloride reacts with methanamine to form N-methyl propanamide in the reaction below. 'N-methyl' means the methyl group, **CH**$_3$, is bonded to the N atom.

$$CH_3CH_2COCl(l) + 2CH_3NH_2(aq) \longrightarrow CH_3CH_2CONHCH_3(aq) + CH_3NH_3^+(aq) + Cl^-(aq)$$

- Amides are neutral; they are neither acidic nor basic.

Esters

- Esters are formed by the reaction of a carboxylic acid and an alcohol with an inorganic acid catalyst, or by the reaction of an acyl chloride and an alcohol with no catalyst required.
- Esters are responsible for many fruit flavours and floral perfumes. For example, octyl methanoate is orange flavour and ethyl heptanoate is apple flavour. Octyl methanoate would be formed from octanol and methanoic acid, and ethyl heptanoate from ethanol and heptanoic acid.

ISBN: 9780170355544 PHOTOCOPYING OF THIS PAGE IS RESTRICTED UNDER LAW.

- An important group of esters are the triglycerides. These are esters of 1,2,3-trihydroxypropane (glycerol) and fatty acids.

  For example, in most natural fats and oils, there are two or three different fatty acids esterified to glycerol.

- These esters are called **fats** if they are solid at room temperature, and **oils** if they are liquids. If the fatty acids in the ester are liquid at room temperature, then the triglyceride tends also to be liquid at room temperature.

- If just one of the –OH groups is esterified, a monoglyceride is formed. If two –OH groups are esterified, the molecule is a diglyceride.

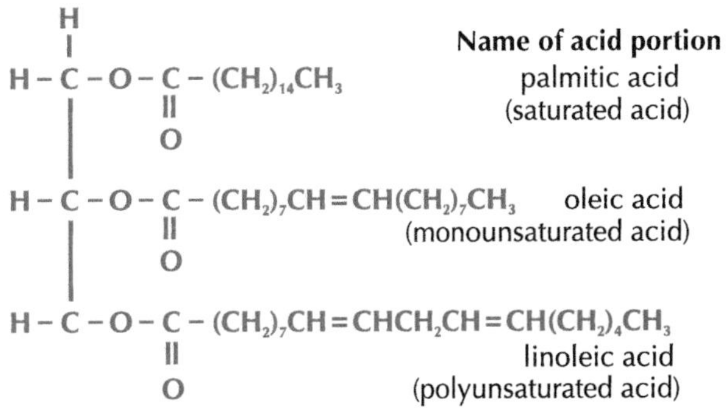

	Name of acid portion
$H-C-O-C-(CH_2)_{14}CH_3$ $\parallel$ O	palmitic acid (saturated acid)
$H-C-O-C-(CH_2)_7CH=CH(CH_2)_7CH_3$ $\parallel$ O	oleic acid (monounsaturated acid)
$H-C-O-C-(CH_2)_7CH=CHCH_2CH=CH(CH_2)_4CH_3$ $\parallel$ O	linoleic acid (polyunsaturated acid)

- Esters hydrolyse when warmed with aqueous sodium hydroxide to give the alcohol and the sodium salt of the acid from which it was derived.

$$R-O-\underset{\underset{O}{\parallel}}{C}-R' + NaOH \longrightarrow R-OH + R'-\underset{\underset{O}{\parallel}}{C}-O^-Na^+$$

For example, the triglyceride shown above would give glycerol shown below, and three sodium salts of fatty acids as shown below.

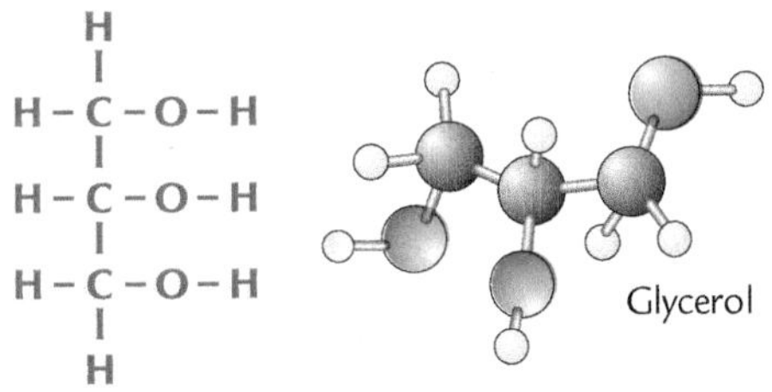

$$
\begin{array}{c}
H \\
| \\
H-C-O-H \\
| \\
H-C-O-H \\
| \\
H-C-O-H \\
| \\
H
\end{array}
$$

Glycerol

Sodium palmitate:

$Na^+\ {}^-O-\underset{\underset{O}{\parallel}}{C}-(CH_2)_{14}CH_3$

Sodium oleate:

$Na^+\ {}^-O-\underset{\underset{O}{\parallel}}{C}-(CH_2)_7CH=CH(CH_2)_7CH_3$

Sodium linoleate:

$Na^+\ {}^-O-\underset{\underset{O}{\parallel}}{C}-(CH_2)_7CH=CHCH_2CH=CH(CH_2)_4CH_3$

- Note that the sodium salts formed have an ionic portion and a hydrocarbon portion. The ionic portion is soluble in water. The hydrocarbon portion will mix with non-polar substances. Thus the non-polar substances can become suspended in water. Sodium salts of fatty acids are soaps. The hydrolysis of triglycerides to form soap and glycerol is called **saponification**.

Amides

- **Amides are the least reactive of the acid derivatives.** They need to be heated to boiling point with aqueous acid or base, often for several hours, before they hydrolyse. In the human body there are enzymes that catalyse the hydrolysis of amides, and hydrolysis occurs rapidly.
- Amides are formed by the reaction of an acyl chloride with ammonia or an amine (see section on acyl chlorides on page 192). The reaction of a carboxylic acid with ammonia or an amine tends to form a salt, that is, an acid-base reaction takes place. If ethanoic acid is reacted with ammonia, the salt ammonium ethanoate forms.

- Amides are acid derivatives.
- The functional group of an amide is $- \overset{\displaystyle}{\underset{\displaystyle \parallel O}{C}} - N -$

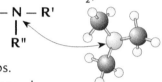
Propanamide

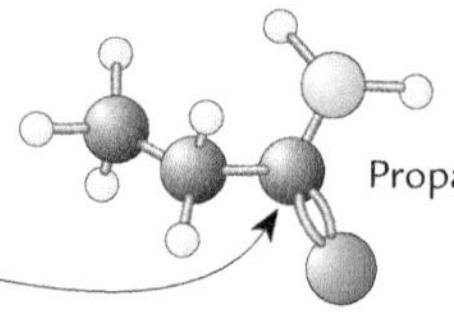

The amide shown above is propanamide, $CH_3CH_2CONH_2$.
- Some common medicines that have the amide functional group are paracetamol and penicillin.

Organic bases: amines

- Amines can be regarded as ammonia (NH_3) in which one or more H atoms have been replaced with alkyl groups. Those soluble in water undergo similar reactions to that of ammonia. Those that are volatile have fishy smells similar to that of ammonia.
- The general formula of a primary amine is **R−NH₂**, that of a secondary amine is **R−NH−R'**, while that of a tertiary amine is $\mathbf{R - \underset{\displaystyle R''}{N} - R'}$

where **R**, **R'** and **R"** are alkyl groups.
- Amines are described as primary, secondary or tertiary depending on the number **of carbon atoms bonded to N. Compare this with alcohols.**

Tertiary alcohol

$$CH_3 - \overset{\displaystyle CH_3}{\underset{\displaystyle OH}{C}} - CH_3$$

2-methyl-2-propanol

The C bonded to OH is bonded to three C atoms.

Primary amine

$$CH_3 - \overset{\displaystyle CH_3}{\underset{\displaystyle NH_2}{C}} - CH_3$$

2-methyl-2-aminopropane

The N is bonded to one carbon atom only.

Naming amines

- There are two methods that can be used to name amides.

Method A: IUPAC
1 Name the longest chain containing the amino group.
2 Number this chain to give the amino group the lowest possible number.
3 Name side chains and amino group with the appropriate number.
  For example, the molecule opposite would be named 2-amino-4-methyl pentane.

2-amino-4-methyl pentane

Method B: CA (Chemical Abstract name)
1 The suffix -amine is added to the largest alkyl group bonded to N.
2 Other alkyl groups bonded to N are added in front with the prefix N–. For example the molecule opposite is N-ethylpropanamine.

N-ethylpropanamine

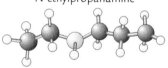

Method C: Common name
1 Name the alkyl groups bonded to the N atom in alphabetical order.
2 Add amine at the end.
  For example, the molecule in Method B could also be named ethylpropylamine.

Ethylpropylamine

- For a given amine, one method of naming is often easier to use than the other.

 ISBN: 9780170355544 PHOTOCOPYING OF THIS PAGE IS RESTRICTED UNDER LAW.

Properties of amines

- Properties of amines are similar to those of ammonia.
- Volatile amines are responsible for fishy odours. They are bases and their odours can be neutralised with acids such as vinegar and lemon juice.
- The properties of ammonia and amines are compared in the following table.

Ammonia	Amines
very soluble in water	most of the common amines are soluble in water
turns litmus blue	turn litmus blue
forms salts in reactions with acids	form salts in reactions with acid

- Amines are weak bases in aqueous solution. For example:

$$CH_3NH_2(aq) + H_2O(l) \rightleftharpoons CH_3NH_3^+(aq) + OH^-(aq)$$

methanamine　　　　　　　　methylammonium ion

Compare this with ammonia:

$$NH_3(aq) + H_2O(l) \rightleftharpoons NH_4^+(aq) + OH^-(aq)$$

ammonia　　　　　　　　ammonium ion

- Amines react with acids to form salts. Volatile amines undergo a similar reaction to ammonia vapour, which forms white clouds of solid ammonium chloride with hydrochloric acid fumes.

$$CH_3NH_2(g) + HCl(g) \longrightarrow CH_3NH_3^+Cl^-(s)$$

methanamine　　　　　　　　methylammonium chloride

Compare this with:

$$NH_3(g) + HCl(g) \longrightarrow NH_4^+Cl^-(s)$$

ammonia　　　　　　　　ammonium chloride

- Amines and ammonia react in similar ways with carboxylic acids.

KEY POINTS SUMMARY

- **Carboxylic acids** have the functional group **–COOH**. Those found in the esters that are fats and oils are called fatty acids. The more volatile acids generally have unpleasant odours.
- Carboxylic acids are said to be unsaturated if they have **C=C** double bonds.
- Acyl chlorides, esters and amides can be prepared from carboxylic acids. They are called **acid derivatives**.
- **Acyl chlorides** have the functional group **–COCl**. **They are very reactive and must be treated with caution.**
- **Esters** have the general formula **RCOOR'**. They have pleasant odours. Fruit flavours are often esters.
- **Esters of glycerol and fatty acids are called triglycerides.**
- **Saponification of triglycerides gives soap and glycerol.**
- **Amides** have the general formula **RCONH$_2$** (**R'** groups may be present instead of the **H** on **–NH$_2$**). They are the least reactive of the acid derivatives.
- **Amines are organic bases with the general formula RNH$_2$ (or RNR'R"). Their properties are similar to that of NH$_3$.**

ASSESSMENT ACTIVITIES

1 Matching terms with descriptions

a	formed from glycerol and three fatty acids		carboxylic acid
b	a carboxylic acid found as an ester in fats and oils		acyl chloride
c	reaction of an ester such as a triglyceride with sodium hydroxide		ester
d	long-chain carboxylic acid with a C=C double bond		amide
e	an organic base		fatty acid
f	contains the functional group – COOH		unsaturated fatty acid
g	contains the functional group – COCl		triglyceride
h	contains the functional group – COOR, where R is an alkyl group		saponification
i	contains the functional group – CONH$_2$		amine

Understanding carboxylic acids and acid derivatives

2 Write the systematic names of the following carboxylic acids.

a

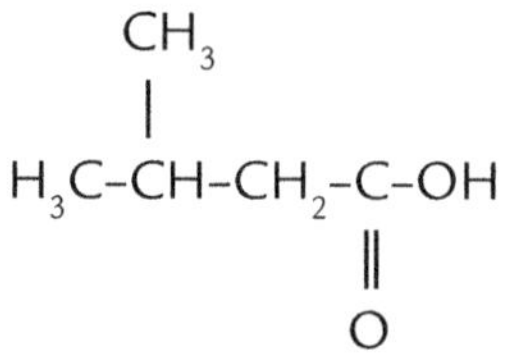

$$CH_3$$
$$|$$
$$H_3C-CH-CH_2-C-OH$$
$$\|$$
$$O$$

b

$$CH_2CH_3$$
$$|$$
$$H_3C- CH - C-OH$$
$$\|$$
$$O$$

c

$$O \qquad CH_2-CH_3$$
$$\| \qquad |$$
$$HO-C-CH_2-CH-CH_2-CH_3$$

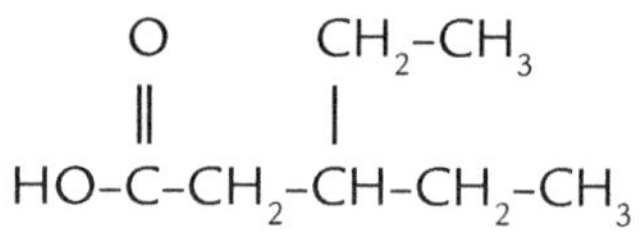

3 Write the systematic names of the following acyl chlorides.

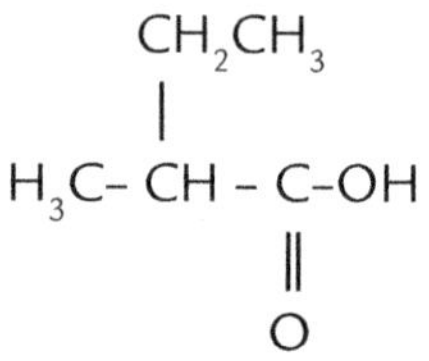

a

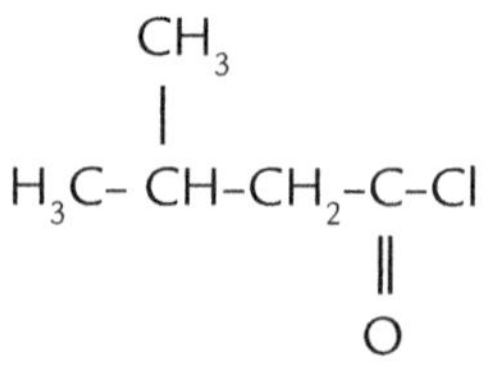

$$CH_3$$
$$|$$
$$H_3C- CH-CH_2-C-Cl$$
$$\|$$
$$O$$

b

$$CH_2CH_2CH_3$$
$$|$$
$$H_3C - C - CH_3$$
$$|$$
$$O=C-Cl$$

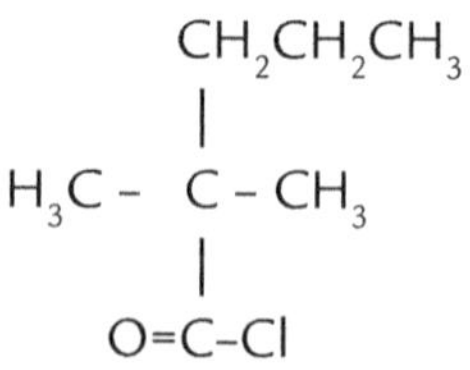

c

$$O$$
$$\|$$
$$Cl-C-CH_2-CH-CH_2-CH_3$$
$$|$$
$$CH_2-CH_2-CH_3$$

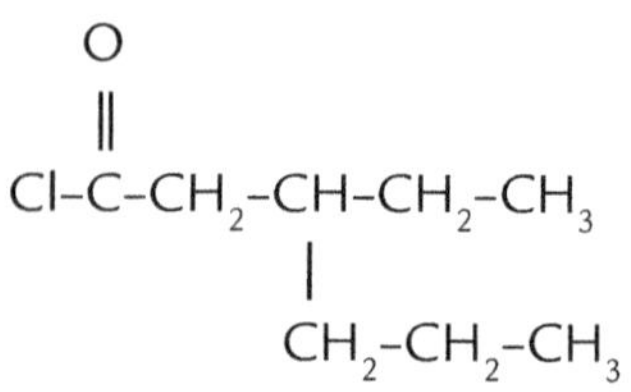

4 Write balanced equations for the following reactions, name the products and describe what you would observe.

  a Ethanoic acid fumes are mixed with ammonia fumes.

  b Ethanoyl chloride is reacted with ammonia.

  c Ethanoyl chloride is reacted with water.

  d Ethanoyl chloride is reacted with ethanol.

ISBN: 9780170355544 PHOTOCOPYING OF THIS PAGE IS RESTRICTED UNDER LAW.

Understanding esters

5 Write the systematic names of the following esters.

a

$H - C - O - CH_2 - CH_3$
$\quad \| $
$\quad O$

b

CH_3
$|$
CH_2
$|$
$O=C-OCH_3$

c

$\qquad\qquad O$
$\qquad\qquad \|$
$H_3C - CH_2 - CH_2 - C - O - CH_2 - CH_3$

6 Name the following esters. Give the names of the alcohol and i the acid and ii the salt that is formed when the compound is hydrolysed by i H_2SO_4 and ii NaOH.

a $CH_3OC - CH_3$
$\qquad\; \|$
$\qquad\; O$

b $CH_3CH_2CH_2O - C - H$
$\qquad\qquad\qquad \|$
$\qquad\qquad\qquad O$

c $CH_3CH_2C - OCH_2CH_3$
$\qquad\quad\; \|$
$\qquad\quad\; O$

7 Aspirin is acetyl salicylate, the ester of ethanoic (acetic) acid and salicylic acid. Aspirin is an analgesic (used to control pain).

Aspirin

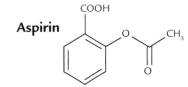

a Copy the diagram and circle the ester functional group.
b Write the formulae of the acid and alcohol formed when aspirin is hydrolysed.
c Name the functional group of salicylic acid that is esterified in aspirin.

8 Another readily available ester of salicylic acid is methyl salicylate. Methyl salicylate is in ointments used to ease the pain of strained muscles and rheumatism. Known also as 'oil of wintergreen', it has a characteristic odour.

Methyl salicylate

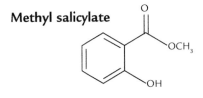

a Copy the diagram and circle the ester functional group.
b Write the formulae of the acid and alcohol formed when methyl salicylate is hydrolysed.
c Name the functional group of salicylic acid that is esterified in methyl salicylate.

Understanding amides

9 Write the systematic names of the following amides.

a

$H_2N - C - CH_2 - CH_2 - CH_3$
$\qquad\; \|$
$\qquad\; O$

b

CH_3
$|$
$N - CH_3$
$|$
$O=C - CH_3$

c

$\qquad\qquad\qquad O$
$\qquad\qquad\qquad \|$
$CH_3 - CH_2 - C - N - CH_3$
$\qquad\qquad\qquad\quad |$
$\qquad\qquad\qquad\; CH_2-CH_3$

10 Amides are the least reactive acid derivatives.
a Draw the group of atoms of the amide functional group.
b Write the formula and name of the amide formed by the reaction of ethanoyl chloride and ammonia.
c Write the formula and name of the amide formed by the reaction of ethanoyl chloride and ethanamine.

ISBN: 9780170355544   PHOTOCOPYING OF THIS PAGE IS RESTRICTED UNDER LAW.

3.5

11 **Paracetamol is an amide which is an analgesic and has the formula shown.**

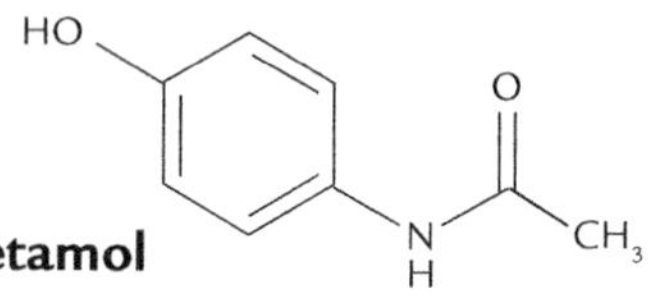

a Copy the diagram and circle the amide functional group.

b Write the formulae of the acid and amine formed when the molecule is hydrolysed.

c Is the amine formed on hydrolysis a primary, secondary or tertiary amine?

Paracetamol

Understanding amines

12 **Name the following amines.**

a

$$CH_3 - CH_2 - \underset{\underset{H}{|}}{\overset{\overset{CH_3}{|}}{N}} - H$$

b

$$H_3C - \underset{\underset{CH_3}{|}}{\overset{\overset{CH_3}{|}}{N}} - CH_3$$

c

$$H_3C - H_2C - CH_2 - \underset{\underset{CH_3-CH_2}{|}}{N} - CH_3$$

13 **The compound shown at right is found in red wine and is believed to be partly responsible for causing headaches. A particular red wine is found to be acidic.**

Write an equation for the equilibrium reaction that this compound undergoes in acidic solution.

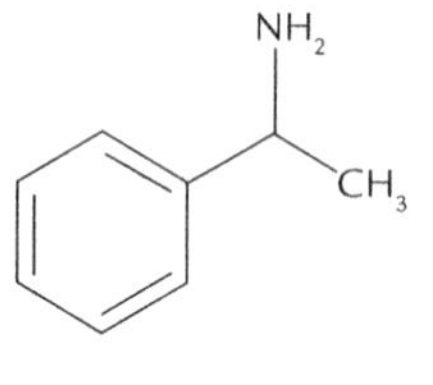

14 **Penicillin has the molecular structure shown.**

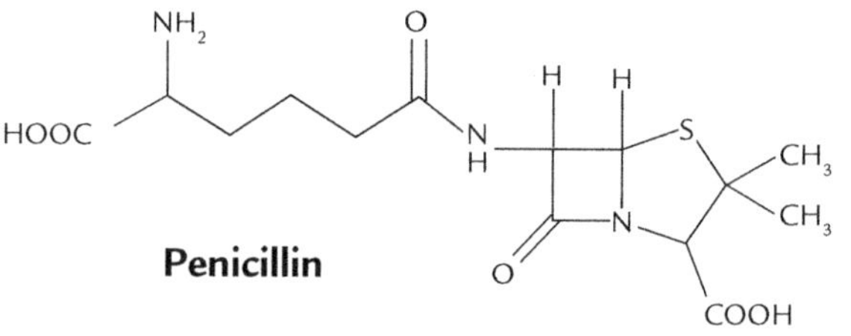

a Draw the fragment of the penicillin molecule that is a primary amine.

b Draw the fragment of the penicillin molecule that is an amide.

c Name another functional group in the molecule.

Penicillin

15 **Histamine is a compound released in breathing passages in sufferers of hayfever, bronchial asthma and other allergies.**

Draw the formula of the cation formed when 1 mole of histamine reacts with 3 moles of hydrochloric acid.

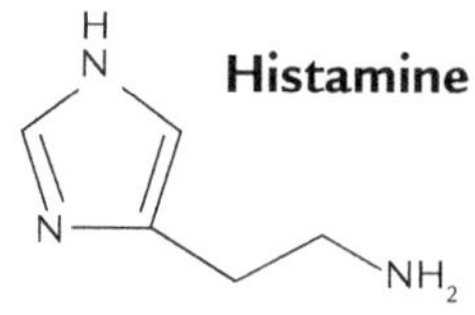

Histamine

16 **Chlorpheniramine is an antihistamine. It blocks those nerve endings that are histamine receptors.**

a Identify an amine functional group. Is chlorpheniramine a primary, secondary or tertiary amine?

b Draw the fragment of the molecule that contains an asymmetric carbon atom.

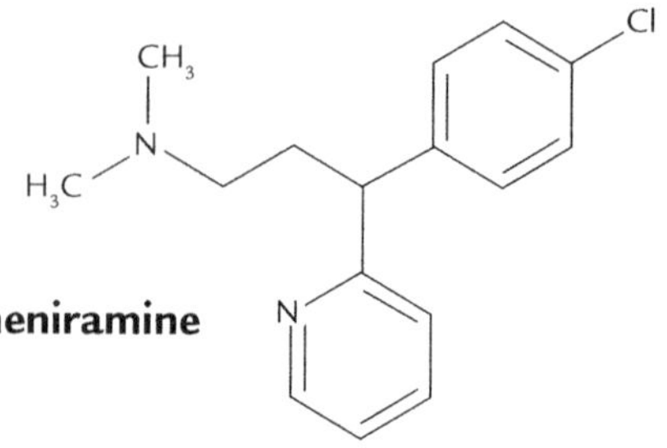

Chlorpheniramine

17 **The adrenalin molecule has the structure shown.**

a Identify an amine functional group. Is adrenalin a primary, secondary or tertiary amine?

b Draw the fragment that contains an asymmetric C atom.

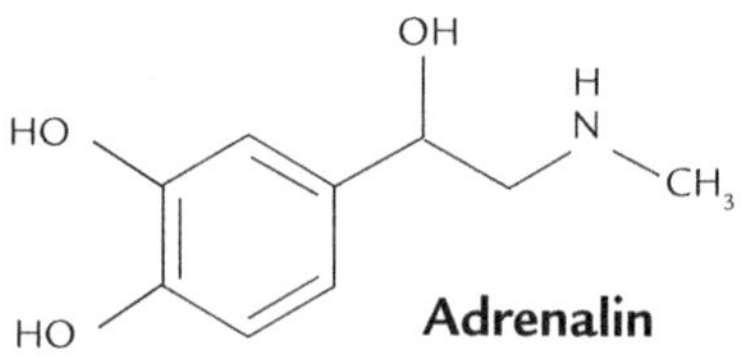

Adrenalin

ISBN: 9780170355544 PHOTOCOPYING OF THIS PAGE IS RESTRICTED UNDER LAW.

Unit 26 | Polymers

Learning Outcomes — on completing this unit you should be able to:

- write structures of addition polymers
- write structures of condensation polymers
- describe the molecular structure and properties of polymeric amides
- describe the molecular structure and properties of polymeric esters
- describe the molecular structure and properties of polypeptides and proteins
- recognise the repeating units in the structural formulae of polymers
- recognise amide and peptide bonds in polyamides and peptides
- recognise ester bonds in polyester
- state the names, structures and properties of natural and synthetic polymers.

Polymers

- Polymers are large molecules with molar masses ranging from tens of thousands to millions. They are formed by **large numbers of molecules bonding together**. A repeating unit can usually be seen in the polymeric structure.
- Naturally occurring polymers include starch and cellulose, which are polymers of different sugar molecules.
- Synthetic polymers are constantly being developed by scientists in materials research. They are being made to order to suit different purposes.
- Recently developed polymers range from strong, heat-resistant materials developed for space exploration, to materials that absorb large amounts of energy which are used for motorway barriers and for bomb disposal, to materials that absorb hundreds of times their weight in water which are used in disposable nappies.
- Polymers are formed by the addition reactions of compounds with double bonds and by condensation reactions.

Addition polymerisation

- An important type of addition reaction of alkenes is **addition polymerisation**. Under condition of high temperature (100–250°C) and high pressure (1000–3000 atmospheres) in the presence of a catalyst, ethene molecules add together to form molecules with molar mass equal to several million.
  Ethene molecules combine to form polythene.

Ethene is the monomer and polythene is the polymer.

- The tetrahedral arrangement around each carbon atom gives the polythene chain a zig-zag structure.
- Whether a straight chain or a branched chain forms depends on the catalyst. The nature and the packing of chains determines the nature of the polythene, for example whether it will be suitable for plastic bags (branched chains which do not pack closely together) or rigid milk containers (straight chains which do pack closely to give a strong dense material).
- Polythene is an example of an addition polymer. Other addition polymers are described below.

Addition polymers

- Ethene molecules add together to form polythene. Other small molecules with **C=C** double bonds add together in silmilar way to form molecules with molar mass equal to several million. The large molecules resulting from smaller molecules adding together are called **addition polymers**.
- **Polypropene** is made by joining propene molecules. Uses for polypropene are similar to those for polythene. Some forms are extremely strong and are used for outdoor carpets and binding around packages. Mixed with an elastic material called lycra, polypropylene is used to make 'thermal' underwear. Both polypropene and polythene can be repeatedly softened by heating and hardened again by cooling. The are said to be **thermoplastic**.

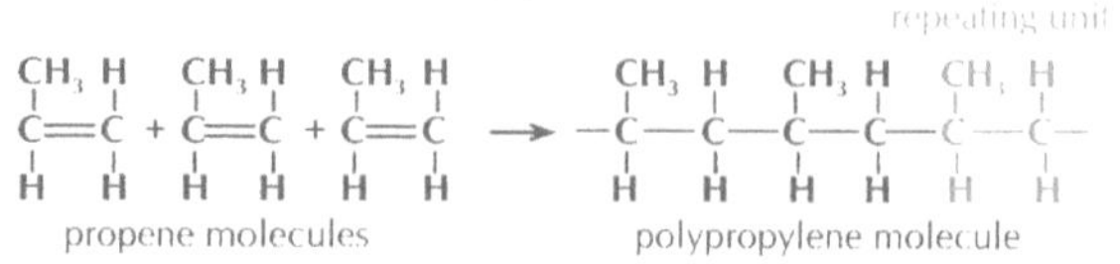

propene molecules → polypropylene molecule

- A small change in the monomer can result in very different polymer. If instead of the –**CH$_3$** in propene there is a –**Cl** in the molecule, **polyvinyl chloride (PVC)** forms. PVC is used for drainpipes and guttering, and as synthetic leather for shoes and furniture.

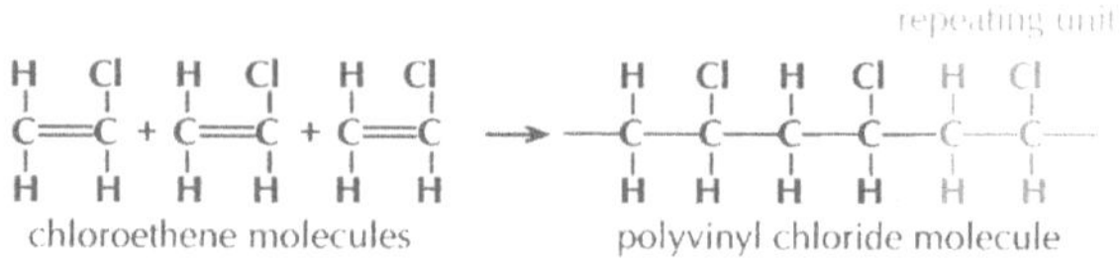

chloroethene molecules → polyvinyl chloride molecule

- If instead of –**CH$_3$** or –**Cl**:
  - there is an ethanoate (acetate) group, the result is **polyvinyl actate**, or **PVA**, which is used as a glue and in latex paints
  - there is a CN group, the result **polyacrylonitrile** or **acrylic**.
- Polymerisation of **CF$_2$=CF$_2$** gives **teflon**, which is used to coat frying pans and to make containers inert to most chemicals.

Condensation polymers

When two large molecules join together with the elimination of a small molecule, the reaction is called a condensation reaction. When molecules with double bonds add together in addition reactions, there is no loss of any substance.

- For example, when an alcohol reacts with a carboxylic acid to give an ester:

alcohol + carboxylic acid ⟶ ester + water

- Water is eliminated when the alcohol condenses with the acid.
- Condensation polymers are produced when large numbers of molecules condense together.

Polyamides

- An amide bond is formed when an acid or an acyl chloride condenses with an amine.

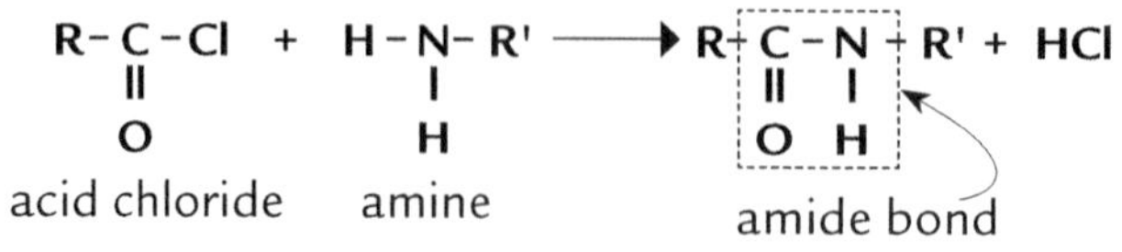

acid chloride   amine   amide bond

- Nylon 6, 6 is a synthetic polyamide formed when hexanedioic acid (or its acyl chloride) condenses with 1,6-diaminohexane. The '6,6' refers to the 6 carbon atoms in the di-acid and the 6 carbon atoms in the diamine.

ISBN: 9780170355544 PHOTOCOPYING OF THIS PAGE IS RESTRICTED UNDER LAW.

- Nylon 6,6 is the most commonly available nylon. Its uses include clothing, parachutes, carpets, clips and fasteners. Other nylons are prepared from starting materials with different chain lengths.

$$Cl-\underset{\underset{O}{\|}}{C}-(CH_2)_4-\underset{\underset{O}{\|}}{C}-Cl + H-\underset{\underset{H}{|}}{N}-(CH_2)_6-\underset{\underset{H}{|}}{N}-H + Cl-\underset{\underset{O}{\|}}{C}-(CH_2)_4-\underset{\underset{O}{\|}}{C}-Cl$$

$$\longrightarrow -\underset{\underset{O}{\|}}{C}-\underset{\underset{H}{|}}{N}-(CH_2)_6-\underset{\underset{H}{|}}{N}-\underset{\underset{O}{\|}}{C}-(CH_2)_4\underset{\underset{O}{\|}}{C}-\underset{\underset{H}{|}}{N}- + nHCl$$

repeating unit ⟶ Nylon 6,6

- The amide bond is highlighted in blue in the repeating unit.
- In nylon 6,6 there are approximately 200 repeating units.

Proteins and polypeptides

- Proteins **and** polypeptides **are naturally occurring polyamides. They are polymers of amino acids, but in these polymers the repeating units are not identical — up to 20 different types of amino acid molecules are used in nature.**
- An **amino acid** has an **amino** group and a **carboxylic acid** group on the same molecule.
- The two simplest amino acids are glycine and alanine.

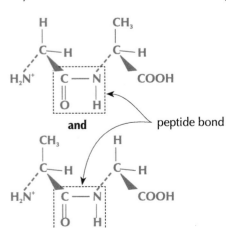

glycine   alanine

Note that alanine will have enantiomers as the central C is asymmetric.
- In the solid state and in solution, amino acids are in the ionic form shown in Structure **I** below. Note that in this form, it has a buffering action. Acid (H⁺) will combine with $-COO^-$ while base (OH⁻) will react with $-NH_3^+$ to give $-NH_2$ plus H_2O. Structure I is an example of a zwitterion, an ion that has positive and negative charges on the same molecule.
- In acidic solution, **II** is the stable form, and **III** is the stable form in basic solution.

Structure I   **Structure II**   **Structure III**

H_3N^+ COO⁻ H_3N^+ COOH H_2N COO⁻

Polypeptides

- Glycine and alanine can form two dipeptides (a molecule formed by two amino acid units).

and   peptide bond

- These dipeptides have an amino group and an acid group and can continue condensing with other amino acids to form polypeptides. The amide bond between amino acids is also called a peptide bond.
- Each polypeptide can be made up of hundreds of amino acid units. Natural polypeptides are formed from 20 different types of amino acids.

Proteins

- Proteins are naturally occurring, highly complex molecules made of one or more large polypeptide chains.
- In the body, proteins have many functions. Enzymes, haemoglobin, insulin, muscle tissue, hair and antibodies are all examples of proteins. Proteins are more abundant in animals than in plants. Some amino acids contain sulfur and give burning protein a characteristic smell. If a material is pure silk or wool, a small amount of it will give the characteristic smell when burnt.
- In proteins and polypeptides, the amino and acid groups are in the zwitterionic form described above.

Polyesters

An ester bond is formed when an acid or acyl chloride condenses with an alcohol.

$$R-\underset{\underset{O}{\|}}{C}-Cl \;+\; H-O-R' \longrightarrow R\dashv\underset{\underset{O}{\|}}{C}-O\vdash R' \;+\; HCl$$

acid chloride   alcohol   ester   ester bond

- Dacron (also known as terylene) is a synthetic polyester formed when 1,4-benzenedicarboxylic acid (terephthalic acid) condenses with 1,2-dihydroxyethane (ethylene glycol).

$$n HO-\underset{\underset{O}{\|}}{C}-\!\!\bigcirc\!\!-\underset{\underset{O}{\|}}{C}-OH \;+\; n HO-CH_2-CH_2-OH \longrightarrow$$

ester bond

$$\dashv \underset{\underset{O}{\|}}{C}-\!\!\bigcirc\!\!-\underset{\underset{O}{\|}}{C}-O-CH_2-CH_2-O\vdash \;+\; n H_2O$$

repeating unit

- Although there are naturally occurring esters such as fruit flavours and floral perfumes, **there are no naturally occurring polyesters**.

KEY POINTS SUMMARY

- **Addition polymers form when alkenes and substituted alkenes add together.**
- **Condensation polymers are formed when many large molecules join together with the elimination of small molecules.**
- **Polyamides form when molecules of carboxylic acid or acyl chloride react with amines. The units are joined by amide bonds.**
- **Polypeptides and proteins are naturally occurring polyamides made by condensation of amino acids. Amino acids have an amino and a carboxylic acid group.** The units are joined by **peptide bonds**.
- **Polyesters form when a large number of molecules of carboxylic acid react with alcohols.** The units are joined by **ester bonds.**

ISBN: 9780170355544 PHOTOCOPYING OF THIS PAGE IS RESTRICTED UNDER LAW.

ASSESSMENT ACTIVITIES

1 Matching terms with descriptions

a	a polymer of a huge number of amino acids	condensation
b	a reaction in which two large molecules combine with the elimination of a small molecule	polymer
c	a large molecule made of small molecules usually with repeating units	amide bond
d	a polymer of a limited number of amino acids	amino acid
e	a molecule with both acid and amine functional groups	peptide bond
f	formed by the reaction of an amine and a carboxylic acid or acyl chloride	polypeptide
g	an amide bond between amino acids	protein

Addition polymers

2 **Draw two repeating units of the polymers formed by addition polymerisation of the following monomers.**

 a $CH_3CH=CH_2$. The polymer is polypropylene.

 b $CH_2=CHCN$. The polymer is acrylic.

 c
$$\begin{array}{ccc} H & & CH_3 \\ \backslash & & / \\ & C = C & \\ / & & \backslash \\ H & & O - C - CH_3 \\ & & \parallel \\ & & O \end{array}$$
The polymer is PVA (polyvinyl acetate) used in adhesives.

Polyamides and amino acids

3 **Polyamides are long chains of condensed amino acid units.**

 a Draw the zwitterionic form of the amino acid alanine.

 b Write an equation for the reaction which occurs when a small amount of acid is added to a solution of alanine.

 c Write an equation for the reaction which occurs when a small amount of base is added to a solution of alanine.

4 **Peptides are long chains of amino acids, but the repeating amino acids are not identical. Peptides are made by the cells of living organisms. A protein may consist of a very long single polypeptide or it may be formed out of several polypeptide chains.**

 a Draw the structure of the amino acid glycine.

 b Draw the structure of a dipeptide of glycine and circle the amide bond.

3.5

c Why is the formation of the dipeptide classed as a condensation reaction?

d The amino acid cysteine has this structure:

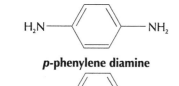

Draw the structure of a tripeptide consisting of cysteine, glycine and alanine in that order.

e How many possible tripeptides of these three amino acids are there? (The tripeptide may have one, two or three different amino acids.)

5 **Many polymers in everyday use were developed by DuPont, a chemical manufacturer in the USA. Among them is a material called kevlar, developed by Stephanie Kwolek, which is used for making bulletproof vests worn by armed offenders squads. Kevlar is also used wherever a high performance material is needed, such as brake linings and underwater cables. Kevlar is a polyamide of p-phenylene diamine and terephthalic acid.**

H_2N——⬡——NH_2

***p*-phenylene diamine**

a Draw a section of the kevlar molecule.
b Draw a box around the repeating unit in your diagram.
c Circle the amide bond.

HOOC ——⬡—— COOH

terephthalic acid

6 **The most famous polyamide, nylon, was also developed by DuPont. Nylon 6,4 is made by polymerising 1,6-diaminohexane and 1,4-butanedioic acid.**

a Write the structural formulae of the monomeric starting materials.
b Draw part of the nylon 6,4 molecule. Draw a box around the repeating unit and circle the amide bond.

Polyesters

7 **Surgical staples are made of a polymer called lactomer, which hydrolyses in the body after six to eight weeks. They are therefore particularly suitable for internal surgical procedures. As staples, they are quick and easy to apply. They allow the wound to heal before reforming the monomers, both of which are compounds that occur naturally in the body. Lactomer is a polymer of lactic acid and glycolic acid.**

$$HO - CH - COOH \qquad\qquad HO - CH_2 - COOH$$
$$\qquad | $$
$$\qquad CH_3$$

lactic acid   glycolic acid

a Write an equation for the formation of lactomer.
b Draw a box around the repeating unit in the product of your equation, and circle the ester bond.

8 **Dacron is a polymer of terephthalic acid and 1, 2 - dihydroxyethane (ethylene glycol).**

a Write the structural formulae of the monomeric starting materials.
b Draw part of the dacron molecule. Draw a box around the repeating unit and circle the ester bond.

ISBN: 9780170355544 PHOTOCOPYING OF THIS PAGE IS RESTRICTED UNDER LAW.

Revision Five

1 Naming compounds

Write the IUPAC systematic names of compounds A, B and C.

A

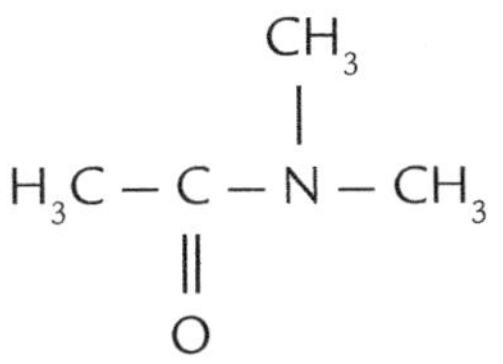

$$CH_3$$
$$|$$
$$H_3C - C - N - CH_3$$
$$\parallel$$
$$O$$

B

$$CH_2 - CH_3$$
$$|$$
$$H - C - CH_2 - CH_3$$
$$|$$
$$OH$$

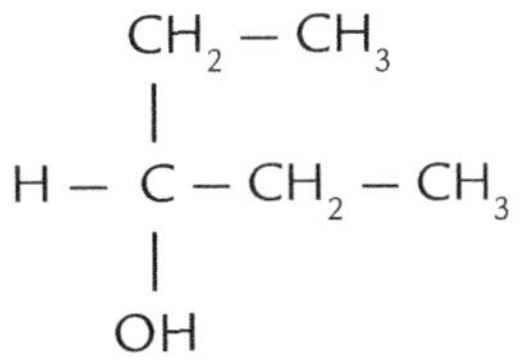

C

$$CH_3 - CH_2 - CH_2 - O - C - H$$
$$\parallel$$
$$O$$

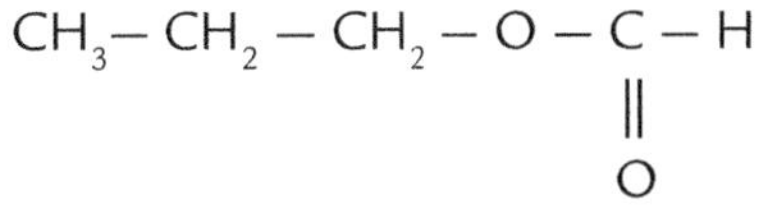

2 Properties of organic compounds

The following three compounds are common flavourings in food.

A Cinnamaldehyde

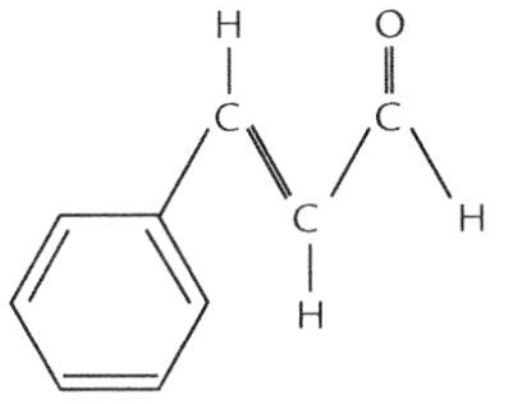

B Vanillin

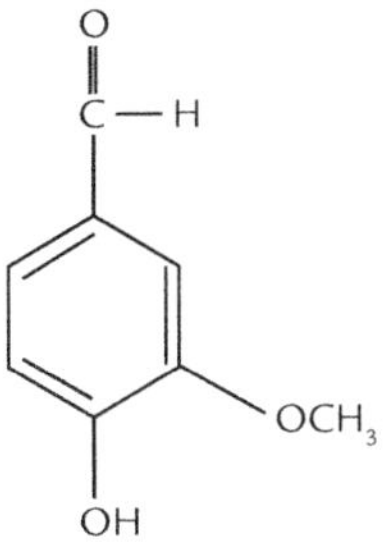

C Malic acid $HOOC - CH_2 - CH - COOH$
                                         $|$
                                         OH

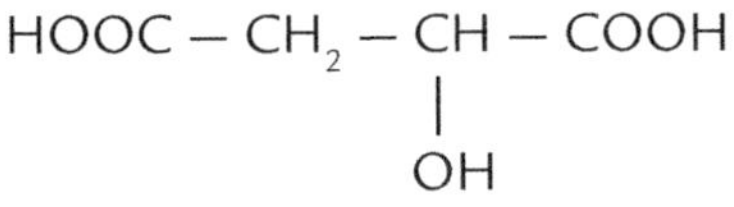

The questions below refer to these three compounds.

a Which compound makes food taste sour? Give a reason for your answer.

b i Which compound has cis and trans isomers?

ii Draw the isomers and label them 'cis' or 'trans'.

iii Describe a property that is different for the cis and trans isomers.

c One mole of cinnamaldehyde reacts with two moles H_2 in the presence of a Pt catalyst.

Assume that the structure ⬡ does not undergo addition with hydrogen.

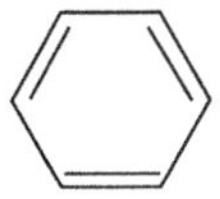

Draw the structure of the product of the reaction.

d i Which of the three compounds exists as enantiomers?

ii Draw 3D structures of the enantiomers of the compound.

iii Describe one *physical* property that is different for the two enantiomers.

e One of the three compounds decolourises acidified potassium permanganate but does not react with Benedict's solution. Which compound is this? Explain your choice.

3.5

3 Chemical reactions

Complete the reaction scheme below by writing *the structural formulae* of compounds A to H, naming reagent X and stating the type of reaction for each of reactions 1 to 5.
B and C have the molecular formula $C_4H_{10}O$. G has the molecular formula $C_5H_{10}O_2$.

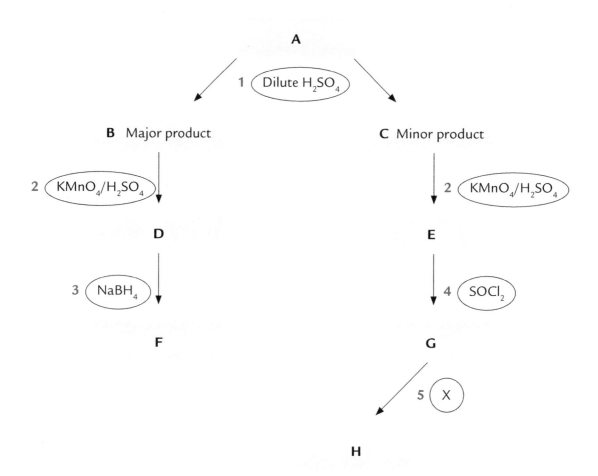

4 Larger molecules and polymers

a PVA glue is a polymer of vinyl acetate $CH_2=CH-O-CO-CH_3$.
   Draw a section of the PVA polymer and underline the repeating unit.

b Leucine is the amino acid $H_3C - CH - CH_2 - CH - COOH$.
$$\qquad\qquad\qquad\qquad\quad |\qquad\qquad\quad |$$
$$\qquad\qquad\qquad\qquad CH_3\qquad\quad NH_2$$

It is one of the essential amino acids.

i Write the systematic name for leucine.

ii Draw a dipeptide that leucine would form with alanine.

iii Name the bond formed between the two amino acids and circle it in the dipeptide that you have drawn.

ISBN: 9780170355544
PHOTOCOPYING OF THIS PAGE IS RESTRICTED UNDER LAW.

1 **Naming compounds and drawing enantiomers**

a Write the systematic names of compounds A, B and C.

A

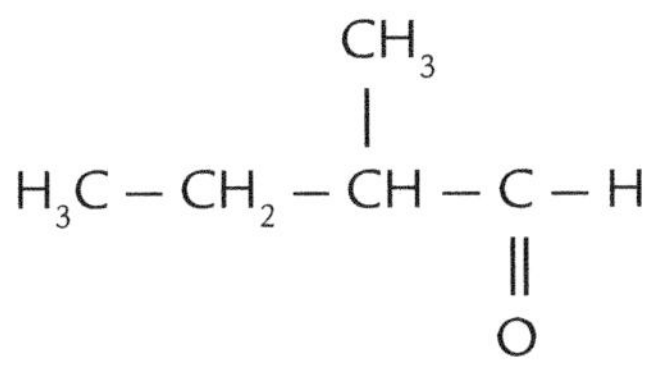

$$CH_3$$
$$|$$
$$H_3C - CH_2 - CH - C - H$$
$$||$$
$$O$$

B

$$CH_3$$
$$|$$
$$H_3C - N - CH_2 - CH_3$$

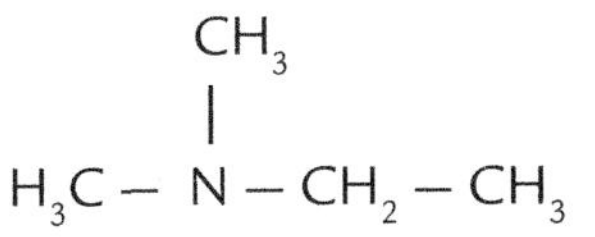

C

$$CH_3 - CH_2 - O - C - H$$
$$||$$
$$O$$

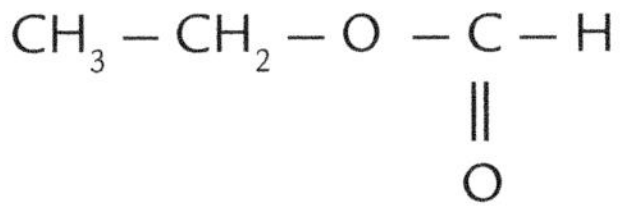

b *One* of the above compounds A, B or C exists as enantiomers (optical isomers).

  i Draw the 3D structures of the enantiomers of the compound.

  ii On one of the structures that you have drawn, mark the asymmetric C atom with an asterisk.

  iii Describe one *physical* property that is different for the two enantiomers.

2 **Primary, secondary and tertiary alcohols**

The compound C_4H_9OH can exist as a primary, secondary or tertiary alcohol.

a Draw the structures and write the names of these alcohols.

b Each of the alcohols above is reacted with an acidified solution of $KMnO_4$. Write the structures of the products of the reactions in each case. If there is no reaction, state 'no reaction'.

3 **Cis and trans isomers**

The alkene pent-2-ene can exist as *cis-trans* isomers.

a Draw the structures of the two isomers and label one as *cis* and the other as *trans*.

b Give one physical property that is different for the *cis-trans* isomers.

4 **Chemical reaction**

For this question the following information could be useful.

Primary haloalkanes tend to undergo substitution reactions, while secondary haloalkanes undergo elimination with KOH.

Complete the following reaction scheme by writing structural formulae for compounds A, B, C, D, E and F, name the reagent X and state the type of reaction for reactions 1 to 4.

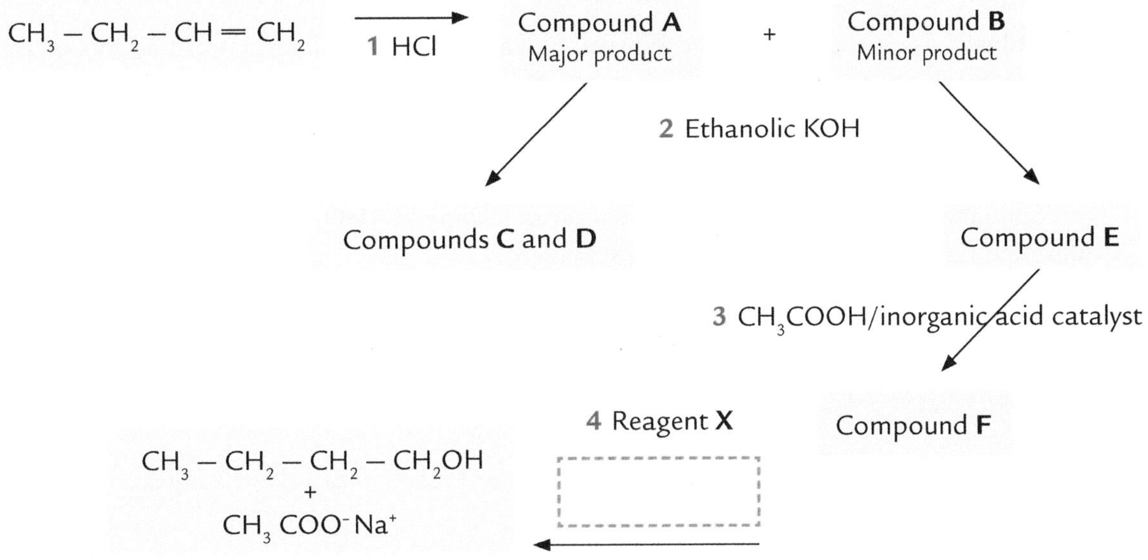

$$CH_3 - CH_2 - CH = CH_2 \xrightarrow{\text{1 HCl}}$$ Compound **A** (Major product) + Compound **B** (Minor product)

2 Ethanolic KOH

Compounds **C** and **D**

Compound **E**

3 CH_3COOH/inorganic acid catalyst

4 Reagent X

Compound **F**

$$CH_3 - CH_2 - CH_2 - CH_2OH$$
$$+$$
$$CH_3 COO^- Na^+$$

3.5

5 **Distinguishing between compounds**

Describe chemical reactions that would distinguish between the following pairs of compounds. Name the reagents you would need, and describe the expected results.

a $CH_3CH_2CH_2Cl$ and CH_3CH_2COCl

b CH_3CH_2CHO and CH_3COCH_3

6 **Polymers**

a i Propene polymerises to give a very useful polymer called polypropylene. This polymer is used for making banknotes. Draw a section of the polymer showing three repeating units.

   ii What name is given to this type of polymerisation?

b Draw the structure of a dipeptide formed by the amino acid alanine and circle the peptide bond.

c A nylon is formed by the reaction of 1,4-diaminobutane and 1,6-hexanedioic acid.

   i Draw the structure of a section of the polymer and underline the repeating unit.

   ii The reaction is rapid if the acid chloride of 1,6-hexanedioic acid is used instead of the acid. Give the formula of a reagent that could be used to prepare the acid chloride from the acid.

   iii What name is given to the bond formed between 1,4-diaminobutane and 1,6-hexanedioic acid?

Question III

1 **Naming compounds**

Write the names of compounds A, B and C.

A

$$H_3C - \underset{\underset{O}{\overset{||}{\underset{|}{C}}}}{\overset{\overset{CH_3}{|}}{CH}} - C - NH - CH_3$$

B

$$H_3C - \underset{\underset{O}{\overset{||}{C}}}{C} - \overset{\overset{CH_3}{|}}{CH} - CH_3$$

C

$$CH_3 - CH_2 - O - \underset{\underset{O}{\overset{||}{C}}}{C} - CH_2 - CH_3$$

2 **Enantiomers**

The compound $CH_3CH_2CH(Br)CH_3$ exists as enantiomers (optical isomers).

a Draw the 3D structures of the enantiomers of the compound.

b On one of the structures that you have drawn, mark the asymmetric C atom with an asterisk.

c Describe how you could determine that you had a solution containing equal amounts of the two enantiomers.

3 **Chemical reactions**

Compound A, $C_4H_{10}O$, is an alcohol that does not react with acidified $KMnO_4$ solution.

Compound B is an isomer of A that does react with acidified $KMnO_4$ to produce compound C.

Compound C gives no reaction with Tollens' reagent.

Compound D is an isomer of C that does react with Tollens' reagent to form compound E.

a Identify compounds A, B, C, D and E by writing the name and structural formula of each compound.

b Describe what you would observe when B reacts with acidified $KMnO_4$. State the type of reaction.

c Describe what you would observe when compound D reacts with Tollens' reagent. State the type of reaction.

ISBN: 9780170355544 PHOTOCOPYING OF THIS PAGE IS RESTRICTED UNDER LAW.

4 A reaction scheme

Complete the following reaction scheme by writing structural formulae for the organic compounds I, II, III, IV and V. Compounds IV and V decolourise bromine water. State what type of reaction 1, 2 and 3 are.

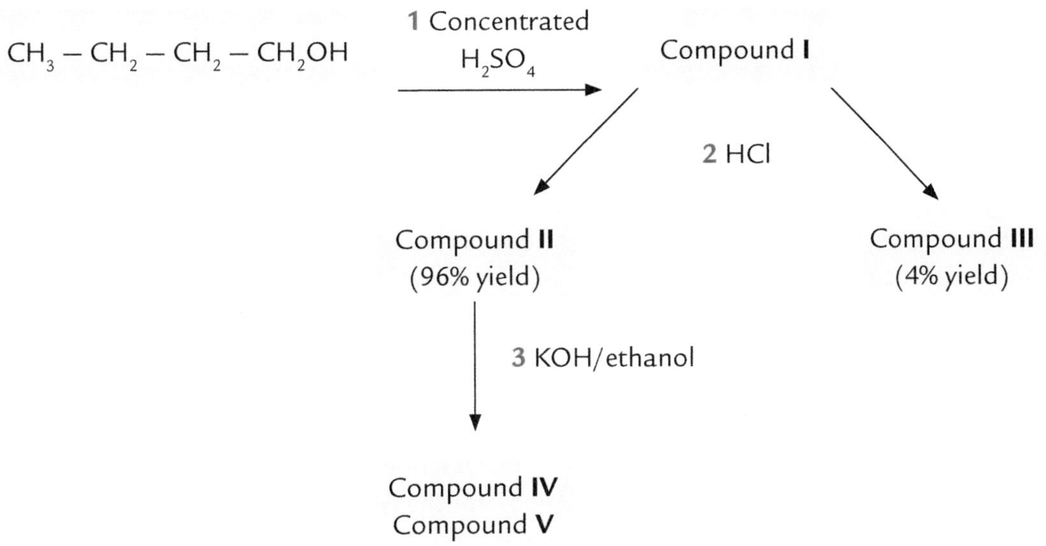

$CH_3 - CH_2 - CH_2 - CH_2OH$

1 Concentrated H_2SO_4 → Compound **I**

2 HCl

Compound **II**
(96% yield)

Compound **III**
(4% yield)

3 KOH/ethanol

Compound **IV**
Compound **V**

5 Reaction products

Write the structural formula of the organic product formed by the reaction of the following pairs of compounds. State the type of reaction in each case.

a $CH_3CH_2OH + CH_3CH_2COCl \longrightarrow$

b $CH_3CH_2COOH + NH_3 \longrightarrow$

c $CH_3CH_2COCl + NH_3 \longrightarrow$

6 Amino acids

a Draw the structure that the amino acid glycine would have in neutral aqueous solution.

b Cysteine is an amino acid with the formula $HS - CH_2 - CH - COOH$.
$|$
NH_2

Draw the structure of a dipeptide formed by two molecules of cysteine.

7 Nylon 6,6

Nylon 6,6 is a polymer of $H_2N-(CH_2)_6-NH_2$ and $HOOC-(CH_2)_4-COOH$.

a Draw the structure of part of the nylon polymer and underline the repeating unit.

b What is the name given to the bond that is formed between the molecules? Circle this bond in the structure that you have drawn.

c What name is given to this type of polymerisation?

Question IV

1 Naming compounds

In the spaces provided, write the IUPAC systematic names of compounds A, B and C.

A

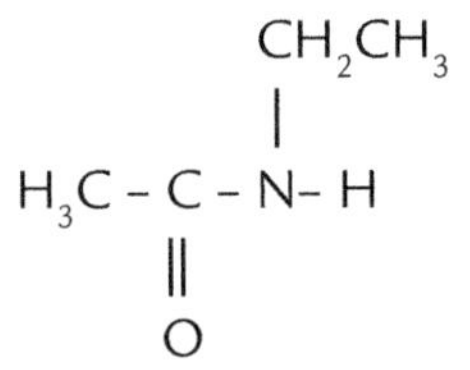

$H_3C - C - N - H$
with CH_2CH_3 attached to C, and $\parallel O$ below C

B

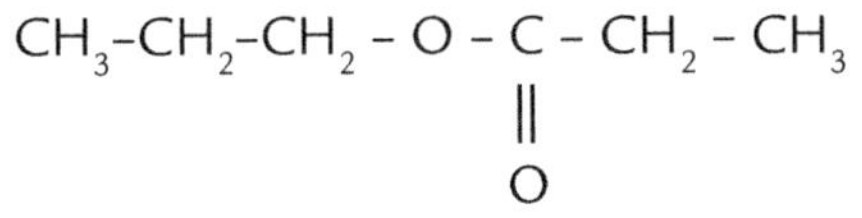

$CH_3-CH_2-CH_2 - O - C - CH_2 - CH_3$
with $\parallel O$ below C

C

$CH_3 - CH - CH_2 - C - Cl$
with CH_3 below the second C and $\parallel O$ below the fourth C

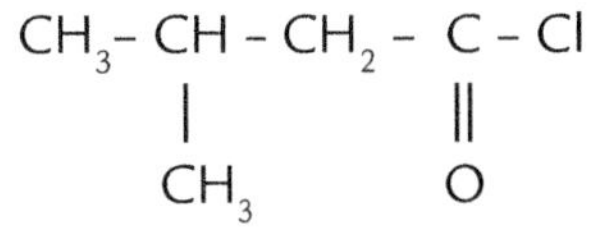

3.5

2 Writing formulae

Draw the structural formula of each of the following compounds.

a 2-methylpentan-3-one

b 3-hydroxy-2-aminobutanoic acid

3 Isomers

a The following questions refer to this compound: $CH_3-CH_2-CH-C-Cl$

$$\begin{array}{cc} | & || \\ CH_3 & O \end{array}$$

  i Draw the 3D structures that clearly show the enantiomers of this compound and write an asterisk beside the asymmetric C atom.

  ii How are the structures of the two enantiomers related?

b Palmitoleic acid, also known as Omega 7, is a beneficial fatty acid. It has the formula:

$$CH_3(CH_2)_5CH=CH(CH_2)_7COOH$$

  i Draw **and** label the cis and trans isomers of $CH_3(CH_2)_5CH=CH(CH_2)_7COOH$.

  ii Describe a physical property that is different for the two isomers.

4 Distinguishing between compounds

Describe a chemical test that would distinguish between the following compounds. In each case, name the reagents used and describe the expected results.

a $CH_3CH=CH_2$ and $CH_3CH_2CH_3$

b CH_3CONH_2 and $CH_3CH_2CH_2NH_2$

c $CH_3CH(OH)CH_2CH_3$ and $CH_3-C-CH_2CH_3$

$$\begin{array}{c} || \\ O \end{array}$$

d $CH_3CH_2CH_2OH$ and CH_3CH_2CHO

Macadamia oil is a source of palmitoleic acid.

5 Chemical reactions

Complete the following reaction scheme that starts with 2-chloropropane by writing the structural formulae of the organic products A, B, C and D. State the type of reaction for reactions 1, 2 and 3.

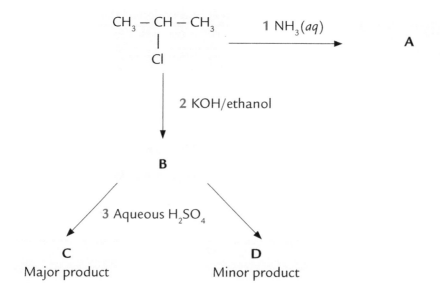

ISBN: 9780170355544   PHOTOCOPYING OF THIS PAGE IS RESTRICTED UNDER LAW.

6   **Polymerisation of methylpropene CH$_3$ — C = CH$_2$ produces an artificial rubber.**

$$CH_3 - \underset{\underset{CH_3}{|}}{C} = CH_2$$

    a Draw a section of the polymer and underline *two* repeating units.

    b What name is given to this type of polymerisation?

7   **Silk is a fibrous protein in which the monomer units are mainly glycine and alanine.**

    a Draw a section of a silk polymer showing glycine bonded to alanine.

Alanine is   Glycine is the most simple amino acid.

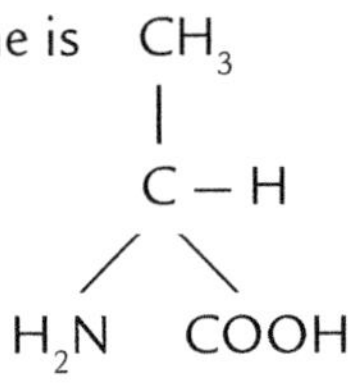

$$CH_3$$
$$|$$
$$C - H$$
$$\diagup \quad \diagdown$$
$$H_2N \quad\quad COOH$$

    b Name the bond formed between glycine and alanine.

    c What name is given to this type of polymerisation?

Silk cocoons. The fibre is a protein made up mainly of units of glycine and alanine.

New Zealand banknotes are made of polypropylene (polypropene).

ISBN: 9780170355544 PHOTOCOPYING OF THIS PAGE IS RESTRICTED UNDER LAW.

3.5

3.2

Demonstrate an understanding of spectroscopic data in chemistry

Internal assessment   **3 credits**

Unit	Content	Aim
27	**Spectroscopic data in organic chemistry** • **Infra red** absorption spectra • **C-13 NMR** • **Mass spectra**	To use spectra of compounds to: • identify the **functional group** using **IR spectra** • look at the **carbon environment** (the atoms bonded to each carbon atom) with **C-13 NMR spectra** • deduce the **molar mass** and **structure** of molecules from **mass spectra** • identify an **organic compound** (in units 21 to 26) using a **combination of spectral data**

Organic molecules will be from this list:
• Alkanes
• Alkenes
• Alcohols
• Aldehydes
• Ketones
• Carboxylic acids
• Esters
• Amines
• Amides
• Acid chlorides
• Haloalkanes

ISBN: 9780170355544   PHOTOCOPYING OF THIS PAGE IS RESTRICTED UNDER LAW.

Unit 27 | Spectroscopic data in organic chemistry

Learning Outcomes — on completing this unit you should be able to:

- identify functional groups from infra-red spectra
- recognise the environment of a carbon atom from carbon-13 NMR spectra
- deduce the molar mass and some of the structure of a molecule from mass spectral data.

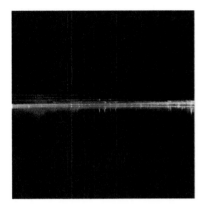

Molecular spectroscopy

The structures of organic molecules can be identified by **spectroscopic methods** as well as by their chemical reactions.

We can often determine the molecular structure of an organic compound if we can identify the following three characteristics.

- The functional group. We do this by looking at its **IR spectrum**.
- The environment of the carbon atoms. We do this by looking at **carbon-13 NMR** data. For example, the environments of the C atoms in $-C=O$, $-CH_2-$ and $-CH_3$ are all different. The environments of the two atoms in H_3C-CH_3 are the same.
- The molar mass. This information is obtained by **mass spectrometry**.

Type of spectroscopy	How it works	The information it gives us
Infra red, IR	Energy is absorbed by bonds that are stretching, bending and vibrating. The energy absorbed is in the infra-red part of the electromagnetic spectrum. The wavelength of the energy absorbed is measured.	The functional groups in the molecule, for example whether it is $-OH$, $-COOH$ or $-CHO$, etc. Different functional groups absorb different wavelengths. The wavelength of the energy absorbed depends on the atoms that are bonded together.
Nuclear magnetic resonance, NMR	Energy is absorbed and released by spinning carbon 13 nuclei placed in a magnetic field. This energy is affected by the electrons of surrounding atoms. They shield the nucleus from the effects of the field. The spinning nucleus must have an odd number of protons + neutrons. C-13 has 6 protons and 7 neutrons. Radio waves (low energy) are absorbed and released.	The environment of the carbon atom. For example, the C atoms in CH_3-, $-CH_2-$, $-C=O-$ are in different environments. The atoms that are bonded to a carbon atom affect the energy that its spinning nucleus absorbs when placed in a magnetic field.
Mass spectrometry	Molecules are ionised (usually +1) by a stream of electrons. Then the unstable molecular ions fragment. Molecular ions and fragment ions are deflected in a magnetic field. The amount of deflection depends on the mass and the charge.	The molar mass of the molecule and its structure. For example $$CH_3\ CH\ CH_3$$ $$|$$ $$CH_3$$ will give CH_3^+ and $CH_3CHCH_3^+$ particles as well as the molecular ions. The way a molecule fragments (shatters) depends on its structure.

The diagram below relates infra-red and radio waves to the visible part of the spectrum.

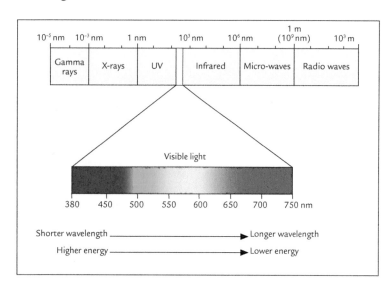

Many **functional groups** of organic molecules absorb energy from the **infra-red** part of the spectrum.

In **NMR spectra**, **radio waves** are absorbed.

Infra-red spectroscopy

Infra-red (IR) spectra usually show absorbances at different wave numbers. The wave number is the number of wavelengths in 1 cm. A wave number of 3000 cm^{-1} means there are 3000 waves in 1 cm. So the wavelength is 0.01/3000 m = 3333 nm.

IR spectra of different compounds are shown below. The peaks attributed to each functional group are indicated by arrows as shown. Alkane – C–H gives a peak at 2850 to 3000 cm^{-1} and will be in most spectra.

$$H H$$
$$| \quad |$$
Alcohol $\quad$ H — C — C — O — H
$$| \quad |$$
$$H \quad H$$

–O–H strong broad peak at 3200 to 3600 cm^{-1}; note the shape of the peak
–C–O strong peak at 1050 to 1150 cm^{-1}

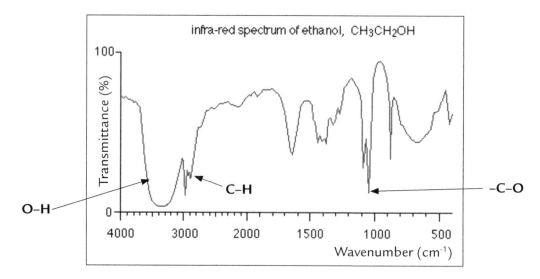

 ISBN: 9780170355544 PHOTOCOPYING OF THIS PAGE IS RESTRICTED UNDER LAW.

Aldehyde

– **C=O** (carbonyl) strong peak 1670 to 1820 cm^{-1}

Aldehyde also has two medium peaks around 2720 to 2750 cm^{-1} and 2820 to 2850 cm^{-1}

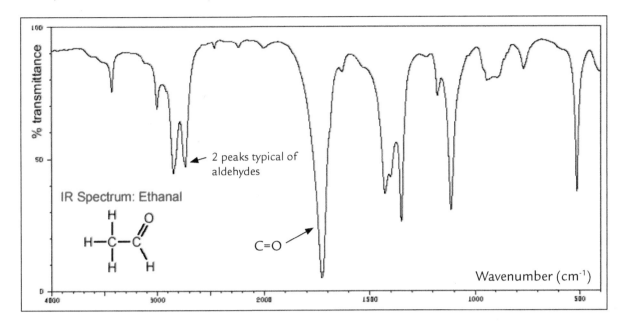

Ketone

– **C=O** (carbonyl) strong peak 1670 to 1820

Ketone also has a strong peak at 1705 to 1725 but does not have the two peaks that an aldehyde has.

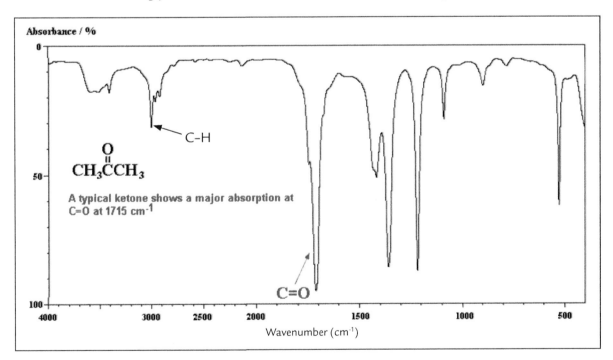

ISBN: 9780170355544   PHOTOCOPYING OF THIS PAGE IS RESTRICTED UNDER LAW.

Acid

$$- C = O$$
$$\quad | $$
$$\quad O - H$$

– C=O   (carbonyl) strong peak 1670 to 1820 cm^{-1}

– O–H   strong broad peak at 3200 to 3600 cm^{-1}; compare the shape of this peak with the one in the alcohol at the same wave number

– C – O   strong peak at 1210 to 1320 cm^{-1}

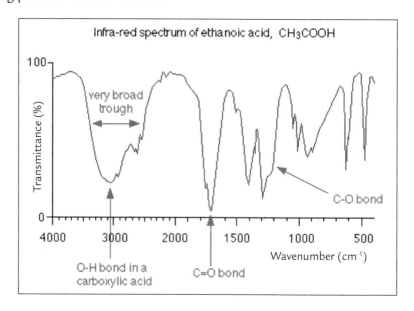

Ester

$$- C = O$$
$$\quad | \qquad\qquad \text{Note there is no O–H peak}$$
$$\quad O - C -$$

– C=O   (carbonyl) strong peak 1670 to 1820 cm^{-1}

– C – O   strong peak at 1000 to 1300 cm^{-1}

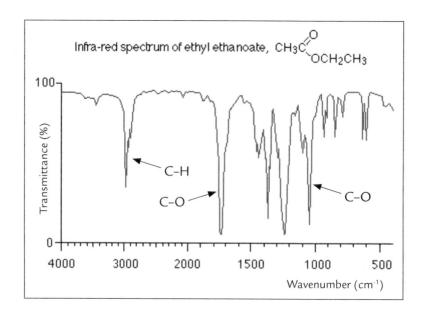

ISBN: 9780170355544   PHOTOCOPYING OF THIS PAGE IS RESTRICTED UNDER LAW.

Haloalkanes

– **C–F**   strong peak 1100 to 1400 cm^{-1}
– **C–Cl**   strong broad peak at 540 to 785 cm^{-1}
– **C–Br**   strong peak at 510 to 650 cm^{-1}
– **C–I**   strong peak at 485 to 600 cm^{-1}

IR spectrum of bromoethane

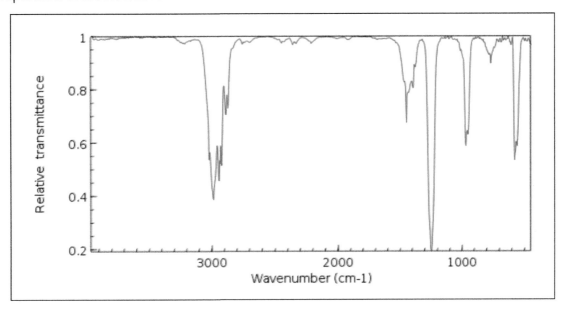

A summary of IR absorption frequencies is shown in the diagram below.

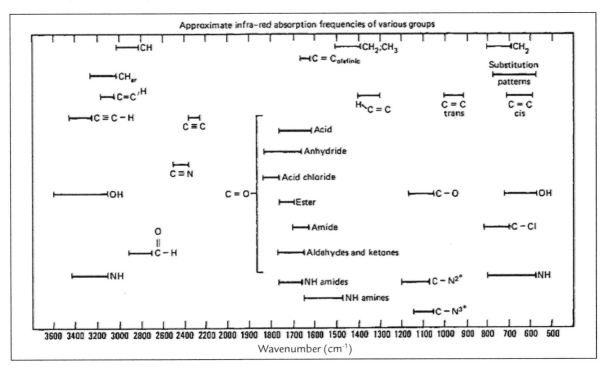

A list giving intensities of absorptions are in Appendix 4.

C-13 nuclear magnetic resonance spectroscopy

A spinning nucleus with an odd number of protons + neutrons generates a magnetic field. So a C-12 nucleus does not generate a magnetic field but a C-13 nucleus does. About 1% of C atoms are C-13. If the spinning nuclei are in an external magnetic field, some will line up with the field (north N and south S in the same direction as the field). Some will oppose the magnetic field (N of the nucleus's magnetic field pointing in the S direction of the external field). More nuclei line up with the field. These have less energy than those that oppose it.

Electromagnetic radiation gives energy for excess aligned nuclei to be against the field. When the energy is removed, the energised nuclei relax back to the aligned state. Fluctuations in the magnetic field associated with the relaxation process is called resonance. The fluctuating field generates an electric current.

Different nuclei absorb and release different energies. The differences arise from the nuclei being in different electron environments. Electrons 'shelter' the nuclei from the applied magnetic field. The applied field must be large for the differences to be significant. Resonance frequencies are measured relative to a standard. Usually the standard is tetramethylsilane, TMS.

$$\begin{array}{c} CH_3 \\ | \\ H_3C - Si - CH_3 \\ | \\ CH_3 \end{array}$$

The difference in resonance is called a chemical shift, which is given by

$$\frac{\text{difference between a resonance frequency and that of a reference substance}}{\text{operating frequency of the spectrometer}}$$

The numerator is in hertz, Hz, and the denominator is in megahertz (10^6 Hz), so the unit for is 10^{-6} or ppm (parts per million).

C-13 NMR spectra give information about the environment of carbon atoms.

Carbon atoms have the same environment if they are bonded to exactly the same atoms and groups.

- In butane, $H_3C-H_2C-CH_2CH_3$, carbon-1 and carbon-4 (C) are in the same environment, This is different from the environment of carbon-2 and carbon-3. Carbon-2 and carbon-3 (C) have the same environment. So the C-13 NMR would show only two peaks.
- Pentane $CH_3-CH_2-CH_2-CH_2-CH_3$ has three different carbon environments.
- 2-bromobutane CH_3 $CH(Br)CH_2CH_3$ has all four C atoms in different environments. The C-13 NMR spectrum would show four peaks.
- This spectrum of ethanol tells us that the two carbon atoms are in two different environments. These are CH_3– and $-CH_2O-$. O is electronegative. The C bonded to it has a smaller share of the electrons in the bond. Its nucleus is less shielded. This C is at a higher ppm than the other C.

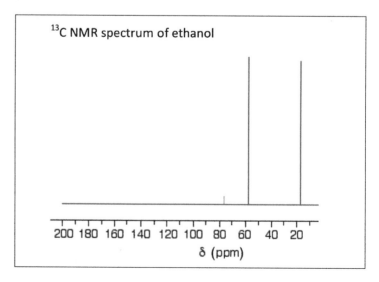

^{13}C NMR spectrum of ethanol

δ (ppm)

ISBN: 9780170355544 PHOTOCOPYING OF THIS PAGE IS RESTRICTED UNDER LAW.

- In **ethyl ethanoate**, the four carbon atoms are in four different environments. So the NMR spectrum has four peaks. $CDCl_3$ is the solvent. Note that C of C=O (C_2 in the diagram) is at a high ppm. A C atom bonded to an electronegative atom, such as O, will give a peak at a much higher ppm. C in C=O is less shielded than the C in C–O.

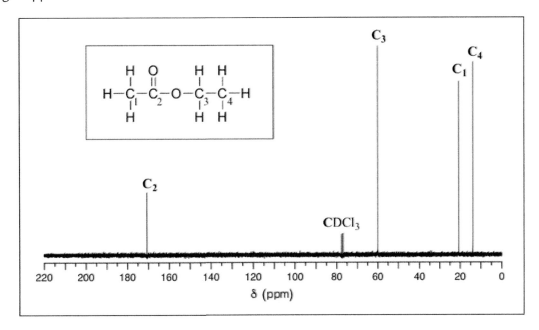

- In the compound below there are six carbon atoms in five different carbon environments. The two carbon atoms on the right are bonded to exactly the same things.

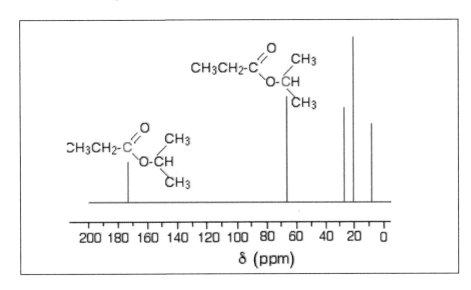

Which peak shows the carbon in C=O?

It is the one at 172ppm. C-13 atoms bonded to electronegative atoms are shielded from the magnetic field. They give larger chemical shifts (higher ppm).

3.2
3.3

This table gives a summary of chemical shifts.

carbon environment	δ (ppm)
C=O (in ketones)	205–220
C=O (in aldehydes)	190–200
C=O (in acids and esters)	170–185
C=C (in alkenes)	115–140
C≡C (in alkynes)	67–85
RCH_2OH	50–65
RCH_2Cl	40–45
RCH_2NH_2	37–45
R_3CH	25–35
CH_3CO	20–30
R_2CH_2	16–25
RCH_3	10–15

Mass spectra

- Masses of molecules are measured in a mass spectrometer. Particles are bombarded with electrons so that they are fragmented (broken up) and ionised. Those with a + charge can be accelerated to a constant velocity through the poles of a magnet by attracting them to a negatively charged screen. The amount of deflection by the magnetic field depends on the mass and charge of the ion. The mass per unit charge, m/z, is measured. For ions with the same charge, for example +1, the deflection depends on the mass. Lighter particles will be deflected more.

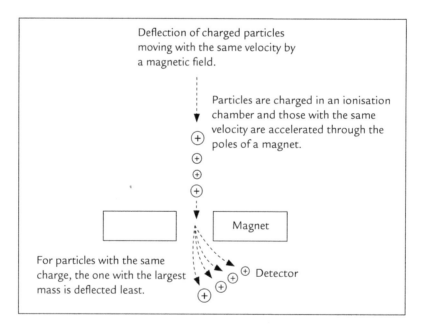

Deflection of charged particles moving with the same velocity by a magnetic field.

Particles are charged in an ionisation chamber and those with the same velocity are accelerated through the poles of a magnet.

Magnet

For particles with the same charge, the one with the largest mass is deflected least.

Detector

- The spectra give us information about the molar mass and the molecular structure. The molecules fragment in a way that depends on how atoms are bonded. For example, a straight chain of carbon atoms breaks up in a different way to a branched chain.

ISBN: 9780170355544 PHOTOCOPYING OF THIS PAGE IS RESTRICTED UNDER LAW.

- The mass spectrum of propanone shows how the peaks are attributed to different particles formed when the molecule is fragmented.

M(propanone) = 58 g mol^{-1}

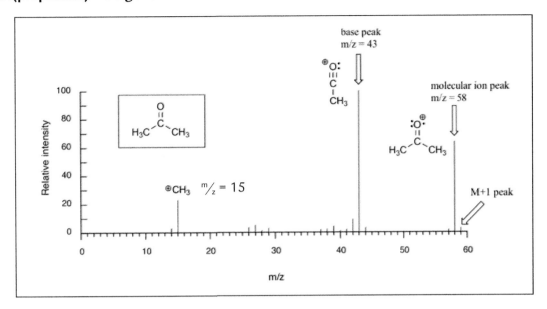

- The peaks in this mass spectrum of ethanol CH_3CH_2OH can be assigned to the molecular ion and possible fragments.

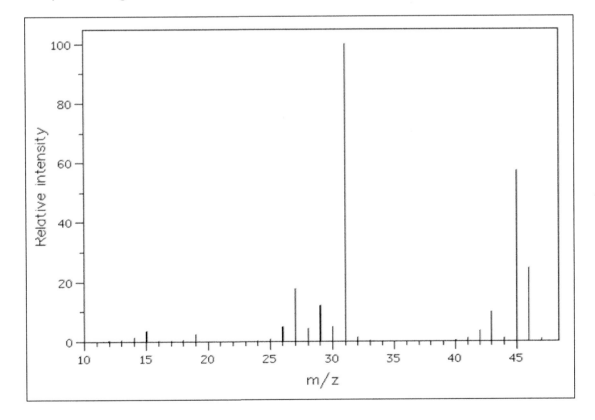

The peak at 46 m/z gives the molecular ion. From this we see that M(ethanol) = 46 g mol^{-1}.

The peak at 15 m/z is due to the particle $[CH_3]^+$.

The peak at 29 m/z is due to the particle $[CH_3CH_2]^+$.

The peak at 31 m/z is due to the particle $[CH_2OH]^+$.

ASSESSMENT ACTIVITIES

1 Interpreting IR spectra

In each of the following, determine the functional group. Give your reasons for each.

a Describe the two peaks that identify the functional group in this spectrum.

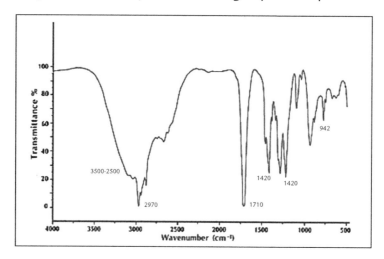

b Describe the peaks that identify the functional group in this spectrum.

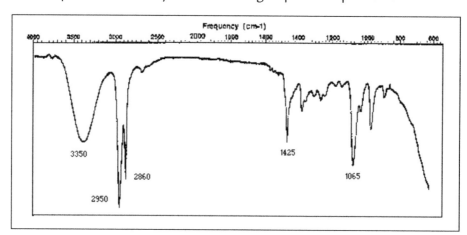

c Describe the peak that identifies the functional group in this spectrum.

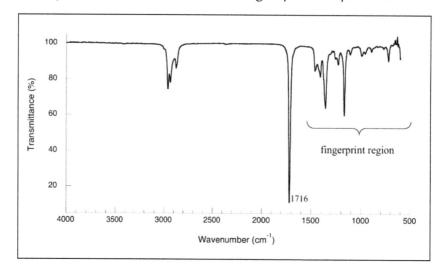

ISBN: 9780170355544 PHOTOCOPYING OF THIS PAGE IS RESTRICTED UNDER LAW.

d Describe the peaks that identify the functional group in this spectrum.

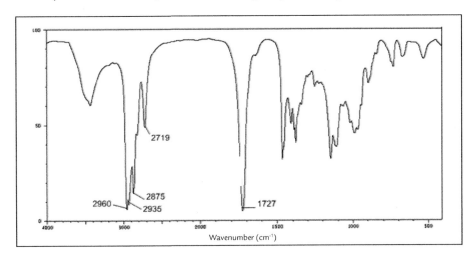

2 Interpreting NMR spectra

a The C-13 NMR spectrum of ethyl ethanoate is shown below.

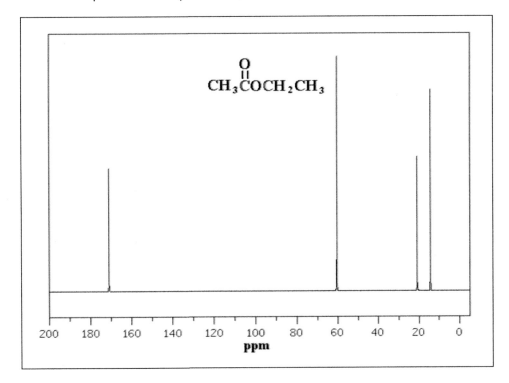

In terms of the peaks in the spectrum, state how the spectrum of methyl ethanoate would differ from that of ethyl ethanoate.

b i Identify the number of carbon environments in each of these compounds:

A CH_3CHCH_3
       |
      OH

B $CH_3CH_2CH_2OH$

C $CH_3CH_2CCH_2CH_3$
        ||
        O

D $CH_3CH_2CH_2CH_2COOH$

ii Which of the compounds **A** to **D** in part **b i** would show a peak at a high ppm (greater than 150 ppm) in their C-13 NMR spectrum?

ISBN: 9780170355544   PHOTOCOPYING OF THIS PAGE IS RESTRICTED UNDER LAW.

3.2

c Copy and complete the table. For each of the compounds, draw the structural formula, state the number of peaks you would expect to see in the C-13 NMR spectrum of the compound and indicate whether the spectrum would show a peak at a high ppm (greater than 150 ppm).

Compound	Structural formula	No. of peaks in C-13 NMR = no. of C environments	Peak at > 100 ppm: Y/N
Butanol			
Methyl propanoate			
Butene			
Butanal			
Methyl propanol			
Butanone			
Methyl propan-2-ol			

3 Interpreting mass spectra

a This is the mass spectrum of 1-butene.
   i   Draw the structural formula of 1-butene. Calculate the molar mass of 1-butene.
   ii Circle the peak that gives the molar mass.
   iii Write the structure of the particle with
       m/z = 27 _____
       m/z = 29 _____
       m/z = 41 _____

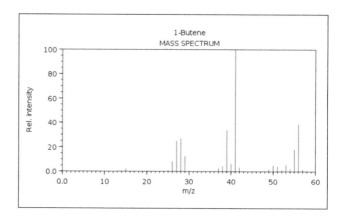

b In the following spectra, the m/z values are given for a number of peaks. Identify the peak that corresponds to the molecular ion (m/z = molar mass). Assign structures of the fragments corresponding to the m/z values given for the other peaks.

i   Pentane

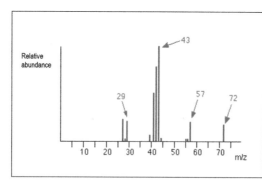

ii Methanol

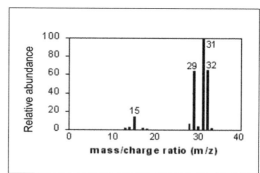

ISBN: 9780170355544 PHOTOCOPYING OF THIS PAGE IS RESTRICTED UNDER LAW.

Combining information from spectra

1 The IR, NMR and mass spectra are given for compound X.

  a Identify the functional group from this IR spectrum.

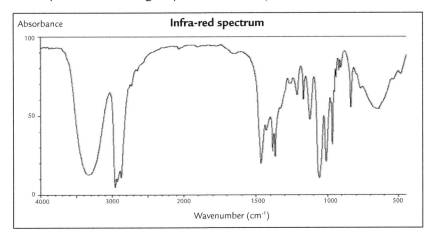

  b From the NMR spectrum, how many different environments do the carbon atoms have?

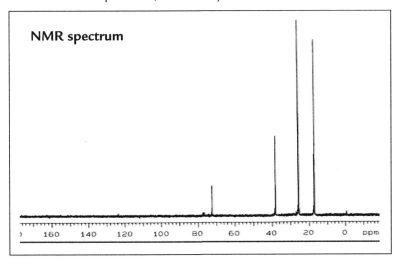

  c The mass spectrum is given below.

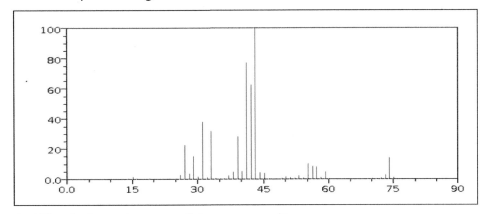

  i What is the molar mass of the compound?

  ii Write a possible structure for it.

2 Compounds A, B and C are related by oxidation.

  a Identify the functional group in compound A from its IR spectrum.

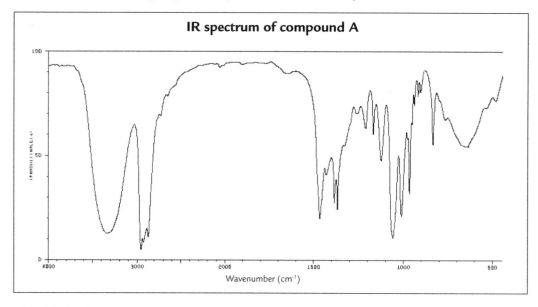

  b Identify the functional group in compound B from its IR spectrum.

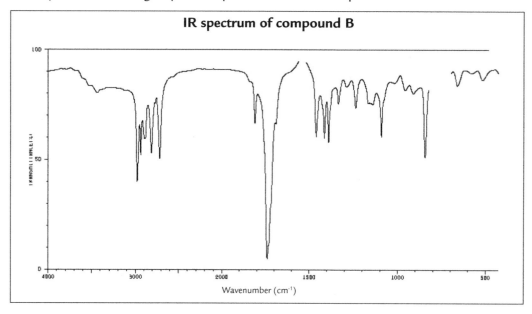

  c Identify the functional group in compound C from its IR spectrum.

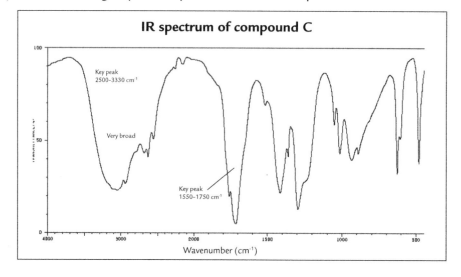

ISBN: 9780170355544 PHOTOCOPYING OF THIS PAGE IS RESTRICTED UNDER LAW.

d The mass spectrum of compound A is shown below.

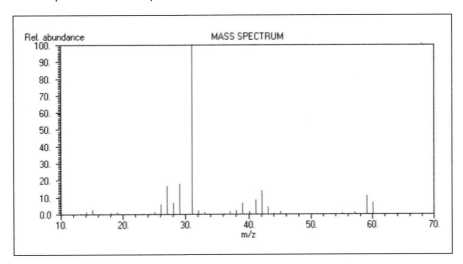

i What is the molar mass of compound A?

ii Identify compounds A, B and C.

iii Give the structure of particles in the mass spectrum of A with m/z = 31 and m/z = 29.

3 Two students reacted ethanol with acidified potassium dichromate.

a The IR spectra of the products of the two students were different. They are shown below.

IR spectrum of compound 1

IR spectrum of compound 2

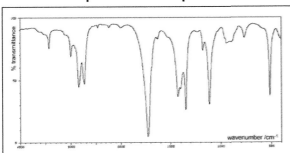
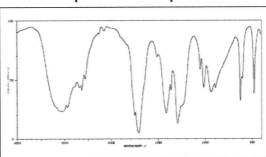

Identify the two products.

Compound 1 is _____

Compound 2 is _____

b Mass spectra of the two products are shown below.

Mass spectrum 1

Mass spectrum 2

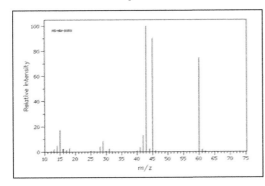
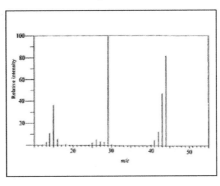

i Explain, by reference to the peaks in the spectra, which compound has mass spectrum 1.

ii Explain, by reference to the peaks in the spectra, which compound has mass spectrum 2.

4 a The IR spectrum of propanamide is shown below.

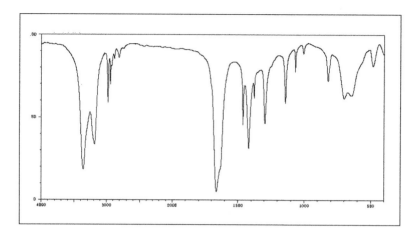

  i   Write the wave number of the peak given by the C=O bond.

  ii   Write the wave numbers of the two bands given by the N-H bond.

b   The C-13 NMR spectrum of propanamide is shown below.

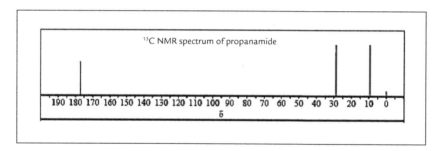

Assign the three peaks to the three carbon atoms in propanamide.

c   The mass spectrum of propanamide is shown below.

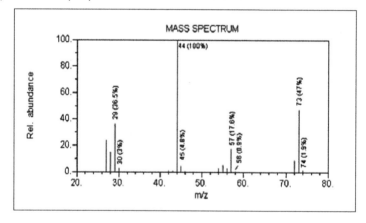

Draw the structure of particles with m/z equal to

  i   73

  ii   44

  iii   29

  iv   57

ISBN: 9780170355544 PHOTOCOPYING OF THIS PAGE IS RESTRICTED UNDER LAW.

5 a Compounds A, B and C all have the empirical formula C_2H_4O. Their IR, NMR and mass spectra are given below. Identify each compound, describing the features of each spectrum used in the identification.

    i Compound A

IR spectrum	NMR spectrum

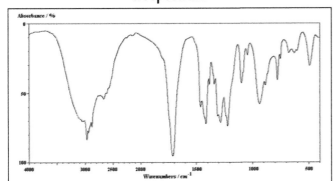
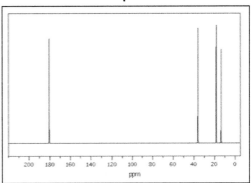

Mass spectrum

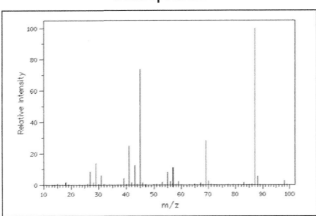

    ii Compound B

IR spectrum	C-13 NMR spectrum

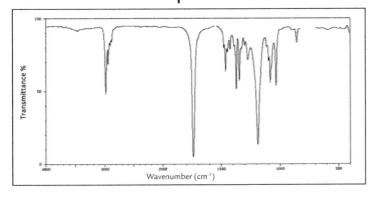
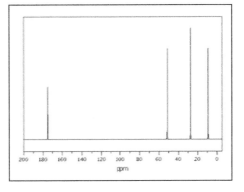

Mass spectrum

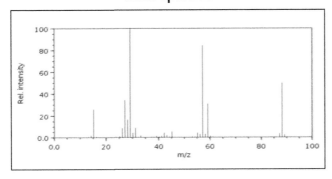

ISBN: 9780170355544   PHOTOCOPYING OF THIS PAGE IS RESTRICTED UNDER LAW.

iii Compound C

IR spectrum

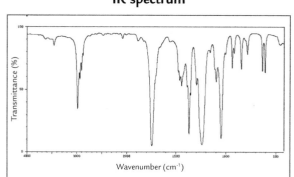

NMR spectrum

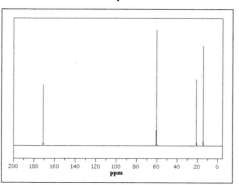

Mass spectrum

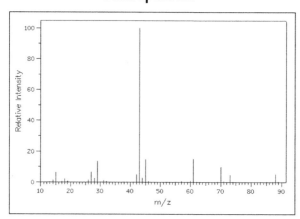

b If only one type of spectral data was available in this case, which one would be the most useful?

ISBN: 9780170355544   PHOTOCOPYING OF THIS PAGE IS RESTRICTED UNDER LAW.

Appendices

Appendix 1

Standard* reduction potentials

These reduction potentials are measured at 25°C and 101.3 kPa with all dissolved reactants 1 molL^{-1}.

Strongest reducing agent

Reduction reaction	Reduction potential E°/V
$Li^+(aq) + e \longrightarrow Li(s)$	–3.02
$K^+(aq) + e \longrightarrow K(s)$	–2.92
$Ca^{2+}(aq) + 2e \longrightarrow Ca(s)$	–2.76
$Na^+(aq) + e \longrightarrow Na(s)$	–2.71
$Mg^{2+}(aq) + 2e \longrightarrow Mg(s)$	–2.37
$Al^{3+}(aq) + 3e \longrightarrow Al(s)$	–1.68
$Zn^{2+}(aq) + 2e \longrightarrow Zn(s)$	–0.76
$Cr^{3+}(aq) + 3e \longrightarrow Cr(s)$	–0.74
$Fe^{2+}(aq) + 2e \longrightarrow Fe(s)$	–0.44
$Ni^{2+}(aq) + 2e \longrightarrow Ni(s)$	–0.23
$Sn^{2+}(aq) + 2e \longrightarrow Sn(s)$	–0.14
$Pb^{2+}(aq) + 2e \longrightarrow Pb(s)$	–0.13
$Fe^{3+}(aq) + 3e \longrightarrow Fe(s)$	–0.04
$2H^+(aq) + 2e \longrightarrow H_2(g)$	0.00
$SO_4^{2-}(aq) + 2H^+(aq) + 2e \longrightarrow SO_3^{2-}(aq) + H_2O(l)$	0.17
$Cu^{2+}(aq) + 2e \longrightarrow Cu(s)$	0.34
$I_2(s) + 2e \longrightarrow 2I^-(aq)$	0.54
$Fe^{3+}(aq) + e \longrightarrow Fe^{2+}(s)$	0.77
$Ag^+(aq) + e \longrightarrow Ag(s)$	0.80
$Hg^{2+}(aq) + 2e \longrightarrow Hg(l)$	0.85
$Br_2(l) + 2e \longrightarrow 2Br^-(aq)$	1.07
$O_2(g) + 4H^+(aq) + 4e \longrightarrow 2H_2O(l)$	1.23
$Cr_2O_7^{2-}(aq) + 14H^+(aq) + 6e \longrightarrow 2Cr^{3+}(aq) + 7H_2O(l)$	1.33
$Cl_2(g) + 2e \longrightarrow 2Cl^-(aq)$	1.36
$MnO_4^-(aq) + 8H^+(aq) + 5e \longrightarrow Mn^{2+}(aq) + 4H_2O(l)$	1.51
$F_2(g) + 2e \longrightarrow 2F^-(aq)$	2.87

Strongest oxidising agent

*In chemistry, IUPAC established **standard temperature and pressure** (informally abbreviated as **STP**) as a temperature of 273.15 K (0°C) and a pressure of 101.3 kPa. An unofficial, but commonly used, standard is a temperature of 298.15 K (25°C) and a pressure of 101.3 kPa. Other conditions are also considered 'standard' by other disciplines.

Appendix 2

Naming organic compounds

No. of C atoms	Prefix	Alkane	Alkene	Alkyl group	Alcohol	Aldehyde	Ketone	Acid
1	Meth-	-ane		-yl	-anol	-anal		-anoic acid
2	Eth-	-ane	-ene	-yl	-anol	-anal		-anoic acid
3	Prop-	-ane	-ene	-yl	-anol	-anal	-anone	-anoic acid
4	But-	-ane	-ene	-yl	-anol	-anal	-anone	-anoic acid
5	Pent-	-ane	-ene	-yl	-anol	-anal	-anone	-anoic acid
6	Hex-	-ane	-ene	-yl	-anol	-anal	-anone	-anoic acid
7	Hept-	-ane	-ene	-yl	-anol	-anal	-anone	-anoic acid
8	Oct-	-ane	-ene	-yl	-anol	-anal	-anone	-anoic acid

No. of C atoms	Prefix	Unsubstituted amide	Acyl chloride
1	Meth-	-anamide	-anoyl chloride
2	Eth-	-anamide	-anoyl chloride
3	Prop-	-anamide	-anoyl chloride
4	But-	-anamide	-anoyl chloride
5	Pent-	-anamide	-anoyl chloride
6	Hex-	-anamide	-anoyl chloride
7	Hept-	-anamide	-anoyl chloride
8	Oct-	-anamide	-anoyl chloride

Other

No. of C atoms	Chloroalkanes
1	Chloromethane
2	Chloroethane
3	Chloropropane
4	Chlorobutane
5	Chloropentane
6	Chlorohexane
7	Chloroheptane
8	Chloro-octane

Naming substituent groups

If the groups are different, they are written alphabetically, e.g. bromochloromethane (not chlorobromomethane) and 3-ethyl-2-methylhexanal (not 2-methyl-3-ethylhexanal).

If the groups are the same, the numbers are given in order, e.g. 2,3-dimethylhexanal.

For N substituted compounds such as substituted amines and amides, the name gives the group(s) on N, e.g. N-methylpropanamide.

Naming amines

Are described on page 194. The IUPAC method is similar to that for haloalkanes.

Appendix 3 Characteristic ^{13}C NMR Chemical Shifts

Carbon environment	δ (ppm)
C=O (in ketones)	205–220
C=O (in aldehydes)	190–200
C=O (in acids and esters)	170–185
C=C (in alkenes)	115–140
C≡C (in alkynes)	67–85
RCH_2OH	50–65

Carbon environment	δ (ppm)
RCH_2Cl	40–45
RCH_2NH_2	37–45
R_3CH	25–35
CH_3CO	20–30
R_2CH_2	16–25
RCH_3	10–15

ISBN: 9780170355544
PHOTOCOPYING OF THIS PAGE IS RESTRICTED UNDER LAW.

Appendix 4

Characteristic IR absorption frequencies of organic functional groups

All compounds
C–H peak at 2850–3000cm^{-1}

Alkene
=C–H medium peak 3010–6100 cm^{-1}, strong peak 675–1000 cm^{-1}

Alcohol
–O–H strong broad peak 3200–3600 cm^{-1}
–C–O strong peak 1050–1150 cm^{-1}

Aldehyde and ketone

Aldehyde
–C=O (carbonyl) strong peak 1670–1820 cm^{-1}
Aldehyde also has two medium peaks around 2720-2750 cm^{-1} and 2820-2850 cm^{-1}

Ketone
–C=O (carbonyl) strong peak 1670–1820 cm^{-1}
Ketone also has a strong peak at 1705–1725 cm^{-1} (often masked by the carbonyl peak)

Acid
–C=O (carbonyl) strong peak 1670–1820 cm^{-1}
–O–H strong broad peak 3200–3600 cm^{-1}
–C–O strong peak 1210–1320 cm^{-1}

Ester
–C=O (carbonyl) strong peak 1670–1820 cm^{-1}
–C–O strong peak 1000–1300 cm^{-1}

Amine
N–H medium peaks 3300–3500 cm^{-1} and 1600 cm^{-1}. Primary amines (–C–NH$_2$) have two bands
C–N medium peak 1080–1360 cm^{-1}

Amide
C=O strong peak 1640–1690 cm^{-1}
N–H see Amine above. Unsubstituted amides (–CONH$_2$) have two bands

Haloalkanes
–C–F strong peak 1100–1400 cm^{-1}
–C–Cl strong broad peak 540–785 cm^{-1}
–C–Br strong peak 510–650 cm^{-1}
–C–I strong peak 485–600 cm^{-1}

PHOTOCOPYING OF THIS PAGE IS RESTRICTED UNDER LAW.

ISBN: 9780170355544

Answers

Unit 1 Atomic properties and electron configuration

1 **a** ground state; **b** orbital; **c** electron configuration; **d** valance electrons; **e** electron shell; **f** electrons in atoms

2 **a** He: $1s^2$
   b B: $1s^22s^22p^1$ or $[He]2s^22p^1$
   c C: $1s^22s^22p^2$ or $[He]2s^22p^2$
   d O: $1s^22s^22p^4$ or $[He]2s^22p^4$
   e K: $1s^22s^22p^63s^23p^64s^1$ or $[Ar]4s^1$
   f Al: $1s^22s^22p^63s^23p^1$ or $[Ne]3s^23p^1$
   g P: $1s^22s^22p^63s^23p^3$ or $[Ne]3s^23p^3$
   h Cl: $1s^22s^22p^63s^23p^5$ or $[Ne]3s^23p^5$
   i Cr: $1s^22s^22p^63s^23p^63d^54s^1$ or $[Ar]3d^54s^1$
   j Fe: $1s^22s^22p^63s^23p^63d^64s^2$ or $[Ar]3d^64s^2$
   k Cu: $1s^22s^22p^63s^23p^63d^{10}4s^1$ or $[Ar]3d^{10}4s^1$
   l Zn: $1s^22s^22p^63s^23p^63d^{10}4s^2$ or $[Ar]3d^{10}4s^2$
   m Br: $1s^22s^22p^63s^23p^63d^{10}4s^24p^5$ or $[Ar]3d^{10}4s^24p^5$
   n Ar: $1s^22s^22p^63s^23p^6$ or $[Ne]3s^23p^6$

3 **a** H^+: no electrons
   b Mg^{2+}: $1s^22s^22p^6$ or $[Ne]$
   c Na^+: $1s^22s^22p^6$ or $[Ne]$
   d O^{2-}: $1s^22s^22p^6$ or $[Ne]$
   e K^+: $1s^22s^22p^63s^23p^6$ or $[Ar]$
   f Al^{3+}: $1s^22s^22p^6$ or $[Ne]$
   g P^{3-}: $1s^22s^22p^63s^23p^6$ or $[Ar]$
   h Cl^-: $1s^22s^22p^63s^23p^6$ or $[Ar]$
   i Cr^{3+}: $1s^22s^22p^63s^23p^63d^3$ or $[Ar]3d^3$
   j Fe^{2+}: $1s^22s^22p^63s^23p^63d^6$ or $[Ar]3d^6$
   k Cu^{2+}: $1s^22s^22p^63s^23p^63d^9$ or $[Ar]3d^9$
   l S^{2-}: $1s^22s^22p^63s^23p^6$ or $[Ar]$

4

At. no.	Charge	Configuration	Symbol
5	0	$[He]2s^22p^1$	B
6	0	$[He]2s^22p^2$	C
7	0	$[He]2s^22p^3$	N
8	0	$[He]2s^22p^4$	O
8	−2	$[Ne]$	O^{2-}
1	0	$1s^1$	H
1	+1	no electrons	H^+
33	0	$[Ar]3d^{10}4s^24p^3$	As
10	0	$[Ne]$ or $1s^22s^22p^6$	Ne
7	−3	$[Ne]$	N^{3-}
9	−1	$[Ne]$	F^-
10	0	$[Ne]$	Ne
12	0	$[Ne]3s^2$	Mg
11	+1	$[Ne]$	Na^+
8	−2	$[Ne]$	O^{2-}
13	+3	$[Ne]$	Al^{3+}
24	+3	$[Ar]3d^3$	Cr^{3+}
30	0	$[Ar]3d^{10}4s^2$	Zn
16	−2	$[Ar]$	S^{2-}
26	+3	$[Ar]3d^5$	Fe^{3+}

At. no.	Charge	Configuration	Symbol
28	+2	$[Ar]3d^8$	Ni^{2+}
33	0	$[Ar]3d^{10}4s^24p^3$	As
35	−1	$[Kr]$	Br^-

Unit 2 Periodic properties

1 **a** electronegativity; **b** oxidation number; **c** group of elements; **d** period of elements; **e** first ionisation enthalpy; **f** second ionisation enthalpy

2 **a** P, Al, Br. Br has four electron shells. Therefore it is larger than Al or P, which have three. Al attracts electrons with 13 protons while P has 15 protons. Al has weaker attraction for electrons and is therefore larger than P.
   b Ca is larger. It has 20 electrons in four electron shells while Ca^{2+} has 18 electrons in only three electron shells. Both have 20 protons so the attraction force of the nucleus is the same.
   c K^+, Cl^-, S^{2-}. All have the same electron configuration $[Ar]$. K^+ has 19 protons, Cl^- has 17 and S^{2-} 16. The attractive force of the nucleus for electrons is greatest for K^+ and least for S^{2-}. K^+ is therefore smallest and S^{2-} the largest.

3 **a** The first ionisation energy of an atom is the energy required to remove an electron from the isolated atom to give a +1 ion. The second ionisation energy for any atom is the energy required to remove an electron from the isolated +1 ion. The third ionisation energy for any atom is the energy required to remove an electron from the isolated +2 ion.
   b Mg with 12 protons has a stronger attraction for electrons than Na with 11 protons. It is therefore more difficult to remove an electron from Mg than Na.
   c The electron to be removed from Mg ($[Ne]3s^2$) in an s sublevel is closer to the nucleus than the electron to be removed from Al ($[Ne]3s^23p^1$), which is in a p sublevel. It therefore is more strongly attracted to the nucleus.
   d The first electron is removed from a neutral atom. The second electron is removed from a +1 ion and experiences a greater effective nuclear charge.
   e The second electron is removed from the second electron shell, which is closer to the nucleus than the first electron that is in the third electron shell. Also the second electron is removed from a +1 ion, while the first electron is removed from a neutral atom.
   f The third electron is in the second electron shell. It is closer to the nucleus than the first two electrons, which are in the third electron shell. An electron in a +2 ion is more difficult to remove than an electron in a +1 ion, which is more difficult to remove than an electron in a neutral atom.
   g The fourth electron is in a different electron energy

ISBN: 9780170355544 PHOTOCOPYING OF THIS PAGE IS RESTRICTED UNDER LAW.

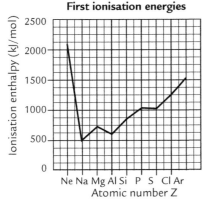

level. It is in the second energy level and closer to the nucleus than the first three electrons, which were in the third level. Also the electron is removed from an ion with a +3 charge.

4 **a**

First ionisation energies

y-axis: Ionisation enthalpy (kJ/mol), from 0 to 2500

x-axis: Ne Na Mg Al Si P S Cl Ar — Atomic number Z

b Across a period, the increasing number of protons means increasing attraction for electrons. Therefore it is more difficult to remove electrons. Ionisation energies increase.

c Al $(...3s^2 3p^1)$ has an electron in a p sublevel, which is further from the nucleus than the s electrons in Na $(...3s^1)$ and Mg $(...3s^2)$. This electron experiences a weaker nuclear force. S has four electrons in three $3p$ orbitals $(...3s^2 3p^4)$. Two electrons occupy a p orbital. Repulsion between these two electrons makes it easier to remove one. P $(...3s^2 3p^3)$ has singly occupied $3p$ orbitals and does not have this electron repulsion.

d Because of the greater electron shielding of nuclear charge in Ar. There are three electron shells in Ar, two in Ne.

5 Electrons in the outer shell of fluorine are shielded by only one electron shell and therefore experience a strong nuclear attraction. In its compounds, fluorine has a strong attraction for bonding electrons. The outer electrons of caesium are well shielded by the electrons in the five lower electron shells and experience weak nuclear attraction. Caesium has little attraction for bonding electrons and tends to lose its outer electron to form a +1 ion.

6 **a** Ionisation energy, electron affinity and electronegativity all increase across a period because of increasing nuclear charge. These properties decrease down a group due to progressively better nuclear shielding by electrons in lower energy levels.

b +1 and +2 ions, and -1 and -2 ions are quite stable whereas ions with higher charges are less stable. Compounds that require a loss or gain of three to four electrons to gain a stable outer shell tend to do so by sharing electrons to avoid high ionic charges.

Unit 3 Bonding and molecular shape

1 **a** Lewis structure; **b** anion; **c** ionic bond; **d** polar molecule; **e** polar covalent bond; **f** covalent bond; **g** cation; **h** VSEPR theory; **i** polyatomic ion; **j** non-polar covalent bond; **k** electronegativity

2 **a** ionic; **b** polar covalent; **c** non-polar covalent; **d** ionic; **e** ionic; **f** non-polar covalent; **g** polar covalent; **h** polar covalent; **i** ionic; **j** polar covalent

3 **a**
$$:\!\overset{..}{F}\!-\!\overset{..}{N}\!-\!\overset{..}{F}\!:$$
$$\underset{..}{:\!\overset{..}{F}\!:}$$
3 x 7 + 5 = 26 valence electrons

b
$$:\!\overset{..}{O}\!-\!\overset{..}{F}\!:$$
$$\underset{..}{:\!\overset{..}{F}\!:}$$
2 x 7 + 6 = 20 valence electrons

c
$$\underset{..}{:\!\overset{..}{F}\!:}$$
$$:\!\overset{..}{F}\!-\!\overset{..}{Br}\!\cdot$$
$$\underset{..}{:\!\overset{..}{F}\!:}$$
4 x 7 = 28 valence electrons

d
$$:\!\overset{..}{O}\!=\!\overset{..}{O}\!-\!\overset{..}{O}\!:$$
3 x 6 = 18 valence electrons

e
$$:\!\overset{..}{Cl}\!-\!\overset{..}{S}\!:$$
$$\underset{..}{:\!\overset{..}{Cl}\!:}$$
2 x 7 + 6 = 20 valence electrons

f
$$:\!\overset{..}{O}\!=\!\overset{..}{S}\!-\!\overset{..}{O}\!:$$
3 x 6 = 18 valence electrons

g
$$:\!\overset{..}{O}\!=\!C\!=\!\overset{..}{O}\!:$$
2 x 6 + 4 = 16 valence electrons

h
$$
\begin{array}{c}
H \\
| \\
H - Si - H \\
| \\
H
\end{array}
$$
4 x 1 + 4 = 8 valence electrons

i
$$
\begin{array}{c}
\quad\ :\!\overset{..}{F}\!: \\
:\!\overset{..}{F}\diagdown\ | \\
\qquad S: \\
:\!\overset{..}{F}\diagup\ | \\
\quad\ :\!\overset{..}{F}\!:
\end{array}
$$
4 x 7 + 6 = 34 valence electrons

j
$$
\begin{array}{c}
:\!\overset{..}{F}\!: \\
| \\
:\!\overset{..}{F}\!-\!\overset{..}{I}\!: \\
| \\
:\!\overset{..}{F}\!:
\end{array}
$$
4 x 7 = 28 valence electrons

k
$$
\begin{array}{c}
:\!\overset{..}{F}\diagdown\ \ :\ \diagup\overset{..}{F}\!: \\
\qquad Xe \\
:\!\overset{..}{F}\diagup\ \ \diagdown\overset{..}{F}\!:
\end{array}
$$
4 x 7 + 8 = 36 valence electrons

l
$$
\left[
\begin{array}{c}
H \\
| \\
H - N - H \\
| \\
H
\end{array}
\right]^{+}
$$
$$\underset{\text{valence e's}\ \ +1\ \text{charge}}{\underline{4 \times 1 + 5 - 1}} = 8$$

m
$$
\left[
\begin{array}{c}
:\!\overset{..}{O}\!-\!\overset{..}{N}\!=\!\overset{..}{O}\!: \\
| \\
:\!\overset{..}{O}\!: \\
\end{array}
\right]^{-}
$$
$$\underset{\text{valence e's}\ \ -1\ \text{charge}}{\underline{3 \times 6 + 5 + 1}} = 24$$

n
$$
\left[
\begin{array}{c}
:\!\overset{..}{O}\!: \\
| \\
:\!\overset{..}{O}\!-\!\overset{..}{S}\!-\!\overset{..}{O}\!: \\
| \\
:\!\overset{..}{O}\!: \\
\end{array}
\right]^{2-}
$$
$$\underset{\text{valence e's}\ \ -2\ \text{charge}}{\underline{4 \times 6 + 6 + 2}} = 32$$

o
$$
\left[
\begin{array}{c}
:\!\overset{..}{O}\!: \\
| \\
:\!\overset{..}{O}\!-\!\overset{..}{P}\!-\!\overset{..}{O}\!: \\
| \\
:\!\overset{..}{O}\!: \\
\end{array}
\right]^{3-}
$$
$$\underset{\text{valence e's}\ \ -3\ \text{charge}}{\underline{4 \times 6 + 5 + 3}} = 32$$

p
$$
\left[
\begin{array}{c}
:\!\overset{..}{O}\!: \\
| \\
:\!\overset{..}{O}\!=\!C\!-\!\overset{..}{O}\!: \\
\end{array}
\right]^{2-}
$$
$$\underset{\text{valence e's}\ \ -2\ \text{charge}}{\underline{3 \times 6 + 4 + 2}} = 24$$

q
$$
\left[
\begin{array}{c}
\quad\ :\!\overset{..}{F}\!: \\
:\!\overset{..}{F}\diagdown\ |\ \diagup\overset{..}{F}\!: \\
\qquad P \\
:\!\overset{..}{F}\diagup\ |\ \diagdown\overset{..}{F}\!: \\
\quad\ :\!\overset{..}{F}\!:
\end{array}
\right]^{-}
$$
$$\underset{\text{valence e's}\ \ -1\ \text{charge}}{\underline{6 \times 7 + 5 + 1}} = 48$$

4 **a** tetrahedral; **b** tetrahedral; **c** trigonal bipyramidal;
d trigonal planar; **e** tetrahedral; **f** trigonal planar;
g linear; **h** tetrahedral; **i** trigonal bipyramidal; **j** trigonal
bipyramidal; **k** octahedral; **l** tetrahedral; **m** trigonal
planar; **n** tetrahedral; **o** tetrahedral; **p** trigonal planar;
q octahedral

5 **a** trigonal pyramid; **b** bent (V-shaped); **c** T-shaped
(distorted tertahedral); **d** bent (V-shaped); **e** bent
(V-shaped); **f** bent (V-shaped); **g** linear; **h** tetrahedral;
i seesaw (distorted tetrahedral); **j** T-shaped; **k** square
planar; **l** tetrahedral; **m** trigonal planar; **n** tetrahedral;
o tetrahedral; **p** trigonal planar; **q** octahedral

6 **a** polar; **b** polar; **c** polar; **d** non-polar; **e** polar; **f** polar;
g non-polar; **h** non-polar; **i** polar; **j** polar; **k** non-polar

Unit 4 More molecular structures

1 **a** expanded octet **b** oxyacid **c** dative bond
d double bond **e** single bond

2 :C ≡ O: :C ≡ N : H
:C ≣ O: :C ≣ N : H

3 **a**
```
          ..O:
          ‖
H – O – N
          \
           O:
```
b H – O – N = O

c
```
       :O:
       |
H – O – S – O:
       |
      :O:
       |
       H
```
d
```
H – O – S – O:
       |
      :O:
       |
       H
```

e
```
        H
        |
       :O:
        |
H – O – P – O – H
        |
       :O:
        |
        H
```
f H – O – Cl :

g
```
       :O:
        |
H – O – Cl – O:
        |
       :O:
```

h
```
H – O – C – O – H
        ‖
       .O.
```

The bond angle H – O – X for all molecules is
approximately 109°. Approximate bond angles about
the central atom are: **a** 120°; **b** 120°; **c** 109°; **d** 109°;
e 109°; **f** 109°; around O; **g** 109°; **h** 120°.

4 **a**
```
H     H
 \   /
   B
   |
   H
```
b : N = O :

c
```
H – O – B – O – H
        |
       :O:
        |
        H
```

5 **a**
```
   H   H
   |   |
H – C – C – H
   |   |
   H   H
```
Four sets of electrons around each C, bond angles
all 109.5°.

b
```
   H   O:
   |   ‖
H – C – C – O – H
   |
   H
```
H – C – H bond angle is 109.5°/ C–O–H bond
angle is about 109°/ C–C=O bond angle is 120°.

c
```
   H   H
   |   |
H – C – C – O – H
   |   |
   H   H
```
All bond angles are about 109.5°.

6 **a**
```
   H              H
   |    ..        |    ..
H – C – O:     H – C – N – H
   |   |          |   |   |
   H   H          H   H   H
```
Bond angle C–O–H is smaller than C–N–H. Two
non-bonding pairs of electrons on O repel bonding
pairs, while there is only one non-bonding pair on
N to repel bonding pairs.

7 **a**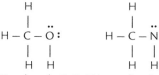

triangular bipyramid

b
```
     .S.
    /   \
 :O      O:
```
bent
(V-shaped)

c

octahedral or square
bipyramid

d
```
    Cl
   / | \
:F   | F:
   :F:
```
T-shaped

e
```
 :Cl   Cl:
    \  /
     S
    /  \
 :Cl   Cl:
```
seesaw or distorted
tetrahedral

f The outer shells of the atoms N and O can
accommodate a maximum of eight electrons. Also,
the atoms N and O are too small to be surrounded
by many other atoms.

8 **a** F has sufficiently strong attraction for electrons
that it can gain a share of electrons from Xe. Xe
can have an expanded octet.

b

The molecule has a square planar shape.

c He can accommodate only two electrons in its
outer shell, Xe in period 5 can accommodate up
to 2 × 50 = 100 electrons in its outermost shell. In
XeF$_4$, Xe has 12 electrons in its outermost shell.

ISBN: 9780170355544
PHOTOCOPYING OF THIS PAGE IS RESTRICTED UNDER LAW.

Revision One

Question I

1 a

	Symbol	Ground state electron configuration in *s, p, d* notation; [Ar] may be used as part of your answer
i	Na	$1s^2 2s^2 2p^6 3s^1$
ii	Fe	$[Ar]3d^6 4s^2$
iii	Fe^{3+} (or Mn^{2+})	$[Ar]3d^5$

b Na atom. The atom has three occupied electron shells, Na^+ has only two.

c **i** $Na(g) \rightarrow Na^+(g) + e$

ii The second electron (in the second electron shell) is closer to the nucleus than the first (in the third electron shell) and experiences a stronger attractive nuclear force. The second electron is removed from a +1 ion, the first was from an uncharged atom.

d Fluorine. S has three electron shells, F has two. Electrons can get closer to the nucleus in F. F therefore has a stronger attraction for bonding electrons.

2 a

Molecule	SF_2	SF_6
Lewis structure		
Name of shape	Bent (or V-shaped)	Octahedral

b S – F. F is more electronegative and has a greater $\delta+$ $\delta-$ share of the bonding electrons. Therefore the F end of the bond is slightly negative.

c The molecule is polar with the polarity shown. The centre of + charge is on S. The centre of – charge is midway between the 2 F's. The two centres are not at the same point. The dipoles of the two S–F bonds do not cancel.

Question II

1 a **i** $1s^2 2s^2 2p^6 3s^2 3p^5$

ii $[Ar]3d^{10}$

iii $[Ar]3d^5$

b P^{3-} (or S^{2-} or K^+ or Ca^{2+}...)

c Cl^- ion is larger. Both Cl and Cl^- have the 17 protons but Cl^- has one more electron. Repulsion between electrons in the same shell forces electrons to move further apart.

d $Ca(g) \rightarrow Ca^+(g) + e$

e The third electron is closer to the nucleus. It is in the third electron shell while the first two are in the fourth shell. Therefore it experiences a stronger attractive force from the nucleus. Also it is more difficult to remove an electron from a +2 ion than from a +1 ion or neutral atom.

f Both have two electron shells. F attracts bonding electrons with 9 protons, a stronger attraction than O, which has 8 protons.

2 a

Molecule	BF_2	NF_3
Lewis structure		
Name of shape	Trigonal planar	Trigonal pyramid

b N – F. F is more electronegative and has a greater $\delta+$ $\delta-$ share of the bonding electrons. Therefore the F end of the bond is slightly negative.

c BF_3 is non-polar. The centres of + and – charges are on B and the dipoles cancel. NF_3 is polar. The centre of + charge is on N, the centre of – charge is at the centre of the triangle formed by the three F atoms. The centres are not at the same point. The dipoles of the three N–F bonds do not cancel.

Question III

1 a

	Symbol	Ground state electron configuration in *s, p, d* notation; [Ar] may be used as part of your answer
i	As	$[Ar]3d^{10}4s^2 4p^3$
ii	Cr	$[Ar]3d^5 4s^1$
iii	Co^{3+}	$[Ar]3d^6$

b Fe^{2+} ion (No atom has the same electron configuration.)

c Cr has four occupied electron shells, Co^{3+} has only three and is therefore smaller.

d $Fe(g) \rightarrow Fe^+(g) + e$

e The first electron is removed from a neutral atom with 26 protons. This takes less energy than removing an electron from a +1 ion with the same number of protons.

f Energy must be supplied to overcome the attraction between a negatively charged electron and a positively charged nucleus.

g P is more electronegative. P has three electron shells, As has four. The nucleus of As is more shielded. As has less attraction for bonding electrons in its compounds.

2 a

Molecule	NCl_3	PCl_5
Lewis structure		
Name of shape	Trigonal pyramid	Trigonal bipyramid

b Cl is more electronegative than P and has a greater share of the bonding electrons. Polarity of the bond is therefore P – Cl.
$\delta+$ $\delta-$

c PCl_5 is non polar. The centres of + and – charge are at the same point, at P. The polar bonds are symmetrically placed around P in such a way that the dipoles cancel.

d The valence shell (n = 3) of P can accommodate up to 18 (2×3^2) electrons. In PCl_3 it has a share of 8 electrons. In PCl_5 it has a share of 10. The valence shell (n = 2) of N can accommodate a maximum of 8 (2×2^2) electrons, so only NCl_3 can form.

e Cl has three electron shells and N has two. Therefore the attraction of Cl's nucleus for bonding electrons is more shielded. However, the nucleus of Cl has 17 protons, which will attract bonding electrons more strongly than N, which has only 7 protons. The two opposing factors result in similar electronegativity values for the two atoms.

Question IV

1 a

	Symbol	Ground state electron configuration in *s, p, d* notation; [Ar] may be used as part of your answer
i	K	$[Ar]4s^1$
ii	Fe^{3+}	$[Ar]3d^5$
iii	Cu	$[Ar]3d^{10}4s^1$

b Mn^{2+}

c Fe^{3+}, Fe, K. Fe^{3+} has only three occupied electron shells while Fe and K have four. Therefore it is the smallest. Electrons in K and Fe have similar shielding of nuclear attraction by electron shells. Fe attracts electrons with 26 protons and is smaller than K, which attracts electrons with only 19 protons.

d $K(g) \longrightarrow K^+(g) + e$

e The second electron is in the third electron shell and so is closer to the nucleus than the first electron. Also it is more difficult to remove an electron from a +1 Ion than from a neutral atom.

f Cl is more electronegative than S. Both have the same number of electron shells and therefore have similar shielding of the nucleus. However, Cl has 17 protons to attract bonding electrons. S has 16 protons.

2 a

Molecule	CF_4	SF_4
Lewis structure	(Lewis structure of CF_4)	(Lewis structure of SF_4)
Name of shape	Tetrahedron	Seesaw or distorted tetrahedron

b The bond is polar covalent. The polarity is S – F.
$\delta+$ $\delta-$

c i Non-polar. Centres of + and – charges are at the same point, on C. The polar C–F bonds are placed in such a way that the dipoles cancel.

ii Polar. The centre of + charge is on S. The centre of – charge is between the two Cl's that are not in line with S. The two centres are not at the same point. (Although the dipoles of the two S–F bonds that are at 180° cancel, the dipoles of the other two S–F bonds do not.)

Unit 5 Bond enthalpy

1 a C=C double bond
b bond enthalpy
c bond forming
d bond order
e free radical

2 a $H–H(g) \longrightarrow H(g) + H(g)$ $\Delta_r H° = 436$ kJmol^{-1}
b $H–Cl(g) \longrightarrow H(g) + Cl(g)$ $\Delta_r H° = 431$ kJmol^{-1}
c $H–O–H(g) \longrightarrow 2H(g) + O(g)$ $\Delta_r H° = 926$ kJmol^{-1}
d $H–N–H \longrightarrow 3H(g) + N(g)$ $\Delta_r H° = 1173$ kJmol^{-1}
   |
   H

3 a The bond in H_2 is a single bond, in O_2 the bond is a double bond, and in N_2 the bond is a triple bond. The bonds become progressively stronger, but as the two bonds in a double bond are not the same, a double bond is not twice as strong as a single bond. Similarly, a triple bond is not three times as strong as a single bond.

b Electronegativity of F > Cl > I. Bonding electrons in HF are strongly attracted to F and the HF is the most polar of the three. More energy is required to return H's share of electrons before bond breaking.

c H is the smallest atom. Bonding electrons are not so shielded from the nucleus and are held strongly, making the bond more difficult to break. As the atoms become larger (I is largest), the bonding electrons are shielded from nuclear attraction and the bonds are weaker.

4 a $:N\equiv N:$
b $[:C\equiv N:]^-$
c $:C\equiv O:$
They all have triple bonds.

5

Molecule	Structural formula	Bonds broken	Energy required
H_2O	(structural formula)	all bonds	926 kJmol^{-1}
CH_4	(structural formula)	2 C–H bonds	826 kJmol^{-1}
C_2H_6	(structural formula)	all bonds	2826 kJmol^{-1}
C_2H_4	(structural formula)	the C=C bond	614 kJmol^{-1}
CH_3Cl	(structural formula)	a C–Cl and a C–H bond	752 kJmol^{-1}
NH_3	(structural formula)	all bonds	1173 kJmol^{-1}

Unit 6 Intermolecular attractions

1 a hydrogen bond
b dipole
c dispersion forces

ISBN: 9780170355544 PHOTOCOPYING OF THIS PAGE IS RESTRICTED UNDER LAW.

2 a A temporary dipole resulting from a momentary concentration of moving electrons induces dipoles on neighbouring molecules. The molecules attract.

b The polar water molecule (a permanent dipole) induces dipoles on nearby oxygen molecules.

c Van der Waals forces between particles increase with particle mass and size. Larger molecules are more easily polarised.

d Argon is non-polar, hydrogen chloride is polar. Dipole attraction plus dispersion forces between hydrogen chloride molecules are greater than the dispersion forces between argon atoms.

3 a CH_4 is non-polar. The other molecules are polar. The only forces between CH_4 molecules are dispersion forces (weak van der Waals forces between temporary dipoles). In other molecules there are dipole-dipole attractions and hydrogen bonds between molecules.

b i

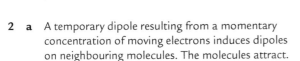

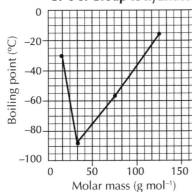

BP's of Group 15 hydrides

ii Hydrogen bonds can form between NH_3 molecules, but not in any of the other Group 15 hydrides. All the compounds have dipole-dipole attractions and dispersion forces between the molecules.

iii As the molecular size increases, dispersion forces between the molecules become stronger.

c Skin, hair, feathers, wood, wooden bridges, piles holding up wharves, pallets for concrete blocks and bricks, power poles.

d All have dispersion forces. The strength of these forces increase with particle size. All are polar, so all have dipole-dipole attractions between molecules. H_2O also has hydrogen bonds so it has the highest melting point. H_2Se has a higher melting point than H_2S as it has stronger dispersion forces.

e A is polar, B is non-polar. A is larger than B. Both have dispersion forces. These are stronger in A. A has dipole-dipole attractions and hydrogen bonds between molecules that B does not have. A has a higher boiling point.
C is large and non-polar. D is small and polar. Both have dispersion forces and these are stronger for C. D also had dipole-dipole attractions but C's stronger dispersion forces result in it have the higher boiling point.
E and F are polar. Both have dispersion forces, dipole-dipole attractions and hydrogen bonds between the molecules. The branched structure of F makes it difficult for molecules to pack closely together. F has a slightly lower boiling point.

f i Water is polar and has dispersion forces, dipole-dipole attractions and hydrogen bonds between its molecules. Petrol is non-polar and has only dispersion forces. However, the water molecule is small and the petrol molecule is large. Dispersion forces are stronger in petrol. These forces result in both substances being liquid at room temperature.

ii Both are non-polar. Both have dispersion forces between the molecules. The branched structure of methyl propane makes it difficult for molecules to pack closely together so that has a lower boiling point.

g N_2 is small and non-polar. There are weak dispersion forces between molecules. NH_3 is small but is polar. There are weak dispersion forces, stronger diploe-dipole attractions and hydrogen bonds between molecules. $CH_3CH_2NH_2$ is larger than NH_3 and also has all 3 types of intermolecular attraction. Therefore boiling point of N_2 is lowest, then NH_3, the $CH_3CH_2NH_2$.

h Wooden piles under houses and wharves, timber framing, floors, buildings.

4 a Only a small amount of energy is needed to break the weak dispersion forces in dry ice.

b Non-polar grease molecules with weak dispersion forces dissolve through the non-polar solvent to which they are attracted by the same weak dispersion forces.

c Non-polar oxygen is attracted to water by permanent dipole-induced dipole forces. The polar ammonia molecule is strongly attracted to polar water molecules.

Unit 7 Thermochemical principles

1 a The heat released when one mole of SO_2 gas at 298 K and 101.3 kPa pressure is formed from its elements in their standard states is 297 kJ.

b The heat released when one mole of carbon is burned in oxygen to form carbon dioxide and water, with all substances in their standard states, at standard pressure is 393 kJ.

2 a $C(s) + O_2(g) \longrightarrow CO_2(g)$ $\Delta_r H° = -393$ kJmol^{-1}
b $C(s) + O_2(g) \longrightarrow CO_2(g)$ $\Delta_r H° = -393$ kJmol^{-1}
c $C(s) + 2H_2(g) \longrightarrow CH_4(g)$ $\Delta_r H° = -75$ kJmol^{-1}
d $CH_4(g) + 2O_2(g) \longrightarrow CO_2(g) + 2H_2O(l)$
$\Delta_r H° = -889$ kJmol^{-1}
e $CO(g) + \frac{1}{2}O_2(g) \longrightarrow CO_2(g)$ $\Delta_r H° = -282$ kJmol^{-1}
f $\frac{1}{2}N_2(g) + 1\frac{1}{2}H_2(g) \longrightarrow NH_3(g)$ $\Delta_r H° = -46$ kJmol^{-1}
g $C_8H_{18}(l) + 12\frac{1}{2}O_2(g) \longrightarrow 8CO_2(g) + 9H_2O(l)$
$\Delta_r H° = -5464$ kJmol^{-1}
3 a $\frac{1}{2}H_2(g) + \frac{1}{2}N_2(g) + 1\frac{1}{2}O_2(g) \longrightarrow HNO_3(l)$
$\Delta_r H° = -174$ kJmol^{-1}
b $H_2(g) + S(s) + 2O_2(g) \longrightarrow H_2SO_4(l)$
$\Delta_r H° = -814$ kJmol^{-1}
c $Na(s) + \frac{1}{2}Cl_2(g) \longrightarrow NaCl(s)$ $\Delta_r H° = -411$ kJmol^{-1}
d $Na(s) + \frac{1}{2}Br_2(l) \longrightarrow NaBr(s)$ $\Delta_r H° = -360$ kJmol^{-1}
e $Na(s) + \frac{1}{2}I_2(s) \longrightarrow NaI(s)$ $\Delta_r H° = -288$ kJmol^{-1}
f $Na(s) + \frac{1}{2}I_2(s) + 2H_2(g) + O_2(g) \longrightarrow NaI.2H_2O(s)$
$\Delta_r H° = -883$ kJmol^{-1}
4 -888 kJmol^{-1}

ISBN: 9780170355544   PHOTOCOPYING OF THIS PAGE IS RESTRICTED UNDER LAW.

5 a −128 kJmol⁻¹ **b** exothermic
   The enthalpy change is negative.
6 a $2CH_3OH(l) \rightarrow C_2H_4(g) + 2H_2O(l)$
   $\Delta_rH° = -40$ kJmol⁻¹
   $4C_2H_4(g) + H_2(g) \rightarrow C_8H_{18}(l) \; \Delta_rH° = -416$ kJmol⁻¹
 b $8CH_3OH(l) + H_2(g) \rightarrow 8H_2O(l) + C_8H_{18}(l)$
   $\Delta_rH° = -576$ kJmol⁻¹
   The reaction is exothermic.
7 a −1526 kJmol⁻¹
 b Energy barrier for the reaction is very high.
8 a 13 kJmol⁻¹
 b Endothermic reaction. 1 mole of liquid particles
   has formed 1 mole of gas and 1 mole of liquid.
   The products in this reaction have more freedom
   of movement than the reactant. Energy has been
   absorbed to give kinetic energy to the particles.
9 i octane
 ii methane
10 a $CH_3OH(l) + 1\frac{1}{2}O_2(g) \rightarrow CO_2(g) + 2H_2O(l)$
   $\Delta_cH°$ (CH₃OH, l, 298 K) = −724 kJmol⁻¹
   $C_2H_5OH(l) + 3O_2(g) \rightarrow 2CO_2(g) + 3H_2O(l)$
   $\Delta_cH°$ (C₂H₅OH, l, 298 K) = −1365 kJmol⁻¹
 b CH_3OH, 22.6 kJg⁻¹; C_2H_5OH, 29.7 kJg⁻¹
   Ethanol is the better fuel on a per gram basis.
11 a −125 kJmol⁻¹
 b $2H_2(g) + O_2(g) \rightarrow 2H_2O(l)$
   $\Delta_cH° = -570$ kJmol⁻¹
   $CO(g) + \frac{1}{2}O_2(g) \rightarrow CO_2(g)$
   $\Delta_cH° = -282$ kJmol⁻¹
   $CH_3OH(l) + 1\frac{1}{2}O_2(g) \rightarrow CO_2(g) + 2H_2O(l)$
   $\Delta_cH° = -727$ kJmol⁻¹
   The reactants release more energy (852 kJ) than the
   products (727 kJ).
 c Gases tend to react quickly as they expand to fill the
   available space. A liquid needs to be aspirated, i.e.
   sprayed into small droplets to get a similar effect.
12 a 613.5 kJmol⁻¹
 b Some of the heat given off is used to heat the
   beaker, the air surrounding the experiment, the
   tripod holding the water container and the spirit
   burner. It is also given off as light. Sometimes,
   combustion may not be complete. In this
   experiment water is formed in the gas phase. In
   data book values, water is returned to its standard
   state, which is liquid. In this process, a further
   amount of heat energy is released.
 c The mass continuously decreases while the burner
   is being weighed. This is because ethanol is
   continuously evaporating.
13 a 125.3 J = 0.1253 kJ
 b 100.8 kJ
 c 100.9(3) kJ
 d −2225 kJmol⁻¹
14 a +
 b 0
 c +
 d −
 e −
 f −
15 a ΔH +, ΔS +
 b ΔH −, ΔS −
 c ΔH +, ΔS +
16 a The drive to minimum energy favours NaOH
   dissolving. Energy is given out in the form of heat.

The drive to maximum disorder also favours NaOH
dissolving. Dissolved ions have more freedom to
move than ions in a solid.
 b The drive to minimum energy favours NH_4NO_3
   remaining a solid. Energy is supplied by the
   surroundings for it to dissolve. The drive to maximum
   disorder favours NH_4NO_3 dissolving. Dissolved ions
   have more freedom to move than ions in a solid.

Revision Two

Question I
a i $C(s) + O_2(g) \rightarrow CO_2(g)$ $\quad \Delta H = -393$ kJmol⁻¹
 ii $C(s) + O_2(g) \rightarrow CO_2(g)$ $\quad \Delta H = -393$ kJmol⁻¹
 iii The reactions are the same.
b Fewer bonds are formed. Bond formation releases
   energy. Forming CO is therefore less exothermic.
c −724 kJmol⁻¹ **d** −75 kJmol⁻¹ **e** −888 kJmol⁻¹

Question II
a i $C(s) + 2H_2(g) \rightarrow CH_4(g)$ $\quad$ H = −75 kJmol⁻¹
 ii $CH_4(g) + 2O_2(g) \rightarrow CO_2(g) + 2H_2O(l)$
   H = −882 kJmol⁻¹
 iii Both are exothermic, ΔH values are negative. Heat
   is given out.
b i $C_4H_{10}(g) + 6\frac{1}{2}O_2(g) \rightarrow 4CO_2(g) + 5H_2O(g)$.
 ii −2876 kJmol⁻¹
 iii Combustion of butane is highly exothermic. The
   drive to minimum energy favours combustion.
   When butane burns, water is formed in the gas
   phase. 7½ moles of gaseous reactants have less
   freedom of movement than 9 moles of gaseous
   products. The drive to maximum entropy also
   favours combustion of butane.
 iv −907 kJmol⁻¹
 v Only some of the heat was used to raise the
   temperature of the water. Some was used to heat
   the surrounding air, the tripod, beaker and part
   of the lighter. Heat was also used to vaporise
   the water. The data book value includes the heat
   released when the water vapour formed is returned
   to its standard state (liquid). This would not have
   happened to water vapour formed in the experiment.
   To minimise losses, a foil top on the beaker would
   reduce vaporisation, the lighter could be placed
   closer to the beaker, the tripod could be removed
   and the beaker held with a clamp, the system could
   be surrounded by a screen to reduce air currents.
c i 124 kJmol⁻¹
 ii The entropy change is positive. 1 mole of gaseous
   reactants forms 3 moles of gaseous products. The
   products have more freedom of movement and can
   become more disordered.

Question III
1 a i $H_2(g) + \frac{1}{2}O_2(g) \rightarrow H_2O(l)$ **ii** $H_2O(l) \rightarrow H_2O(g)$
 b Energy must be supplied to give liquid particles
   enough kinetic energy to move into the gas phase.
 c −242 kJmol⁻¹
2 a By definition the enthalpy change is for the element
   to be formed from its elements in their standard
   states. Therefore there is no reaction.
 b −1368 kJmol⁻¹
 c i −106 kJmol⁻¹ **ii** 0.967 g **iii** Heat from the burning
   propane heated the beaker, the surrounding air
   and any equipment that was placed near the
   flame. Heat was also used to vaporise some
   water. Only the heat used to raise the temperature

ISBN: 9780170355544 PHOTOCOPYING OF THIS PAGE IS RESTRICTED UNDER LAW.

of the water was used in this calculation.

d i -1172 kJmol^{-1}

ii Negative. In the reaction, 9 gas moles form 4 gas and 6 liquid moles. Gases have more freedom of movement.

iii $\Delta_c H^\circ$ would be more negative. In combustion, NO_2 would be formed. Formation of NO shows incomplete combustion. A further amount of heat is released when NO forms NO_2.

e i ΔH is negative, heat is given out. ΔS is negative, particles in solid ice have less freedom to move.

ii ΔH is negative. All combustion reactions are exothermic. ΔS is positive. 7 gas moles have greater freedom of movement to become more disordered than 6 gas moles.

Unit 8 Redox reactions

1 a Reduction (CO has gained hydrogen or C has decreased its oxidation number from +2 to −2).

b Oxidation ($S_2O_3^{2-}$ has lost electrons, S has increased its oxidation number).

c Neither (no ion has changed its oxidation number).

d Reduction (As has reduced its oxidation number from +5 to +3 or AsO_4^{3-} has lost oxygen).

e Neither (no changes in oxidation number).

f Reduction (N has reduced its oxidation number from +5 to +4).

2 a $Cl_2(aq) + 2e \longrightarrow 2Cl^-(aq)$

b $2H^+(aq) + H_2O_2(aq) + 2e \longrightarrow 2H_2O(l)$

c $O_3(g) + 2H^+(aq) + 2e \longrightarrow O_2(g) + H_2O(l)$

d $S_2O_8^{2-}(aq) + 2e \longrightarrow 2SO_4^{2-}(aq)$

3 a $Fe^{2+}(aq) \longrightarrow Fe^{3+}(aq) + e$ oxidation
$14H^+(aq) + Cr_2O_7^{2-}(aq) + 6e \longrightarrow 2Cr^{3+}(aq) + 7H_2O(l)$ reduction
$14H^+(aq) + Cr_2O_7^{2-}(aq) + 6Fe^{2+}(aq) \longrightarrow 2Cr^{3+}(aq) + 6Fe^{3+}(aq) + 7H_2O(l)$

b $2I^-(aq) \longrightarrow I_2(aq) + 2e$ oxidation
$12H^+(aq) + 2IO_3^-(aq) + 10e \longrightarrow I_2(aq) + 6H_2O(l)$
reduction
$12H^+(aq) + 2IO_3^-(aq) + 10I^-(aq) \longrightarrow 6I_2(aq) + 6H_2O(l)$
Or: $6H^+(aq) + IO_3^-(aq) + 5I^-(aq) \longrightarrow 3I_2(aq) + 3H_2O(l)$

c $C_2H_5OH(aq) + H_2O(l) \longrightarrow CH_3COOH(aq) + 4H^+(aq) + 4e$ oxidation
$14H^+(aq) + Cr_2O_7^{2-}(aq) + 6e \longrightarrow 2Cr^{3+}(aq) + 7H_2O(l)$ reduction
$16H^+(aq) + 2Cr_2O_7^{2-}(aq) + 3C_2H_5OH(aq) \longrightarrow 4Cr^{3+}(aq) + 3CH_3COOH(aq) + 11H_2O(l)$

d $2I^-(aq) \longrightarrow I_2(aq) + 2e$ oxidation
$6H^+(aq) + BrO_3^-(aq) + 6e \longrightarrow Br^-(aq) + 3H_2O(l)$
reduction
$6H^+(aq) + BrO_3^-(aq) + 6I^-(aq) \longrightarrow Br^-(aq) + 3I_2(aq) + 3H_2O(l)$

e $Fe^{2+}(aq) \longrightarrow Fe^{3+}(aq) + e$ oxidation
$8H^+(aq) + MnO_4^-(aq) + 5e \longrightarrow Mn^{2+}(aq) + 4H_2O(l)$
reduction
$8H^+(aq) + MnO_4^-(aq) + 5Fe^{2+}(aq) \longrightarrow Mn^{2+}(aq) + 5Fe^{3+}(aq) + 4H_2O(l)$

4 a $C_2H_5OH(aq) + H_2O(l) \longrightarrow CH_3COOH(aq) + 4H^+(aq) + 4e$
$O_2(g) + 4H^+(aq) + 4e \longrightarrow 2H_2O(l)$

$C_2H_5OH(aq) + O_2(g) \longrightarrow CH_3COOH(aq) + H_2O(l)$

b i $2Cl^-(aq) \longrightarrow Cl_2(g) + 2e$
$MnO_2(s) + 4H^+(aq) + 2e \longrightarrow Mn^{2+}(aq) + 2H_2O(l)$

$4H^+(aq) + 2Cl^-(aq) + MnO_2(s) \longrightarrow Cl_2(g) + Mn^{2+}(aq) + 2H_2O(l)$

ii Chlorine gas escaping from the system.

iii Acid must be added carefully as the reaction is highly exothermic. Because chlorine is poisonous and corrosive, the reaction must be carried out in a fume hood.

c $[Ag_2S(s) + 2e \longrightarrow 2Ag(s) + S^{2-}(aq)] \times 3$
$[Al(s) + 4OH^-(aq) \longrightarrow Al(OH)_4^-(aq) + 3e] \times 2$

$3Ag_2S(s) + 2Al(s) + 8OH^-(aq) \longrightarrow 2Al(OH)_4^-(aq) + 6Ag(s) + 3S^{2-}(aq)$

d $[Fe(s) + 3OH^-(aq) \longrightarrow Fe(OH)_3(s) + 3e] \times 4$
$[O_2(g) + 2H_2O(l) + 4e \longrightarrow 4OH^-(aq)] \times 3$

$4Fe(s) + 3O_2(g) + 6H_2O(l) \longrightarrow 4Fe(OH)_3(s)$

e i acid solution
$2MnO_4^-(aq) + 5SO_2(g) + 2H_2O(l) \longrightarrow 2Mn^{2+}(aq) + 5SO_4^{2-}(aq) + 4H^+(aq)$

ii slightly basic solution
$4OH^-(aq) + 2MnO_4^-(aq) + 3SO_2(g) \longrightarrow 2MnO_2(s) + 3SO_4^{2-}(aq) + 2H_2O(l)$

f i +2 to +4

ii S has been oxidised.

iii −3 to −5

Unit 9 Electrochemical cells

1

Cell	a	b	c	d
zinc and copper	copper	zinc	+1.10 V	Zn to Cu
silver and iron	silver	iron	+1.24 V	Fe to Ag
copper and silver	silver	copper	+0.46 V	Cu to Ag
zinc and nickel	nickel	zinc	+0.51 V	Zn to Ni
nickel and silver	silver	nickel	+1.05 V	Ni to Ag
iron and zinc	iron	zinc	+0.32 V	Zn to Fe

Cell	e	f
zinc and copper	Zn(s) / Zn^{2+}(aq) // Cu^{2+}(aq) / Cu(s)	Zn(s) + Cu^{2+}(aq) $\longrightarrow$ Zn^{2+}(aq) + Cu(s)
silver and iron	Fe(s) / Fe^{2+}(aq) // Ag$^+$(aq) / Ag(s)	Fe(s) + 2Ag$^+$(aq) $\longrightarrow$ Fe^{2+}(aq) + 2Ag(s)
copper and silver	Cu(s) / Cu^{2+}(aq) // Ag$^+$(aq) / Ag(s)	Cu(s) + 2Ag$^+$(aq) $\longrightarrow$ Cu^{2+}(aq) + 2Ag(s)
zinc and nickel	Zn(s) / Zn^{2+}(aq) // Ni^{2+}(aq) / Ni(s)	Zn(s) + Ni^{2+}(aq) $\longrightarrow$ Zn^{2+}(aq) + Ni(s)
nickel and silver	Ni(s) / Ni^{2+}(aq) // Ag$^+$(aq) / Ag(s)	Ni(s) + 2Ag$^+$(aq) $\longrightarrow$ Ni^{2+}(aq) + 2Ag(s)
iron and zinc	Zn(s) / Zn^{2+}(aq) // Fe^{2+}(aq) / Fe(s)	Zn(s) + Fe^{2+}(aq) $\longrightarrow$ Zn^{2+}(aq) + Fe(s)

2 a silver and zinc
$Zn(s) / Zn^{2+}(aq) // Ag^+(aq) / Ag(s)$

b $E°_{cell} = 1.56\ V$

3 a $MnO_4^-(aq) + 8H^+(aq) + 5e \rightarrow Mn^{2+}(aq) + 4H_2O(l)$

b $2Cl^-(aq) \rightarrow Cl_2(aq) + 2e$

c $2MnO_4^-(aq) + 16H^+(aq) + 10Cl^-(aq) \rightarrow 2Mn^{2+}(aq) + 8H_2O(l) + 5Cl_2(aq)$

d From the chloride/chlorine half-cell to the permanganate/manganous half-cell.

e $0.11\ V$

f The smell of chlorine gas would become apparent at one electrode. The fading of the purple permanganate solution may also be noticeable.

4 a $NO_3^-(aq) + 2H^+(aq) + e \rightarrow NO_2(g) + H_2O(l)$ reduction

$Fe^{2+}(aq) \rightarrow Fe^{3+}(aq) + e$ oxidation

$NO_3^-(aq) + Fe^{2+}(aq) + 2H^+(aq) \rightarrow NO_2(g) + Fe^{3+}(aq) + H_2O(l)$

b $0.03\ V$

c NO_3^- oxidises Fe^{2+} to Fe^{3+}.

d Yes.

5 a

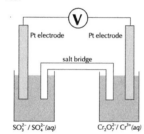

b $+1.16\ V$

c $Cr_2O_7^{2-}$ is stable only in acid solution and H^+ ions are required for the reaction.

d $Cr_2O_7^{2-}(aq) + 14H^+(aq) + 6e \rightarrow 2Cr^{3+}(aq) + 7H_2O(l)$

e $SO_3^{2-}(aq) + H_2O(l) \rightarrow SO_4^{2-}(aq) + 2H^+(aq) + 2e$

f $Cr_2O_7^{2-}(aq) + 8H^+(aq) + 3SO_3^{2-}(aq) \rightarrow 2Cr^{3+}(aq) + 4H_2O(l) + 3SO_4^{2-}(aq)$

g The orange solution at the cathode fades as orange dichromate forms green chromic ions.

6 a $8H^+(aq) + MnO_4^-(aq) + 5e \rightarrow Mn^{2+}(aq) + 4H_2O(l)$

b $3I^-(aq) \rightarrow I_3^-(aq) + 2e$

c It provides a pathway for the movement of ions between the two half-cells thus completing the circuit. Salt bridge. Contains a concentrated solution of a strong electrolyte, e.g. solution of sodium or potassium cations, and nitrate or sulfate anions. (The ions must not themselves undergo oxidation or reduction under the conditions of the cell, and they must not react with any species in either of the half cells. For example, the ions in the salt bridge should not form precipitates with any ions in the half cells.)

d $16H^+(aq) + 2MnO_4^-(aq) + 15I^-(aq) \rightarrow 2Mn^{2+}(aq) + 5I_3^-(aq) + 8H_2O(l)$

e The purple colour fades slightly in the half-cell containing electrode A and the brown colour becomes more intense in the half-cell containing electrode B.

f From electrode B to electrode A.

g $Pt(s), I^-(aq) / I_3^-(aq) // MnO_4^-(aq) / Mn^{2+}(aq), Pt(s)$

7 a cathode   **b** salt bridge   **c** electrochemical cell
d cell voltage, $E°_{cell}$ **e** standard electrode potential
f anode   **g** half-cell

Unit 10 Spontaneous reactions

1 a spontaneous redox reaction

b strong oxidant

c strong reductant

2 a $Cl_2(aq) + 2e \rightarrow 2Cl^-(aq)$

b $2I^-(aq) \rightarrow I_2(aq) + 2e$

c $Cl_2(aq) + 2I^-(aq) \rightarrow 2Cl^-(aq) + I_2(aq)$

d $E°_{cell} = E°_{\text{for the reduction}} - E°_{\text{for the oxidation}}$
Chlorine is reduced, iodide is oxidised so:
$E°_{cell} = E°_{(Cl_2/Cl^-)} - E°_{(I_2/I^-)} = +1.40\ V - 0.62\ V = 0.78\ V$
Positive, therefore reaction is spontaneous.

e The reaction mixture would become blue-black.

3 a Possible. $E°_{cell} = 1.40 - 0.77 = 0.63\ V > 0$

b Not possible. $E°_{cell} = 0.80 - 1.40 = -0.60\ V < 0$

c Possible. $E°_{cell} = 1.77 - 0.77 = 1.00V > 0$

d $E°_{(Br_2/Br^-)}$ is not as positive as $E°_{(Cl_2/Cl^-)}$

e $Cu(s) + Cl_2(g) \rightarrow CuCl_2(s)$

f $E°_{cell} = 1.40\ V - 0.34\ V = 1.06\ V$

g Fe^{3+} would oxidise I^- to I_2. (I^- would reduce Fe^{3+} to Fe^{2+}.)
$E°_{cell} = 0.15\ V > 0$

h Mg, Zn, Fe

i Mg, Zn, Fe, Pb

4 a $2Cl^-(aq) \rightarrow Cl_2(g) + 2e$

b $MnO_2(s) + 4H^+(aq) + 2e \rightarrow Mn^{2+}(aq) + 2H_2O(l)$

c $-0.17\ V$

d The reaction is performed under non-standard conditions. Concentrated HCl (not 1 molL^{-1}) is used and complete removal of Cl_2 gas ensures that the reaction is driven to the right.

5 a Ag^+

b Mg^{2+}

c Mg

d Ag

e H_2O_2

Unit 11 Electrolytic cells and commercial electrical cells

1 a zinc case, $Zn \rightarrow Zn^{2+} + 2e$

b $MnO_2 + 4NH_4^+ + 2e \rightarrow Mn^{2+} + 4NH_3 + 2H_2O$

c $Zn + MnO_2 + 4NH_4^+ \rightarrow Zn(NH_3)_4^{2+} + Mn^{2+} + 2H_2O$

d The ammonia generated is bound up in a complex ion with Zn^{2+}.

2 a MnO_2 has gained hydrogen and so has been reduced. (By balancing the half-equation, it can be seen that MnO_2 has gained electrons and so has been reduced.)

b $Zn + MnO_2 + 2H_2O \rightarrow Mn(OH)_2 + Zn(OH)_2$

c zinc, negative

d No gases are produced in the reactions and this prolongs the life of the battery.

3 a $Zn + 2OH^- \rightarrow ZnO + H_2O + 2e$

b $HgO + H_2O + 2e \rightarrow Hg + 2OH^-$

c $Zn + HgO \rightarrow ZnO + Hg$ (note that no H_2O or charged ions are formed or used up by the overall reaction)

d zinc

e zinc

4 a $Li \rightarrow Li^+ + e$

b MnO_2

c $MnO_2(s) + 2H_2O(l) + 2e \rightarrow Mn(OH)_2(s) + 2OH(aq)$

d Li

ISBN: 9780170355544   PHOTOCOPYING OF THIS PAGE IS RESTRICTED UNDER LAW.

e Lithium is lighter and less toxic than mercury. It is also a stronger reducing agent.

5 a Pb, the lead plates

b $Pb(s) + SO_4^{2-}(aq) \rightarrow PbSO_4(s) + 2e$

c It decreases. The equation shows both H^+ and SO_4^{2-} ions are removed from the solution.

d The density of the liquid in the cell will decrease.

e $2PbSO_4(s) + 2H_2O(l) \rightarrow Pb(s) + PbO_2(s) + 4H^+(aq) + 2SO_4^{2-}(aq)$

f The increase in density gives a viscous solution. Ions are not able to move so freely and the battery output is decreased.

g Advantages: can be readily recharged and is inexpensive. Disadvantages: heavy and cumbersome, lead compounds are toxic, sulfuric acid is corrosive, lead and sulfuric acid are difficult to dispose of when the battery becomes non-functional.

6 a $Cd + 2OH^- \rightarrow Cd(OH)_2 + 2e$

b $Ni_2O_4 + 4H_2O + 6e \rightarrow 2NiOH + 6OH^-$

c $3Cd + 4H_2O + Ni_2O_4 \rightarrow 3Cd(OH)_2 + 2NiOH$

d $2NiOH + 3Cd(OH)_2 \rightarrow Ni_2O_4 + 4H_2O + 3Cd$

7 a $H_2(g) + 2OH^-(aq) \rightarrow 2H_2O(l) + 2e$

b $O_2(g) + 2H_2(g) \rightarrow 2H_2O(l)$

c Large amounts of energy produced, lightweight H_2 and O_2, low pollution, the water produced can be used for other purposes.

d carbon

e carbon dioxide

f $C + O_2 \rightarrow CO_2$

g Advantages: carbon can be obtained without electrolysis, cell will be safer than having an explosive mixture of gases, carbon is inexpensive. Disadvantages: CO_2 is a pollutant, cell will be heavier.

8 a $Pt(s) / Fe^{3+}(aq), Fe^{2+}(aq) // Zn^{2+}(aq) / Zn(s)$. The cathode is the electrode where reduction occurs. It is the platinum (or other inert material such as C, graphite) electrode immersed in a solution of Fe^{2+} and Fe^{3+}. The anode is the zinc electrode immersed in a solution of Zn^{2+} ions. The half-cells are connected by a salt bridge, containing a redox inert salt such as potassium sulfate.

b 1.5 V batteries are used in everyday equipment, so this battery could be used in place of a dry cell, for example.

9 a fuel cell   b dry cell

c electrolytic cell   d electrochemical cell

e lead-acid cell

Revision Three

Question I

1 a $8H^+(aq) + MnO_4^-(aq) + 5e \rightarrow Mn^{2+}(aq) + 4H_2O(l)$

b $Fe^{2+}(aq) \rightarrow Fe^{3+}(aq) + e$

c $8H^+(aq) + MnO_4^-(aq) + 5Fe^{2+}(aq) \rightarrow Mn^{2+}(aq) + 5Fe^{3+}(aq) + 4H_2O(l)$

d A purple solution reacts with a very pale green solution to form a colourless solution.

e Sulfuric acid (or H_2SO_4)

f $Cr_2O_7^{2-}(aq) + 14H^+(aq) + 6e \rightarrow 2Cr^{3+}(aq) + 7H_2O(l)$

g $SO_3^{2-}(aq) + H_2O(l) \rightarrow SO_4^{2-}(aq) + 2H^+(aq) + 2e$

h $Cr_2O_7^{2-}(aq) + 8H^+(aq) + 3SO_3^{2-}(aq) \rightarrow 2Cr^{3+}(aq) + 4H_2O(l) + 3SO_4^{2-}(aq)$

i An orange solution reacts with a colourless solution to form a green solution.

j $Cu^{2+}(aq) + I^-(aq) + e \rightarrow CuI(s)$

k $2I^-(aq) \rightarrow I_2(aq)$

l $2Cu^{2+}(aq) + 4I^-(aq) \rightarrow 2CuI(s) + I_2(aq)$

m A blue solution reacts with a colourless solution to form a white precipitate in a brown solution (that looks like mustard).

2 a +6   b +3   c +6

3 a $Cu(s) / Cu^{2+}(aq) // Ag^+(aq) / Ag(s)$ b $Ag^+(aq) / Ag(s)$ c $Cu(s) \rightarrow Cu^{2+}(aq) + 2e$ d $Cu(s) + 2Ag^+(aq) \rightarrow 2Ag(s) + Cu^{2+}(aq)$ e $E°_{cell} = 0.80 - 0.34 = 0.46 V$ f from the copper half-cell to the silver half-cell g Cu electrode. Solid, $Cu(s)$, has formed aqueous Cu^{2+} ions. h Sodium nitrate (or potassium nitrate). The salt bridge allows movement of ions between the half-cells, thus completing the circuit.

4 a Br_2 b $S_2O_3^{2-}$ c i If I^- is oxidised, $E°_{cell} = 0.08 - 0.54 = -0.46 V < 0$. No, the reaction cannot occur. ii If Cu is oxidised, $E°_{cell} = 1.09 - 0.34 = 0.75 V > 0$. Copper would be seen to dissolve, the orange bromine colour would fade and the blue colour of $Cu^{2+}(aq)$ would appear. d i $Br_2(aq) + 2I^-(aq) \rightarrow I_2(aq) + 2Br^-(aq)$ ii $E°_{cell} = 1.09 - 0.54 = 0.55 V$ iii The orange colour of bromine would fade in one half-cell, and the brown colour of iodine would intensify in the other half-cell.

5 I a i $2MnO_2(s) + 2NH_4^+(aq) + 2e \rightarrow Mn_2O_3(s) + H_2O(l) + 2NH_3(aq)$

ii $Zn(s) \rightarrow Zn^{2+}(s) + 2e$

b i carbon ii zinc

c i It is portable. ii It cannot be recharged.

II a i $HgO(s) + H_2O(l) + 2e \rightarrow Hg(l) + 2OH^-(aq)$

ii $2Zn(s) + 2OH^-(aq) \rightarrow 2ZnO(s) + H_2O(l)$

b Mercury and its compounds are poisonous and must be contained in the cell.

III a $2PbSO_4(s) + 2H_2O(l) \rightarrow Pb(s) + PbO_2(s) + 4H^+(aq) + 2SO_4^{2-}(aq)$

b sulfuric acid

c Sulfuric acid (dense) has been removed from the solution by the reaction and water (less dense) has been formed. There is not sufficient H^+ present for the cell reaction to take place.

d i The reaction is readily reversed by a car's alternator.

ii Contents are poisonous and corrosive (also battery is heavy).

IV a methane

b $CH_4(g) + 2O_2(g) \rightarrow CO_2(g) + 2H_2O(l)$

c There must be a continuous supply of gas (and they are expensive).

6 a $2H_2O(l) + 4e \rightarrow O_2(g) + 4H^+(aq)$

b $2H_2O(l) + 2e \rightarrow H_2(g) + 2OH^-(aq)$

c $2H_2O(l) \rightarrow 2H_2(g) + O_2(g)$

d Both pieces of litmus will be blue.

Question II

1 a Solution 1, Fe^{2+} and Fe^{3+}. Solution 2, $Cr_2O_7^{2-}$ and Cr^{3+}. Solution 3, Na_2SO_4. Electrode 1, Pt (or C). (Note: Solutions 1 and 2 are interchangeable for the diagram.) b $Pt(s) / Fe^{2+}(aq), Fe^{3+}(aq) // Cr_2O_7^{2-}(aq), Cr^{3+}(aq) / Pt(s)$ c $E°_{cell} = 1.33 - 0.77 = 0.56 V$ d From the Fe^{2+}/Fe^{3+} half-cell to the $Cr_2O_7^{2-}/Cr^{3+}$ half-cell. e The $Cr_2O_7^{2-}/Cr^{3+}$ half-cell. Reduction takes place at the cathode.

2 a $2I^-(aq) \rightarrow I_2(aq)$

   b $Cu^{2+}(aq) + I^-(aq) + e \longrightarrow CuI(s)$

   c $2Cu^{2+}(aq) + 4I^-(aq) \longrightarrow 2CuI(s) + I_2(aq)$

   d Thiosulfate reduces brown I_2 to colourless I^- ions.

3 **a** MnO_2

   b Al

   c $E^\circ_{cell} = 1.23 - (-0.76)$ V $= 1.99$ V

   d MnO_2/Mn^{2+} and Al^{3+}/Al. $E^\circ_{cell} = 1.23 - (-1.66)$ V $= 2.89$ V

   e For the reaction, $E^\circ_{cell} = -1.66 - 0.54$ V $= -2.20$ V < 0. No, Al^{3+} cannot be reduced by I^-.

Question III

1 **a** $Zn(s) / Zn^{2+}(aq) // Fe^{3+}(aq), Fe^{2+}(aq) / Pt(s)$

   b $Fe^{3+}(aq), Fe^{2+}(aq) / Pt(s)$

   c Zn

   d $Zn(s) + 2Fe^{3+}(aq) \longrightarrow Zn^{2+}(aq) + 2Fe^{2+}(aq)$

   e $E^\circ_{cell} = 0.77 - (-0.76) = 1.53$ V

   f From the zinc electrode to the platinum electrode

   g **i** potassium sulfate

       ii It should be a strong electrolyte and must not react with any species in the cell (e.g. nitrate ions could oxidise Fe^{2+} to Fe^{3+} so potassium nitrate is not used).

       iii It allows ions to move between the half-cells completing the circuit.

2 **a** Fe^{3+}

   b Ni

   c $E^\circ_{cell} = 0.77 - 0.17 = 0.60$ V > 0. Yes

   d $Pt(s) / Fe^{3+}(aq), Fe^{2+}(aq) // Ni^{2+}(aq) / Ni(s)$. $E^\circ_{cell} = 0.77 - (-0.23) = 1.00$ V

   e **i** Pt or C

       ii Ni

   f **i** Zn could be oxidised by H^+ to give Zn^{2+} and H_2, or it can be oxidised by NO_3^- to give Zn^{2+} and NO_2. For the first reaction, $E^\circ_{cell} = 0.76$ V. For the second reaction, $E^\circ_{cell} = 1.55$ V.

       ii Cu could not be oxidised by H^+ but it can be oxidised by NO_3^- to give Cu^{2+} and NO_2. For the first reaction, $E^\circ_{cell} = -0.34$ V. For the second reaction, $E^\circ_{cell} = 0.45$ V.

3 **a** anode

   b anode reaction $Zn \longrightarrow Zn^{2+} + 2e$, cathode reaction $MnO_2 + 4NH_4^+ + 2e \longrightarrow Mn^{2+} + 4NH_3 + 2H_2O$, overall reaction $Zn + MnO_2 + 4NH_4^+ \longrightarrow Zn(NH_3)_4^{2+} + Mn^{2+} + 2H_2O$

Question IV

1 **a** +7   **b** +4

   c +6   **d** +3

2 **a** $8H^+(aq) + MnO_4^-(aq) + 5e \longrightarrow Mn^{2+}(aq) + 4H_2O(l)$

   b $(COOH)_2(aq) \longrightarrow 2CO_2(g) + 2H^+(aq) + 2e$

   c $6H^+(aq) + 2MnO_4^-(aq) + 5(COOH)_2(aq) \longrightarrow 2Mn^{2+}(aq) + 10CO_2(g) + 8H_2O(l)$

   d The purple solution (MnO_4^-) reacts with the hot colourless solution ($(COOH)_2$) to form a colourless solution.

3 **a** When a copper half-cell is connected to a hydrogen half-cell under standard conditions (25°C, 101.3 kPa, all metals are pure, all dissolved reactants 1 molL^{-1}), the measured voltage is 0.34 V and reduction is observed in the copper half-cell.

   b $Fe(s) / Fe^{2+}(aq) // Cu(s) / Cu^{2+}(aq)$

   c $Cu(s) / Cu^{2+}(aq)$

   d $Fe(s) \longrightarrow Fe^{2+}(aq) + 2e$

   e Fe

   f $Fe(s) + Cu^{2+}(aq) \longrightarrow Cu(s) + Fe^{2+}(aq)$

   g $E^\circ_{cell} = 0.34 - (-0.44)$ V $= 0.78$ V

   h From the Fe electrode to the Cu electrode

   i In one half-cell, the iron electrode dissolves. Copper metal is deposited in the other half-cell.

   j Sodium sulfate. It allows ions to move between the half-cells thus completing the circuit.

4 **a** MnO_4^-

   b $S_2O_3^{2-}$

   c **i** No. $E^\circ_{cell} = 0.08 - 0.54 = -0.46$ V < 0

       ii Brown pieces of copper would dissolve in the orange bromine, and the blue colour of $Cu^{2+}(aq)$ would appear. $E^\circ_{cell} = 1.09 - 0.34$ V $= 0.75$ V > 0

   d $Fe^{2+}, I^-, Cu, S_2O_3^{2-}$

5 **a** $2Cl^-(aq) \longrightarrow Cl_2(g) + 2e$

   b $2H_2O(l) + 2e \longrightarrow H_2(g) + 2OH^-(aq)$

   c In a solution of Na^+ and Cl^- ions, the Cl^- ions are removed as Cl_2 gas at the anode. OH^- ions are formed at the cathode. Na^+ and OH^- ions are left in the solution.

6 **a** $Zn(s) / Zn^{2+}(aq) // Ag(s) / Ag^+(aq)$

   b Solutions: 1 $Zn(NO_3)_2$, 2 $AgNO_3$, 3 $NaNO_3$. Electrodes: 1 Zn, 2 Ag

   c Ag

   d reduction

   e $Zn(s) + Ag^+(aq) \longrightarrow Ag(s) + Zn^{2+}(aq)$

   f $E^\circ_{cell} = 0.80 - (-0.76)$ V $= 1.56$ V

   g $NaNO_3$ was chosen because it does not form a precipitate with Ag^+ ions. Both Cl^- and SO_4^{2-} form precipitates with Ag^+ ions.

   h From Zn electrode to Ag electrode.

   i From the silver half-cell to the zinc half-cell.

Unit 12 Systems in equilibrium

1 **a** Neither, 2 gas moles $\longrightarrow$ 2 gas moles

   b Reactants, 4 gas moles $\longrightarrow$ 2 gas moles

   c Products, 7 gas moles $\longrightarrow$ 8 gas moles

   d Reactants, 2 gas moles $\longrightarrow$ 0 gas moles

   e Reactants, 2 gas moles $\longrightarrow$ 1 gas mole

2 **a** Remove CO — concentration/backward reaction (that forms reactants) favoured/decreased/increased/decreased/no change

       Increase pressure by decreasing the volume — concentration/forward reaction favoured/increased/increased/increased/no change

       Increase pressure by adding an unreactive gas — neither/no change/no change/ no change/no change/no change

       Heat the system — temperature/backward reaction favoured/increased/increased/decreased/decreased

       Add a catalyst — neither/no change/no change/no change/no change/no change

   b Add CH_3OH — concentration/forward reaction/increased/increased/increased/no change

       Remove CH_3OCH_3 — concentration/forward reaction favoured/decreased/decreased/increased/no change

       Increase pressure by decreasing the volume — concentration/no change/increased/increased/increased/no change

       Increase pressure by adding an unreactive gas — neither/no change/no change/ no change/no change/no change

       Heat the system — temperature/backward reaction favoured/increased/decreased/decreased/decreased

       Add a catalyst — neither/no change/no change/no change/no change/no change

ISBN: 9780170355544
PHOTOCOPYING OF THIS PAGE IS RESTRICTED UNDER LAW.

Unit 13 Solubility of ionic solids

1 a reaction quotient b saturated solution
   c solubility constant d common ion effect
   e ionise

2 a $[Na^+][Cl^-]$
   b $[Ca^{2+}][SO_4^{2-}]$
   c $[Na^+]^2[SO_4^{2-}]$
   d $[Ca^{2+}][OH^-]^2$
   e $[Pb^{2+}][Cl^-]^2$
   f $[Fe^{3+}][OH^-]^3$
   g $[Li^+]^2[CO_3^{2-}]$
   h $[Fe^{3+}]^2[SO_4^{2-}]^3$
   i $[Pb^{2+}][SO_4^{2-}]$
   j $[Ag^+]^2[SO_4^{2-}]$
   k $[Ca^{2+}][CO_3^{2-}]$
   l $[Mg^{2+}][Br^-]^2$
   m $[Pb^{2+}][CrO_4^{2-}]$
   n $[Pb^{2+}][I^-]^2$
   o $[Cu^{2+}][OH^-]^2$

3 a $[Ag^+] = [Cl^-] \gg [H_3O^+] = [OH^-]$
   b $[Ca^{2+}] > [CO_3^{2-}] \gg [OH^-] > [HCO_3^-] > [H_2CO_3] > [H_3O^+]$
   c $[Cl^-] = 2[Pb^{2+}] \gg [H_3O^+] = [OH^-]$
   d $[Ag^+] = 2[SO_4^{2-}] \gg [H_3O^+] = [OH^-]$

4 a 1.96×10^{-10}
   b 4.00×10^{-6}
   c 5.04×10^{-9}
   d 2.84×10^{-11}

5 a 1.00×10^{-5}
   b $[Mg^{2+}] = 1.36 \times 10^{-4}$, $[OH^-] = 2.72 \times 10^{-4}$
   c 0.27 g L⁻¹
   d i 4.76 g L⁻¹
      ii 3.54 g
   e i 3.91×10^{-4} mol L⁻¹, and 7.44×10^{-3} g L⁻¹
      ii Sodium fluoride is soluble (and therefore the fluoride is available), while calcium fluoride is virtually insoluble.
      iii Because calcium fluoride has such a low solubility, teeth will not dissolve.

6 a 0.001 mol L⁻¹ $MgCO_3$, $K_s(MgCO_3) = 1 \times 10^{-5}$, ion product = 1.00×10^{-6}. Ion product < K_s, therefore solution possible.
   b 0.01 mol L⁻¹ $Ca(OH)_2$, $K_s(Ca(OH)_2) = 4 \times 10^{-6}$, ion product = 4.00×10^{-6}. Ion product = K_s, therefore a saturated solution (solution only just possible).
   c 0.001 mol L⁻¹ $PbBr_2$, $K_s(PbBr_2) = 9 \times 10^{-6}$, ion product = 4.00×10^{-9}. Ion product < K_s, therefore solution possible.
   d 0.02 mol L⁻¹ $Ba(OH)_2$, $K_s(Ba(OH)_2) = 5 \times 10^{-3}$, ion product = 3.20×10^{-5}. Ion product < K_s, therefore solution possible.
   e ion product = 1.00×10^{-10} therefore no precipitation
   f ion product = 1.20×10^{-9} therefore no precipitation
   g 1.6×10^{-4} g

   h

Q_s	Y/N
$0.0005 \times 0.0005 = 2.5 \times 10^{-7} > K_s$	Y
1.25×10^{-11}	N
$[(10/50) \times 0.00010]^2 \times [(40/50) \times 0.0010] = 3.2 \times 10^{-15}$	N
1.25×10^{-9}	Y

If $Q = K_s$, the solution is just saturated. When no precipitate forms, $Q < K_s$.

7 a i 1.3×10^{-5} molL⁻¹ ii 1.7×10^{-10} molL⁻¹
      iii The high concentration of Cl^- in 1 mol L⁻¹ NaCl allows only a small amount of Ag^+ to be in solution (common ion effect)
   b i 1.71×10^{-2} molL⁻¹ in water, 2.24×10^{-3} molL⁻¹ in 1 molL⁻¹ H_2SO_4
      ii In 1 molL⁻¹ H_2SO_4, the concentration of SO from the acid is so high that any from the dissociation of Ag_2SO_4 is insignificant. At 1×10^{-3} molL⁻¹ H_2SO_4, the concentration of SO_4^{2-} from ionisation of Ag_2SO_4 becomes significant.

   c

Equilibrium	K_s expression	Value	Value
$BaSO_4(s) \rightleftharpoons Ba^{2+}(aq) + SO_4^{2-}(aq)$	$[Ba^{2+}][SO_4^{2-}]$	3.3×10^{-5} molL⁻¹	1.1×10^{-9} molL⁻¹
$CaF_2(s) \rightleftharpoons Ca^{2+}(aq) + 2F^-(aq)$	$[Ca^{2+}][F^-]^2$	1.1×10^{-3} molL⁻¹	5.3×10^{-7} molL⁻¹
$FeS(s) \rightleftharpoons Fe^{2+}(aq) + S^{2-}(aq)$	$[Fe^{2+}][S^{2-}]$	7.7×10^{-10} molL⁻¹	6.0×10^{-18} molL⁻¹
$Fe(OH)_2(s) \rightleftharpoons Fe^{2+}(aq) + 2OH^-(aq)$	$[Fe^{2+}][OH^-]^2$	5.8×10^{-6} molL⁻¹	8.0×10^{-14} molL⁻¹
$Ag_2SO_4(s) \rightleftharpoons 2Ag^+(aq) + SO_4^{2-}(aq)$	$[Ag^+]^2[SO_4^{2-}]$	1.5×10^{-2} molL⁻¹	1.4×10^{-4} molL⁻¹

8 **Solubility problem 1**
   a $PbI_2(s) \rightleftharpoons Pb^{2+}(aq) + 2I^-(aq)$
   b $K_s = [Pb^{2+}][I^-]^2$
   c i 1.3×10^{-3} mol L⁻¹
      ii 0.59 g
   d 8.0×10^{-7} mol L⁻¹
   e The high concentration of $I^-(aq)$ from KI lowers the amount of $Pb^{2+}(aq)$ that can be in solution (common ion effect).
   f Precipitation will occur (a bright yellow precipitate forms). $[Pb^{2+}][I^-]^2 = 0.010 \times (1.0 \times 10^{-3})^2 = 1.0 \times 10^{-8} > K_s$

Solubility problem 2
   a $Zn(OH)_2(s) \rightleftharpoons Zn^{2+}(aq) + 2OH^-(aq)$
   b $K_s = [Zn^{2+}][OH^-]^2$
   c i 1.44×10^{-6} molL⁻¹
      ii 1.43×10^{-4} gL⁻¹
   d 1.20×10^{-13} molL⁻¹
   e i As the OH^- concentration increases, $Zn(OH)_2(s)$ reacts to form soluble $Zn(OH)_4^{2-}$. $Zn(OH)_2(s) + 2OH^-(aq) \rightarrow Zn(OH)_4^{2-}(aq)$. The solubility of $Zn(OH)_2(s)$ increases.
      ii As the H^+ concentration increases, it reacts with $OH^-(aq)$ from $Zn(OH)_2$ to form water. $H^+(aq) + OH^-(aq) \rightarrow H_2O(l)$. More $Zn(OH)_2(s)$ will dissolve to restore OH^- and re-establish the equilibrium.

Solubility problem 3

a $K_s = [Ag^+]^2[CO_3^{2-}]$

b 1.26×10^{-4} mol L^{-1}

c 4.47×10^{-6} mol L^{-1}

d The high concentration of CO_3^{2-}(aq) from Na_2CO_3 lowers the amount of Ag^+(aq) that can be in solution (common ion effect).

e i Bubbles of gas form and solid silver carbonate is seen to dissolve. The reaction CO_3^{2-}(aq) + $2H^+$(aq) $\longrightarrow$ CO_2(g) + H_2O(l) removes CO_3^{2-} (aq) from the equilibrium Ag_2CO_3(s) $\rightleftharpoons$ $2Ag^+$(aq) + CO_3^{2-}(aq). Ag_2CO_3(s) $\longrightarrow$ $2Ag^+$(aq) + CO_3^{2-}(aq) to restore CO_3^{2-}.

   ii Solid silver carbonate is seen to dissolve. Formation of the complex ion $Ag(NH_3)_2^+$(aq) removes Ag^+(aq) from the equilibrium Ag^+(aq) + $2NH_3$(aq) $\longrightarrow$ $Ag(NH_3)_2^+$(aq). Ag_2CO_3(s) will dissolve to restore Ag^+(aq) and re-establish the equilibrium.

Unit 14 Acids

1 a A strong acid (e.g. 1 mol L^{-1} HCl) completely ionises in solution, while a concentrated acid (e.g. 10 mol L^{-1} HCl, or glacial CH_3COOH) contains a large amount of the acid in a given volume of solution. This large amount may be completely or partially ionised.

   b With a weak acid (e.g. CH_3COOH), only a few molecules are ionised in solution. A dilute acid (e.g. 0.1 mol L^{-1} CH_3COOH or 0.1 mol L^{-1} HCl) contains a small amount of the acid in a given volume of solution. This small amount may be partially or completely ionised.

   c Each molecule of a monoprotic acid (e.g. HCl or CH_3COOH) ionises to give one proton only, while a diprotic acid molecule (e.g. H_2SO_4) can give one or two protons.

   d A molecular acid is made up of molecules (e.g. HCl, CH_3COOH), while an ionic acid is made up of ions (e.g. NH_4^+).

   e An acid loses a proton to form its conjugate base (e.g. the acid CH_3COOH has CH_3COO^- as its conjugate base).

2 a HCl(g) $\longrightarrow$ H^+(aq) + Cl^-(aq)

   b H_2SO_4(l) $\longrightarrow$ H^+(aq) + HSO_4^-(aq)

   c CH_3COOH(aq) $\rightleftharpoons$ CH_3COO^-(aq) + H^+(aq)

   d NH_4^+(aq) $\rightleftharpoons$ NH_3(aq) + H^+(aq)

   e HSO_4^-(aq) $\rightleftharpoons$ H^+(aq) + SO_4^{2-}(aq)

   f H_2CO_3(aq) $\rightleftharpoons$ H^+(aq) + HCO_3^-(aq)
   HCO_3^-(aq) $\rightleftharpoons$ H^+(aq) + CO_3^{2-}(aq)

3 Both H^+(aq) and H_3O^+(aq) are used to represent a hydrated proton.

   a ~1 mol L^{-1} H^+(aq), 1 mol L^{-1} Cl^-(aq), ~10^{-14} mol L^{-1} OH^-(aq)

   b ~2 mol L^{-1} H^+(aq), 1 mol L^{-1} SO_4^{2-}(aq), < 10^{-14} (~5 × 10^{-15}) mol L^{-1} OH^-(aq)

   c ~1 mol L^{-1} CH_3COOH(aq), ~10^{-2} mol L^{-1} H_3O^+(aq), ~10^{-2} mol L^{-1} CH_3COO^-(aq), ~10^{-12} mol L^{-1} OH^-(aq)

   d ~1 mol L^{-1} HCN(aq), ~10^{-4} mol L^{-1} H_3O^+(aq), ~10^{-4} mol L^{-1} CN^-(aq), ~10^{-10} mol L^{-1} OH^-(aq)

   e ~1 mol L^{-1} NH_4^+(aq), ~10^{-5} mol L^{-1} H_3O^+(aq), ~10^{-5} mol L^{-1} NH_3(aq), ~10^{-9} mol L^{-1} OH^-(aq)

4 I a 2

   b 1.46

   c 2.98

   d 3.45

   e 7

   f -0.5

   II For weak acids, assume all H^+ ions are from ionisation of the acid, and that from ionisation of water is negligible. It is also assumed that the amount of acid that is ionised is negligible compared with the total amount of acid present.

   g 3.38   h 3.11   i 4.62

   j 5.42   k 3.09   l 1.55

   m Let x = $[H_3O^+]$ and $[CH_3COO^-]$

   $1.74 \times 10^{-5} = \dfrac{x \cdot x}{(0.01 - x)} = \dfrac{x^2}{(0.01 - x)}$ must be solved.

   $\longrightarrow (1.74 \times 10^{-7}) - (1.74 \times 10^{-5})x = x^2$

   $\longrightarrow x^2 + (1.74 \times 10^{-5})x - (1.74 \times 10^{-7}) = 0$

   If $ax^2 + bx + c = 0$, then

   $x = \dfrac{-1.74 \times 10^{-5} \pm \sqrt{(3.03 \times 10^{-10} + 4 \times 1.74 \times 10^{-7})}}{2}$

   $x = 4.09 \times 10^{-4}$ and pH = 3.39

5 a 0, 1, 2

   b Assume all H^+ comes from ionisation of the acid, and that the amount of ionised acid is negligible compared with the total amount of acid. 1 molL^{-1}, pH = 2.38; 0.1 molL^{-1}, pH = 2.88 0.01 molL^{-1} solution, pH = 3.38.

   c Diluting a weak acid has a less dramatic effect on pH than diluting a strong acid. To distinguish between the two, one would monitor pH change as a function of dilution.

   d As a weak acid becomes more dilute, the proportion of the acid that ionises increases.

6 a Hydrochloric acid is a strong acid and may cause further damage. Heat is released on the reaction of strong acid and strong base, which may also cause damage.

   b To dilute or remove excess base.

7 All answers are in molL^{-1}.

   a 1×10^{-1}

   b 1×10^{-3}

   c 1×10^{-5}

   d 1×10^{-9}

   e 1

   f 3.16×10^{-5}

   g 1.58×10^{-3}

   h 5.62×10^{-1}

   i 3.98×10^{-11}

   j 1.08

8 a Methanoic acid

   b Greater than 1. Methanoic acid is a stronger acid than ethanoic acid and consequently will donate its protons to ethanoate ions readily. There will be more methanoate and ethanoic acid (than of their respective conjugates), so the ratio will be greater than 1.

9 Pyruvate is in higher concentration because it has a greater tendency to ionise. This in turn will suppress the ionisation of ethanoic acid.

ISBN: 9780170355544   PHOTOCOPYING OF THIS PAGE IS RESTRICTED UNDER LAW.

10

Acid	Conjugate base
HCl	Cl^-
H_2SO_4	HSO_4^-
HSO_4^-	SO_4^{2-}
H_2O	OH^-
H_3O^+	H_2O
NH_4^+	NH_3
H_2CO_3	HCO_3^-
HCO_3^-	CO_3^{2-}
CH_3COOH	CH_3COO^-
$HCOOH$	$HCOO^-$

11 HSO_4^-, H_2O, HCO_3^-

12 a H_3PO_4
  b $H_2PO_4^-$
  c PO_4^{3-}
  d An uncharged molecule such as H_3PO_4 loses a proton (+ charge) more readily than a negatively charged ion, $H_2PO_4^-$, which in turn loses a proton more easily than an ion with a –2 charge.

Unit 15 Bases

1 a weak base
  b basic solution
  c pOH
  d strong base

2 a 0.1 mol L^{-1} Na^+(aq), ~0.1 mol L^{-1} OH^-(aq), ~10^{-13} mol L^{-1} H_3O^+(aq)
  b ~2×10^{-4} mol L^{-1} OH^-(aq), 1×10^{-4} mol L^{-1} Ca^{2+}(aq), ~5×10^{-11} mol L^{-1} H_3O^+(aq)
  c 1 mol L^{-1} NH_3(aq), ~10^{-3} mol L^{-1} OH^-(aq), ~10^{-3} mol L^{-1} NH_4^+(aq), ~10^{-11} mol L^{-1} H_3O^+(aq)
  d ~0.25 mol L^{-1} $C_2H_5NH_2$(aq), ~10^{-2} mol L^{-1} OH^-(aq), ~10^{-2} mol L^{-1} $C_2H_5NH_3^+$(aq), ~10^{-12} mol L^{-1} H_3O^+(aq)

3 a i 13
    ii 10.3
  b i 11.6
    ii 10.9
    iii 12.3
    iv 10.5

4 0.01 mol L^{-1} NH_3, 0.01 mol L^{-1} $C_2H_5NH_2$, 0.01 mol L^{-1} NaOH, 0.01 mol L^{-1} $Ca(OH)_2$

5 All the following are in mol L^{-1}.
  a 1×10^{-4}
  b 1×10^{-1}
  c 1×10^{-9}
  d 1×10^{-5}
  e 1
  f 3.16
  g 6.3×10^{-6}
  h 1.8×10^{-14}
  i 10
  j 9.3×10^{-15}

6 a 2.00×10^{-2} molL^{-1}
  b Calcium hydroxide is not completely soluble in water. Once a pH of 12.30 is obtained, no further calcium hydroxide dissolves.
  c 4.00×10^{-6}

7 a Vinegar and lemon juice contain weak acids that react with amines, which have fishy smells. Amines are bases.
  b CO_3^{2-}(aq) + H^+(aq) $\rightleftharpoons$ HCO_3^-(aq)
  HCO_3^-(aq) + H^+(aq) $\rightleftharpoons$ H_2CO_3(aq) [or H_2O(l) + CO_2(g)]

Unit 16 Titration curves

1 a 2.8
  b 5
  c 5, 1×10^{-5}
  d 9
  e 0.20 mol L^{-1}
  f phenolphthalein

2

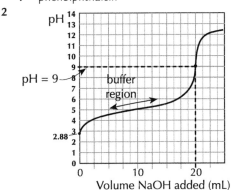

3

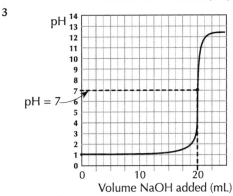

  c As NaOH is added, $[H^+]$ decreases. Conductivity decreases. At the equivalence point, the concentrations of H^+(aq) and OH^-(aq) are at their lowest. These ions have a higher conductivity than Na^+(aq) and Cl^-(aq). After equivalence point, there is an excess of OH^-(aq).

4

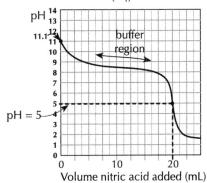

5 a 20 mL 0.10 molL^{-1} HCl

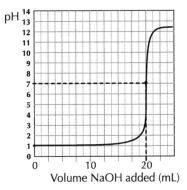

b 20 mL 0.10 mol L⁻¹ CH₃COOH

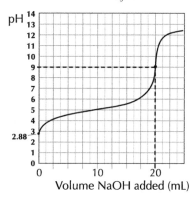

pH vs Volume NaOH added (mL)

c 20 mL water

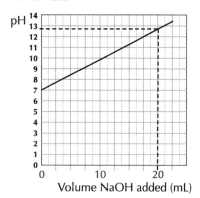

pH vs Volume NaOH added (mL)

6 a phenolphthalein **b** methyl red

7 a equivalence point of an acid-base titration
   b acid-base titration curve
   c end point of an acid-base titration
   d acid-base indicator

Unit 17 pH of buffers and salt solutions

1

	Buffer solution		Salt solution
b	formed by mixing a weak acid and its conjugate base	**a**	formed by reacting equal amounts of acid and base
c	pH of solution changes little with addition of a small amount of strong base	**d**	pH of solution increases rapidly with addition of a small amount of strong base
f	can be used to keep the pH of a solution constant	**e**	can be used to change the pH of a solution

2 a $CH_3COO^-(aq)$, $CH_3COOH(aq)$
   b $HCO_3^-(aq)$, $H_2CO_3(aq)$, or $HCO_3^-(aq)$, $CO_3^{2-}(aq)$
   c $NH_4^+(aq)$, $NH_3(aq)$

3 a $H_2CO_3(aq) + OH^-(aq) \rightleftharpoons HCO_3^-(aq) + H_2O(l)$
   b Conjugate base is $HCO_3^-(aq)$. $HCO_3^-(aq) + H^+(aq) \rightleftharpoons H_2CO_3(aq)$
   c Added base reacts with the weak acid H_2CO_3, while added acid reacts with the conjugate base HCO_3^-, to produce species that do not affect pH.

4 a $NH_4^+(aq) + OH^-(aq) \rightleftharpoons NH_3(aq) + H_2O(l)$
   b $NH_3(aq) + H_3O^+(aq) \rightleftharpoons NH_4^+(aq) + H_2O(l)$

5 a CH_3COOH, CH_3COO^-
   b CH_3COO^-
      $CH_3COO^-(aq) + H_3O^+(aq) \rightleftharpoons CH_3COOH(aq) + H_2O(l)$
   c CH_3COOH
      $CH_3COOH(aq) + OH^-(aq) \rightleftharpoons CH_3COO^-(aq) + H_2O(l)$
   d The weak acid CH_3COOH reacts with any OH^-, the conjugate base reacts with H^+ resulting in little net pH change.

6 a CH_3COOH
   b 0.05 mol L⁻¹
   c CH_3COO^-
   d 0.06 mol L⁻¹
   e $pH = 4.76 + \log \frac{[0.06]}{[0.05]} = 4.84$
   f Because there is more conjugate base.
   g No. When NaOH is added, it reacts with some of the acid to form conjugate base. Less than half of the acid remains. In the buffer made by adding the salt, all of the original acid is present. It contains more weak acid and conjugate base. This buffer has a greater capacity to react with added OH^- and H^+.

7 a NH_4^+
   b 0.06 molL⁻¹
   c NH_3
   d 0.05 molL⁻¹
   e $pH = 9.24 + \log \frac{[0.05]}{[0.06]} = 9.16$
   f Because there is more weak acid.
   g No. When HCl Is added, it reacts with some of the NH_3 to form the weak acid NH_4^+. Less than half of the NH_3 remains. In the buffer made by adding the salt, all of the original NH_3 is present. It contains more NH_3 and NH_4^+. This buffer has a greater capacity to react with added OH^- and H^+.

8 a 4.76
   b yes, pH = 5.06
   c pH = 4.76
   d 9.24
   e 9.32
   f 9.06
   g 4.97
   h no, pH = 4.87
   i 5.05
   j 9.16
   k 9.42

9 a 0.1 molL⁻¹ ethanoic acid solution and solid sodium ethanoate
      $pH = pK_a + \log[\text{conjugate base}]/[\text{acid}]$
      $0.04 = \log[\text{conjugate base}]/[\text{acid}]$
      $1.096 = [\text{conjugate base}]/0.1$
      $[\text{conjugate base}] = 0.1096$ molL⁻¹
      So, to 1 L of 0.1 molL⁻¹ ethanoic acid, need to add 0.1096 mol or 9.0 g of sodium acetate.
   b 0.1 molL⁻¹ ethanoic acid and 0.1 molL⁻¹ sodium hydroxide
      pK_a(ethanoic acid) = 4.76
      $0.04 = \log[\text{conjugate base}]/[\text{acid}]$
      $1.096 = [\text{conjugate base}]/[\text{acid}]$
      So, add 1.096 mL NaOH to 2.096 mL of ethanoic acid (or a similar ratio of the two), giving 1.096 mol of conjugate base to every 1 mol of unreacted acid.

10 a $[Cl^-] = 0.20$, $[Ca^{2+}] = 0.10$, $[H^+] = 1 \times 10^{-7}$, $[OH^-] = 1 \times 10^{-7}$

ISBN: 9780170355544   PHOTOCOPYING OF THIS PAGE IS RESTRICTED UNDER LAW.

b $[Cl^-] = 0.10$, $[Na^+] = 0.10$, $[H^+] = 1 \times 10^{-7}$, $[OH^-] = 1 \times 10^{-7}$

c $[Cl^-] = 0.10$, $[NH_4^+] = 0.10$, $[NH_3] = [H^+] = 1 \times 10^{-5}$, $[OH^-] = 1 \times 10^{-9}$

d $[Na^+] = 0.10$, $[CH_3COO^-] = 0.10$, $[CH_3COOH] = [OH^-] = 1 \times 10^{-5}$, $[H^+] = 1 \times 10^{-9}$

11 $NH_4Cl(aq) \rightarrow NH_4^+(aq) + Cl^-(aq)$, $NH_4^+(aq) \rightleftharpoons NH_3(aq) + H^+(aq)$
$CH_3COONa(aq) \rightarrow CH_3COO^-(aq) + Na^+(aq)$,
$CH_3COO^-(aq) + H_2O(l) \rightleftharpoons CH_3COOH(aq) + OH^-(aq)$

12 a 2.0 b 5.06

Revision Four

Question I

1 a $[CH_3COOH] = 0.200$, $[CH_3COO^-] = [H^+] = 10^{-2.73}$ $= 1.86 \times 10^{-3}$, $[OH^-] = 5.38 \times 10^{-12}$

b $[NH_3] = 0.300$, $[H^+] = 10^{-11.4} = 3.98 \times 10^{-12}$, $[OH^-] = [NH_4^+] = 2.51 \times 10^{-3}$

c $[Ca^{2+}] = 1.00 \times 10^{-4}$, $[OH^-] = 2.00 \times 10^{-4}$, $[H^+] = 5 \times 10^{-11}$

d $[Cl^-] = 0.100$, $[NH_4^+] = 0.100$, $[NH_3] = [H^+] = 10^{-5.12}$ $= 7.59 \times 10^{-6}$, $[OH^-] = 1.32 \times 10^{-9}$

2 List 1: Na_2SO_4, HCOONa, NH_3. Na_2SO_4 is the salt of a strong acid and strong base; it has pH 7. HCOONa is the salt of a strong base and weak acid so it has a basic pH (> 7). $HCOO^-$ is a weaker base than NH_3. NH_3 has pH greater than HCOONa.
List 2: CH_3COOH, NH_4Cl, NaOH. CH_3COOH is a weak acid but stronger than NH_4^+. NaOH is a strong base.

3 The conductivity of a solution depends on the number of dissolved ions.

  a i $CaCl_2$, CH_3NH_3Cl, CH_3COONa, C_2H_5COOH, $C_6H_{12}O_6$

    ii The more ions that are in solution, the higher the solution's conductivity. Each mole of $CaCl_2$ gives 3 moles of ions. Each mole of CH_3NH_3Cl gives 2 moles of ions as does each mole of CH_3COONa. In both cases one ion ($CH_3NH_3^+$ and CH_3COO^-) hydrolyses to a small extent but for every ion that reacts, another is formed (H^+ and OH^- respectively; H^+ is more conductive than OH^-). A mole of C_2H_5COOH remains mainly as molecules and gives only a few ions. A solution $C_6H_{12}O_6$ contains dissolved molecules only.

  b i HCl, NaOH, CH_3COOH, NH_3 **ii** 0.10 mol L^{-1} H^+ has a higher conductivity than 0.10 mol L^{-1} OH^-. The small amount of H^+ from 0.10 mol L^{-1} CH_3COOH has a higher conductivity than the small amount of OH^- from 0.10 mol L^{-1} NH_3.

4 a $Ag_2CrO_4(s) \rightleftharpoons 2Ag^+(aq) + CrO_4^{2-}(aq)$

  b $K_s = [Ag^+]^2[CrO_4^{2-}]$

  c 6.54×10^{-5} mol L^{-1}

  d 5.29×10^{-7} mol L^{-1}

  e The high concentration of CrO_4^{2-} from K_2CrO_4 lowers the amount of Ag^+ that can be in solution (common ion effect). Therefore the value is smaller in **d** than in **c**.

  f Addition of the yellow solution to the colourless solution causes more red solid to form.

  g i The solid dissolves as the soluble complex ion $Ag(NH_3)_2^+$ forms. The reaction $Ag^+(aq) + 2NH_3(aq) \rightarrow Ag(NH_3)_2^+(aq)$ removes $Ag^+(aq)$ from the equilibrium $Ag_2CrO_4(s) \rightleftharpoons 2Ag^+(aq)$

+ $CrO_4^{2-}(aq)$. $Ag_2CrO_4(s)$ will dissolve in an attempt to restore $Ag^+(aq)$.

    ii Unchanged. K_s changes only if the temperature changes.

5 a -0.311

  b 12.7

  c 2.85

  d 11.4

  e $[NH_4^+] = 1.00$, pH = 4.62

6 a To ensure that a solution's pH remains approximately constant.

  b i $CH_3COO^-(aq) + H^+(aq) \rightleftharpoons CH_3COOH(aq)$

    ii $CH_3COOH(aq) + OH^-(aq) \rightleftharpoons CH_3COO^-(aq) + H_2O(l)$

  c 4.75

  d The one prepared by adding solid sodium ethanoate is more effective. All of the original acid is still present. In the second buffer, the conjugate base is obtained by reacting acid with NaOH. Just over half of the original acid is present. This buffer has a lower capacity to react with added base or acid.

7 a HCOONa, 0.0500 molL^{-1}

  b 8.23

  c 3.76

  d 3.94

  e

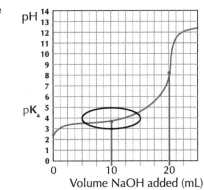

  f In the diagram the region is circled.

  g Phenolphthalein

Question II

1 a $[Cl^-] = [H^+] = 0.25$, $[OH^-] = 4.0 \times 10^{-14}$

  b $[Na^+] = [OH^-] = 2.5$, $[H^+] = 4.0 \times 10^{-15}$

  c $[Na^+] = [CH_3COO^-] = 0.10$, $[H^+] = 1.26 \times 10^{-9}$, $[CH_3COOH] = [OH^-] = 7.94 \times 10^{-6}$

  d $[CH_3NH_3^+] = 0.20$, $[SO_4^{2-}] = 0.10$, $[CH_3NH_3] = [H^+] = 2.0 \times 10^{-6}$, $[OH^-] = 5.01 \times 10^{-9}$

  e $[K^+] = 0.010$, $[SO_4^{2-}] = 0.0050$, $[H^+] = [OH^-] = 1.0 \times 10^{-7}$

2 The conductivity of a solution depends on the number of dissolved ions.

  a List 1: NH_4Cl, CH_3NH_2, C_2H_5OH
List 2: HCl, CH_3COONa, KNO_3, HCOOH

  b Each mole of NH_4Cl gives 2 moles of ions, CH_3NH_2 is a weak base and gives only a small number of ions, C_2H_5OH does not ionise.
1 mole HCl, 1 mole CH_3COONa and 1 mole KNO_3 all give 2 moles of ions, but the conductivity of H^+ is about five times and OH^- about three times greater than other ions. Although CH_3COO^- hydrolyses, for every CH_3COO^- that reacts, one OH^- is formed. Dissolved HCOOH, a weak acid, is largely in the molecular form.

3 **a** $Pb(OH)_2(s) \rightleftharpoons Pb^{2+}(aq) + 2OH^-(aq)$
   b $K_s = [Pb^{2+}][OH^-]^2$
   c 1.53×10^{-7} mol L^{-1}
   d $[OH^-] = 3.06 \times 10^{-7}$, pH = 7.49
   e 1.43×10^{-10} mol L^{-1}
   f The higher concentration of OH$^-$ in the solution of higher pH lowers the amount of Pb^{2+} that can be in solution. The value in **e** is smaller than the value in **c**.
   g $Pb(OH)_2(s)$ forms a soluble complex anion $Pb(OH)_4^{2-}(aq)$.
      $Pb(OH)_2(s) + 2OH^-(aq) \rightarrow Pb(OH)_4^{2-}(aq)$
4 **a** 0.0 **b** 12 **c** 2.73. In **c** to **f**, assume that the amount that hydrolyses is negligible. **d** 8.88 **e** 10.6 **f** 4.53
5 **a** To maintain a constant pH for a solution when small amounts of H$^+$ or OH$^-$ are added to it.
   b **i** $CH_3NH_2(aq) + H^+(aq) \rightarrow CH_3NH_3^+(aq)$
      ii $CH_3NH_3^+(aq) + OH^-(aq) \rightarrow CH_3NH_2(aq) + H_2O(l)$
   c 10.6
   d 10.6
   e 10.5
6 **a** 9.25
   b 8.95
   c

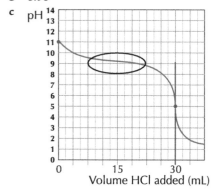

Volume HCl added (mL)

   d The buffer region is circled on the graph.
   e NH$_4$Cl 0.05 mol L^{-1}
   f methyl red

Unit 18 Investigations using volumetric analyses

1 **a** 0.72 mol L^{-1}
   b 43.2 g L^{-1}
   c 4.32% (g in 100 mL)
2 **a** 0.033 mol L^{-1}
   b No. The amount of acid is greater.
   c 0.247% (g in 100 mL)
3 **a** 0.15 mol L^{-1}
   b Yes, there is 5.0% ammonia in the solution.
4 **a** 9.73 mL
   b 9.73×10^{-4} mol
   c $Cl^-(aq) + Ag^+(aq) \rightarrow AgCl(s)$
   d 9.73×10^{-4} mol
   e 4.87×10^{-3} mol, 0.285 g
   f Yes. This sausage has only 0.285% salt (Note: Only a few types of sausages are so low in salt. Most have 10 times this amount.)
5 **a** 4.08×10^{-2} mol L^{-1} **b** 1.08 g **c** 54 % **d** Freshly prepared washing soda crystals contain 37% sodium carbonate. On storage, the crystals lose

water and the percentage of sodium carbonate increases. Because the sample contains 54% sodium carbonate, they are not freshly prepared.
6 **a** 7.80 mL
   b 7.80×10^{-3} moL
   c 2.60×10^{-3} moL
   d 0.500 g (0.4995 g)
   e 100%
7 **a** $5(COOH)_2(aq) + 2MnO_4^-(aq) + 6H^+(aq) \rightarrow 2Mn^{2+}(aq) + 10CO_2(g) + 8H_2O(l)$
      $5Fe^{2+}(aq) + MnO_4^-(aq) + 8H^+(aq) \rightarrow 5Fe^{3+}(aq) + Mn^{2+}(aq) + 4H_2O(l)$
   b 0.1045 molL^{-1} **c** 0.03398 molL^{-1} **d** 0.1199 g **e** yes
8 **a** 14.13 mL
   b 3.53×10^{-4} mol
   c 3.53×10^{-4} mol
   d 0.062 g
   e 0.248 g (248 mg), yes
9 **a** $2Cu^{2+}(aq) + 4I^-(aq) \rightarrow 2CuI(s) + I_2(aq)$
   b $I_2(aq) + 2S_2O_3^{2-}(aq) \rightarrow 2I^-(aq) + S_4O_6^{2-}(aq)$
   c $2Cu^{2+}(aq) + 2S_2O_3^{2-}(aq) + 2I^-(aq) \rightarrow 2CuI(s) + S_4O_6^{2-}(aq)$ (the result of adding two reactions)
   d 5.28×10^{-4} mol
   e 5.28×10^{-4} mol, 0.0335 g
   f 25.4%
   g yes
   h The experiment should be repeated at least twice and the results compared.
10 **a** Water in town supplies often contains chlorine.
   b $Cl_2(aq) + 2I^-(aq) \rightarrow 2Cl^-(aq) + I_2(aq)$
   c $I_2(aq) + 2S_2O_3^{2-}(aq) \rightarrow 2I^-(aq) + S_4O_6^{2-}(aq)$
   d $Cl_2(aq) + 2S_2O_3^{2-}(aq) \rightarrow 2Cl^-(aq) + S_4O_6^{2-}(aq)$
   e 4.88×10^{-3} mol **f** 0.0244 mol **g** Solution contains 34.6 g of Cl$_2$ in 100 mL. The claim is supported.
11 **a** 0.015 mol
   b 0.00955 mol
   c 0.00545 mol
   d 0.00545 mol
   e 0.00273 mol
   f 0.491 g
   g 164 mg aspirin in each tablet, results are high (109% of the manufacturer's claim)

Unit 19 Gravimetric and colorimetric analyses

1 **a**

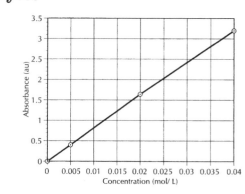

   b 0.033 mol L^{-2}
   c 0.050 mol L^{-1}, 3.2 g L^{-1}

ISBN: 9780170355544 PHOTOCOPYING OF THIS PAGE IS RESTRICTED UNDER LAW.

2 a

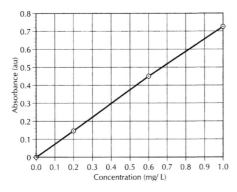

b 0.45 mg L^{-1}

c 0.0075 mg g^{-1}

d It should be diluted with distilled water so that its absorbance is in the same range as of the standards. For example, add 5 mL of the solution to 10 mL of distilled water.

3 a 1.17 g

b 94.4%

4 a 2.675 g

b 1.09 g

c 1.09 %

d Add a little more water to the chips in the filter paper. Test the filtrate with silver nitrate solution. If a precipitate forms, there is still chloride present.

Unit 21 Alkanes, alkenes and addition polymers of alkenes

1 a 2,3-dimethylbutane

b 3-ethylpentane

c 3,4-dimethylheptane

2 a 2,3-dimethylbut-2-ene

b 3-ethylpentene or 3-ethylpent-1-ene

c 4,5-dimethylhept-2-ene

3 a 4,4-dimethylpent-2-ene

b 3-ethyl-3-methylpentane

c 3,3,4-trimethylheptane

4 a Cl_2 substitution

b Cl_2 addition

c Aqueous H_2SO_4 addition

5 a i

```
    H  Br H  H
    |  |  |  |
H – C – C – C – C – H
    |  |  |  |
    H  H  H  H
```

ii

```
    H  OH H  H
    |  |  |  |
H – C – C – C – C – H
    |  |  |  |
    H  H  H  H
```

iii

```
    Br Cl H  H
    |  |  |  |
H – C – C – C – C – H
    |  |  |  |
    H  H  H  H
```

iv

```
    I  Br H  H
    |  |  |  |
H – C – C – C – C – H
    |  |  |  |
    H  H  H  H
```

b i

```
        CH₃
        |
H₃C – C – CH₃
        |
        Br
```

ii

```
        CH₃
        |
H₃C – C – CH₃
        |
        OH
```

iii

```
    Br CH₃
    |  |
H – C – C – CH₃
    |  |
    H  Cl
```

iv

```
    I  CH₃
    |  |
H – C – C – CH₃
    |  |
    H  Br
```

c i

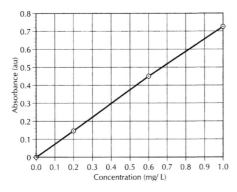

ii

iii

iv

Unit 22 Haloalkanes

1 a freon

b chloroform

c haloalkane

d addition reaction

e tetrachloromethane

f elimination reaction

g substitution reaction

h primary haloalkane

2 a 2-chloro-3-methylbutane

b 3-bromopentane

c 3,4-dimethylchloroheptane or 1-chloro-3,4-dimethylheptane

3 a primary

b secondary

c secondary

d tertiary

e secondary

f tertiary

g primary

4 a Cl_2, HCl

b $3Cl_2$, 3HCl

5 a

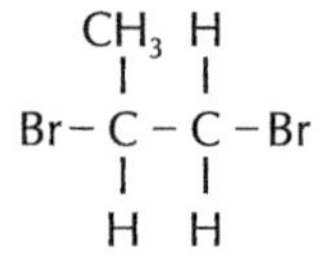

CH₃ H
Br–C–C–Br
H H

b

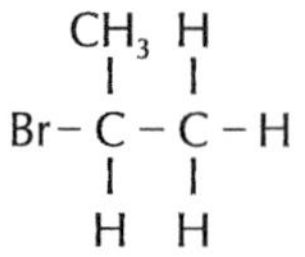

CH₃ H
Br–C–C–H
H H

c

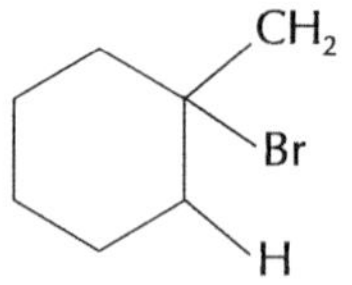

6 a no minor product

b

CH₃ H
H–C–C–Br
H H

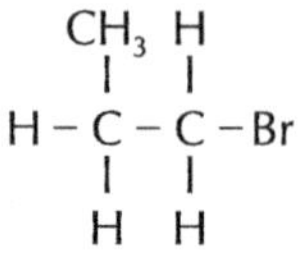

c

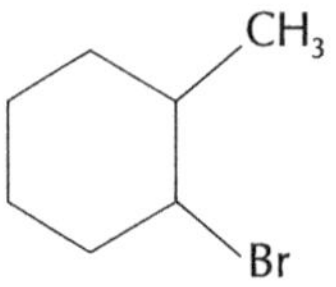

7 Some haloalkanes are gases, all alcohols are liquids. Alcohols are water-soluble, haloalkanes are insoluble in water.

8 a butanol

b 2-butene (major), butene (minor)

c methylpropanol

d methylpropene

e 3-methylbut-2-ene (major), 3-methylbutene (minor)

f cyclohexene

g 2,3-dimethylbut-2-ene (major), 2.3-dimethylbutene (minor)

9 The CFCs need high-energy UV light to split them and form the ozone-depleting fluorine atom. This high-energy UV radiation is largely removed from sunlight (by the existing upper atmospheric ozone) by the time it reaches Earth, so any CFCs near ground level will not react. Consequently ozone near cities is protected from the reaction which forms fluorine atoms. It is only in the outer atmosphere that the fluorine atoms can form from CFCs where there is enough high-energy UV to cause the reaction.

Unit 23 Isomers

1 a cis and trans isomers

b asymetric carbon

c polarised light

d racemic mixture

e constitutional isomers

f enantiomers

g stereoisomers

2 a

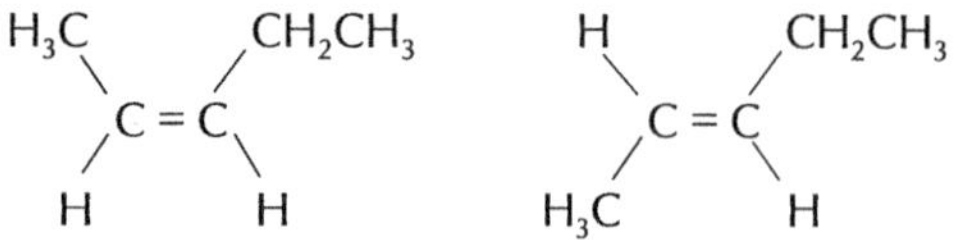

b cis ... trans

3 a no isomers   **b** no isomers

c cis ... trans

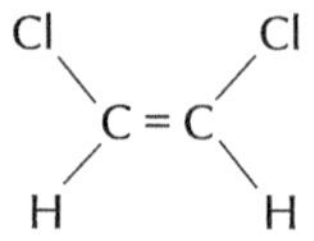

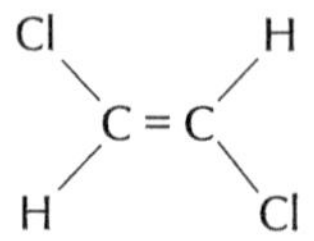

d no isomers

e cis ... trans

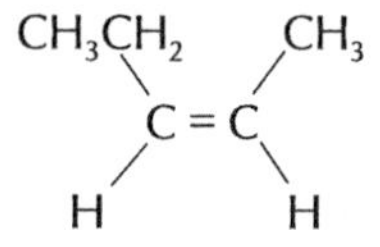

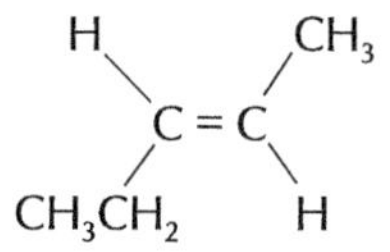

f cis ... trans

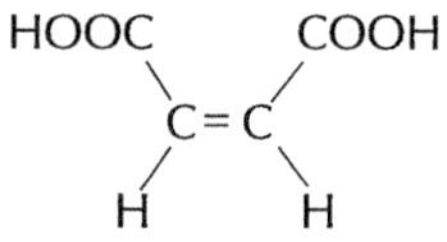

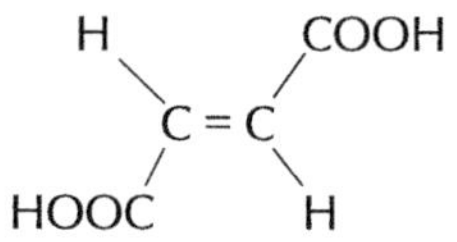

g cis ... trans

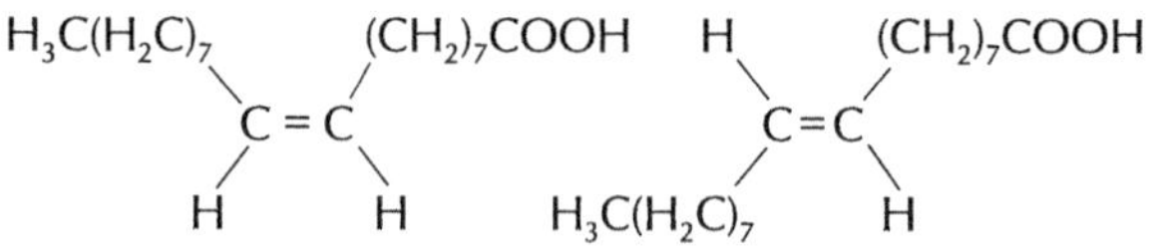

4 a ibuprofen

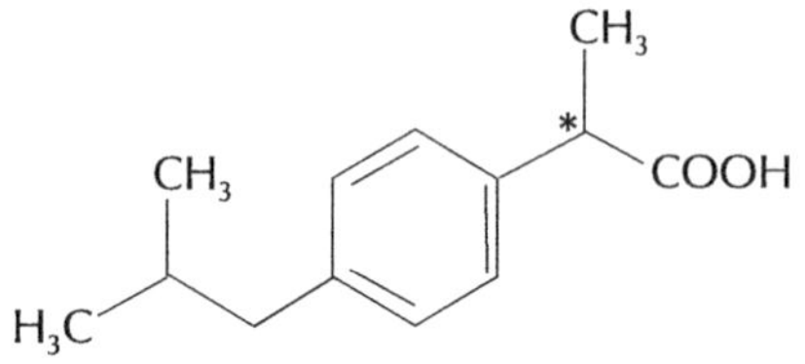

naproxen

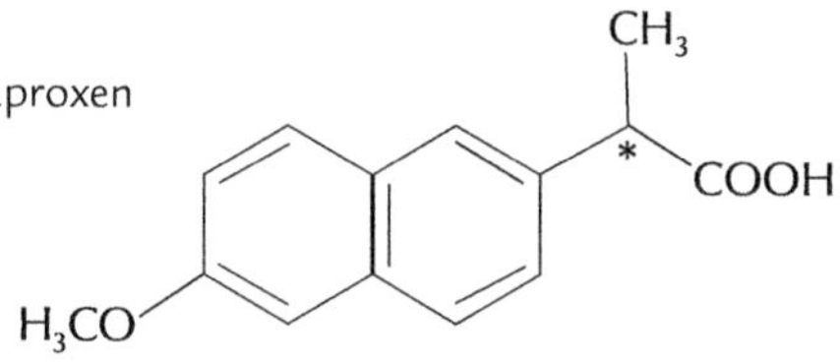

b The racemic mixture would be much easier (and therefore cheaper) to make.

c By seeing the effect that a solution of the drug has on the plane of polarised light. A pure enantiomer will rotate the plane of polarised light, but a racemic mixture will not.

5 a

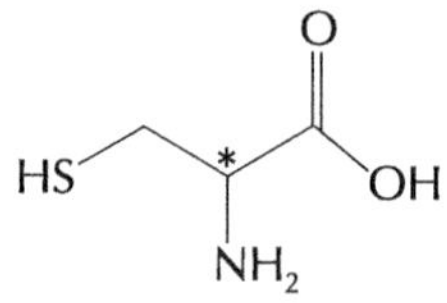

b

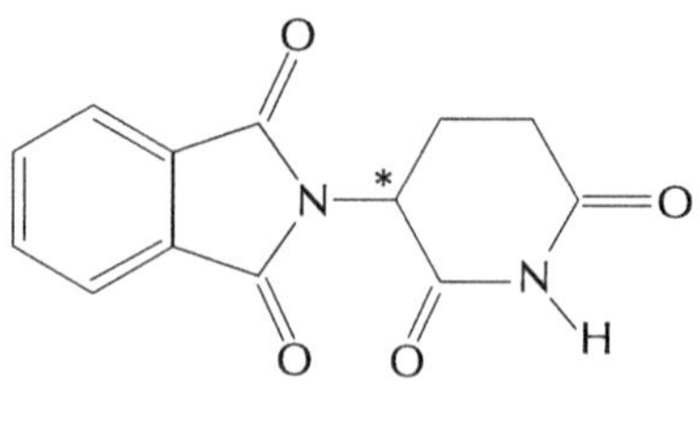

6 a 3-methylhexane

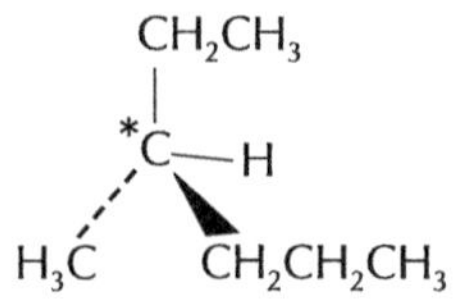

ISBN: 9780170355544   PHOTOCOPYING OF THIS PAGE IS RESTRICTED UNDER LAW.

b 2-pentanol   **c** 1-chloroethanol

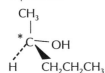

Unit 24 Alcohols, aldehydes and ketones

1 a diol
  b Lucas reagent
  c Tollens' reagent
  d Fehling's solution
  e silver mirror test
  f ketone
  g alcohol
  h tertiary alcohol
  i primary alcohol
  j ester
  k aldehyde
  l triol
  m secondary alcohol

2 a 3,3-dimethylbutanol
  b pentan-3-ol
  c 3,4-dimethylheptan-2-ol

3 a primary
  b primary
  c primary
  d secondary
  e secondary
  f tertiary

4 a to e The colourless alcohol reacts with the purple solution to form a colourless solution. **f** The colourless alcohol does not decolourise the purple solution.

5 a 2-methylpropan-2-ol
  b No reaction will occur. The purple potassium permanganate remains purple.
  c The alcohol –OH will be replaced rapidly by a chloro –Cl group to give a compound that is not soluble in water. The reaction mixture will separate into two layers.
  d 2-methylpropanol
  e The purple potassium permanganate will be decolourised by the colourless primary alcohol. $(CH_3)_2CHCOOH$ is formed.
  f After several hours, the alcohol –OH will be replaced by a chloro –Cl group to give a compound that is not soluble in water. The reaction mixture will separate into two layers. $(CH_3)_2CHCH_2Cl$ is formed.

6 a 2-methylbutanal
  b 2-methylpropanal
  c 4,5-dimethylheptanal

7 a 4-methylpentan-2-one
  b butanone
  c heptan-3-one

8 a propanone, ketone
  b propanal, aldehyde
  c butanal, aldehyde
  d octan-4-one, ketone
  e heptan-4-one, ketone

9 a and b

Compound	Tollens' reagent (colourless)	Fehling's solution (blue)
propanone	no reaction	no reaction
propanal	silver mirror forms	red precipitate forms
butanal	silver mirror forms	red precipitate forms
octan-4-one	no reaction	no reaction
heptan-4-one	no reaction	no reaction

10 a A: ethanol, CH_3CH_2OH; B: ethanoic acid, CH_3COOH; C: ethylethanoate, $CH_3COOCH_2CH_3$.
  b A → B:
$3CH_3CH_2OH + 2Cr_2O_7^{2-} + 16H^+ \rightarrow 3CH_3COOH + 4Cr^{3+} + 11H_2O$
B → C:
$CH_3CH_2OH + CH_3COOH \rightarrow CH_3CH_2OOCCH_3 + H_2O$
  c A → B: Colour changes from orange dichromate to green Cr^{3+}.
B → C: No colour change.

11 a X aldehyde, Y carboxylic acid, Z alcohol, W ester
  b X

```
    H  H  H  O
    |  |  |  ||
H – C– C– C– C – H
    |  |  |
    H  H  H
```

Y

```
    H  H  H  O
    |  |  |  ||
H – C– C– C– C – O – H
    |  |  |
    H  H  H
```

Z

```
    H  H  H  H
    |  |  |  |
H – C– C– C– C – O – H
    |  |  |  |
    H  H  H  H
```

W

```
    H  H  H  H              H  H  H
    |  |  |  |              |  |  |
H – C– C– C– C – O – C – C– C– C – H
    |  |  |  |      ||  |  |  |
    H  H  H  H      O   H  H  H
```

  c X butanal, Y butanoic acid, Z butanol, W butylbutanoate

12 M, 2-butanol

```
    H  OH
    |  |
H₃C – C – C – CH₃
    |  |
    H  H
```

N, 2-methyl-2-propanol

```
     OH
     |
H₃C – C – CH₃
     |
     CH₃
```

Q, butanone

```
    H  O
    |  ||
H₃C – C – C – CH₃
    |
    H
```

13	a	propanol and 2-propanol	reaction with zinc chloride and hydrochloric acid	Lucas reagent (colourless)	Reaction with 2-propanol would slowly produce an oily layer. No change observed for several hours with propanol.
	b	2-methylpropan-2-ol and ethanal	reaction with zinc chloride and hydrochloric acid	Lucas reagent	Reaction with 2-methylpropan-2-ol would produce an oily layer. No change with ethanal.
			mild oxidation	Tollens' reagent (colourless)	Reaction with ethanal would produce a silver mirror. No reaction with 2-methylpropan-2-ol.
				Benedict's solution (blue)	Reaction with ethanal would produce a red precipitate. No reaction with 2-methylpropan-2-ol.
	c	propanal and propanone	mild oxidation	Tollens' reagent	Reaction with propanal would produce a silver mirror. No reaction with propanone.
				Benedict's solution	Reaction with propanal would produce a red precipitate. No reaction with propanone.
	d	propanol and 2-methylpropan-2-ol	reaction with zinc chloride and hydrochloric acid	Lucas reagent	Reaction with 2-methyl-2-propanol would rapidly produce an oily layer.
			oxidation	acidified permanganate (purple)	Propanol will reduce the purple permanganate ion to the near-colourless manganous ion.
				acidified dichromate (orange)	Propanol will reduce the orange dichromate ion to the green chromic ion.

Unit 25 Carboxylic acids and amines

1 a triglyceride
   b fatty acid
   c saponification
   d unsaturated fatty acid
   e amine
   f carboxylic acid
   g acid chloride
   h ester
   i amide
2 a 3-methylbutanoic acid
   b 2-methylbutanoic acid
   c 3-ethylpentanoic acid
3 a 3-methylbutanoyl chloride
   b 2,2-dimethylpentanoyl chloride
   c 3-ethylhexanoyl chloride
4 a $CH_3COOH(g) + NH_3(g) \rightarrow NH_4CH_3COO(s)$
      ammonium ethanoate, particles of white solid would form.
   b $CH_3COCl + 2NH_3 \rightarrow CH_3CONH_2 + NH_4Cl$
      ethanamide ammonium chloride, exothermic reaction
   c $CH_3COCl + H_2O \rightarrow CH_3COOH + HCl$
      ethanoic acid (acetic acid) and hydrochloric acid, vigorous fuming reaction, fumes turn blue litmus paper red.
   d $CH_3COCl + CH_3CH_2OH \rightarrow CH_3COOCH_2CH_3 + HCl$
      ethyl ethanoate and hydrochloric acid, vigorous reaction followed by 'pleasant' smell.
5 a ethyl methanoate
   b methyl propanoate
   c ethyl butanoate

6 a methyl ethanoate, methanol
      i   ethanoic acid
      ii sodium ethanoate
   b propyl methanoate, propanol
      i   methanoic acid
      ii sodium methanoate
   c ethyl propanoate, ethanol
      i   propanoic acid
      ii sodium propanoate
7 a
   b

      and CH_3COOH

   c the alcohol group OH
8 a

ISBN: 9780170355544
PHOTOCOPYING OF THIS PAGE IS RESTRICTED UNDER LAW.

b

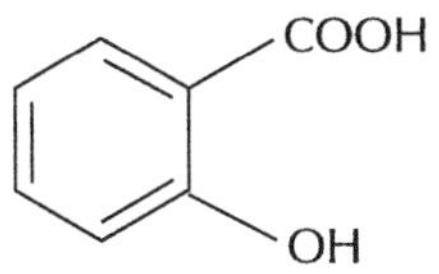

and CH_3OH

c the carboxylic acid group COOH

9 a butanamide
b N,N-dimethylethanamide
c N-ethyl-N-methylpropanamide

10 a

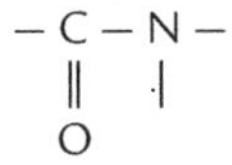

$$-\overset{\parallel}{\underset{O}{C}}-\overset{\mid}{N}-$$

b ethanamide CH_3CONH_2
c ethyl ethanamide $CH_3CONHCH_2CH_3$

11 a

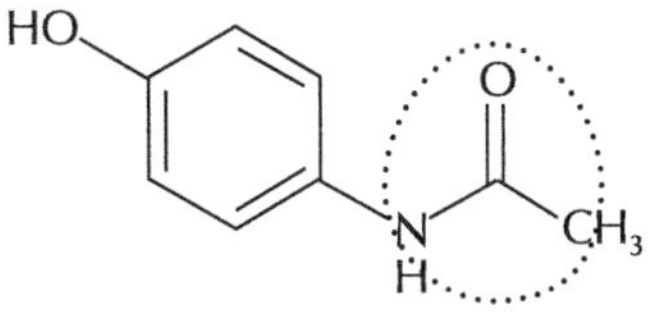

b

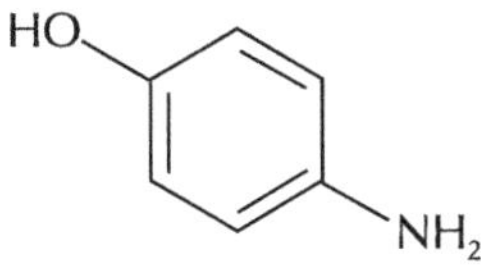

and CH_3COOH

c primary

12 The IUPAC method is less suitable for these secondary and tertiary amines
a N-methylethanamine (CA) or ethylmethylamine (common name)
b N,N-dimethylmethanamine (CA) or trimethylamine (common name)
c N-ethyl-N-methylpropanamine (CA) or ethylmethylpropylamine (common name)

13

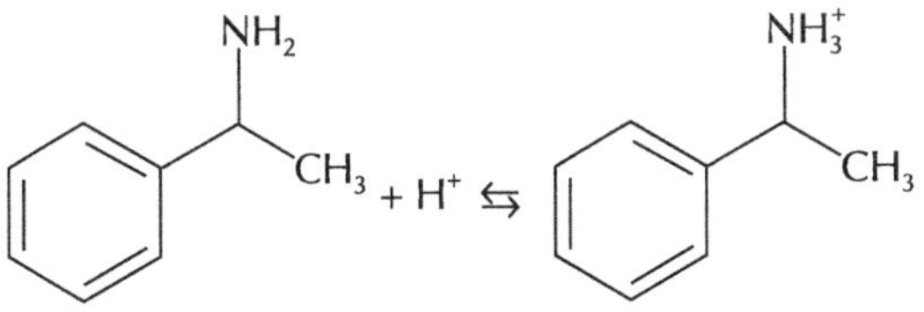

14 a See solid circle below.
b See dotted circle below.

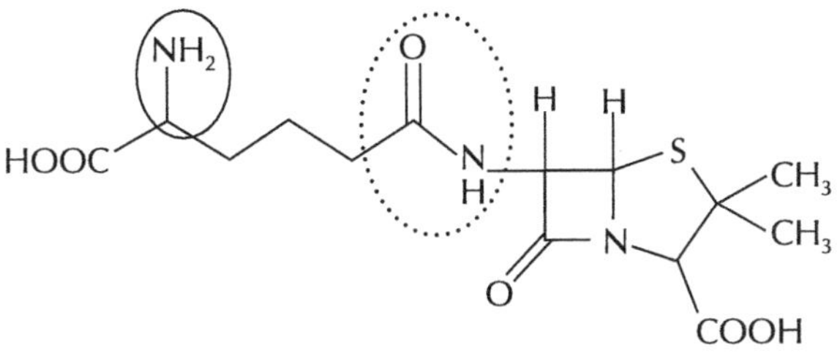

c Another functional group is carboxylic acid (COOH).

15

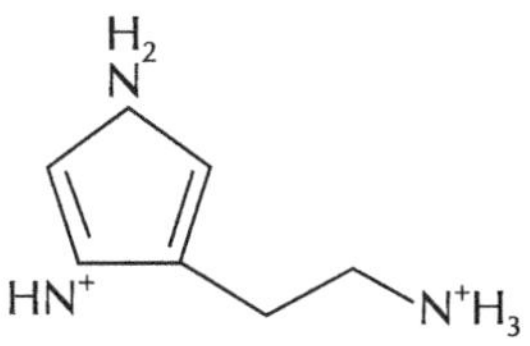

16 a See solid circle below, tertiary amine.
b See dotted circle below.

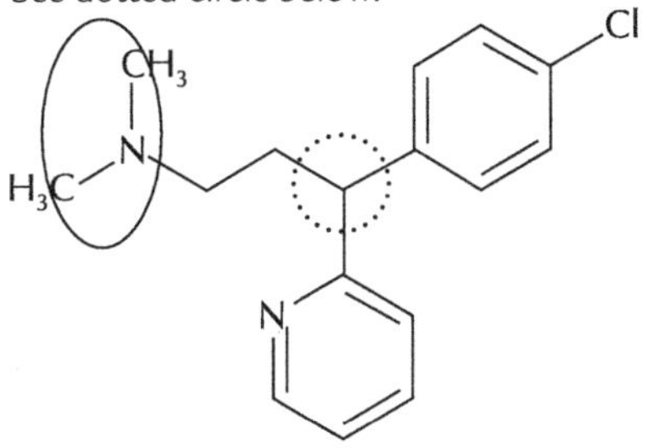

17 a See solid circle below, secondary amine.
b See dotted circle below.

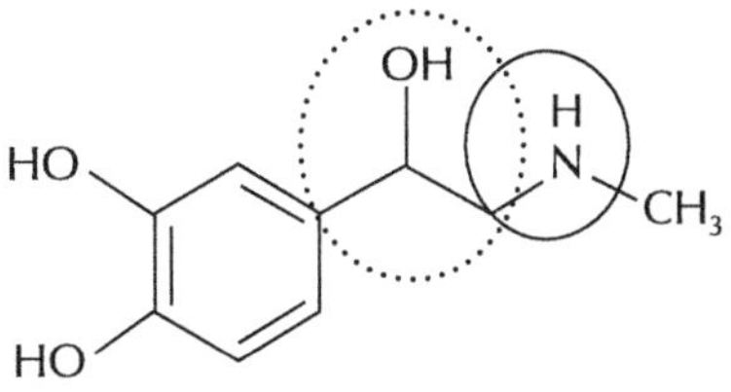

Unit 26 Polymers

1 a protein
b condensation reaction
c polymer
d polypeptide
e amino acid
f amide bond
g peptide bond

2 a

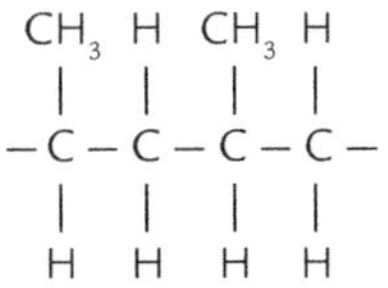

b

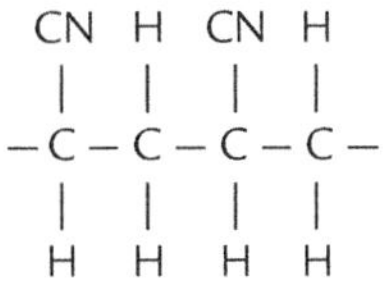

c

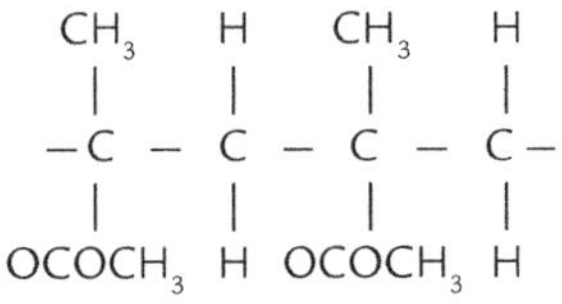

3 a

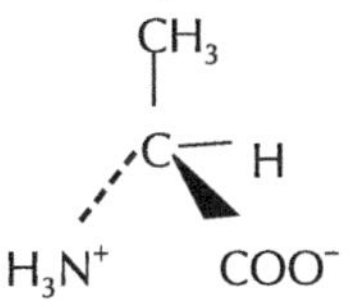

b

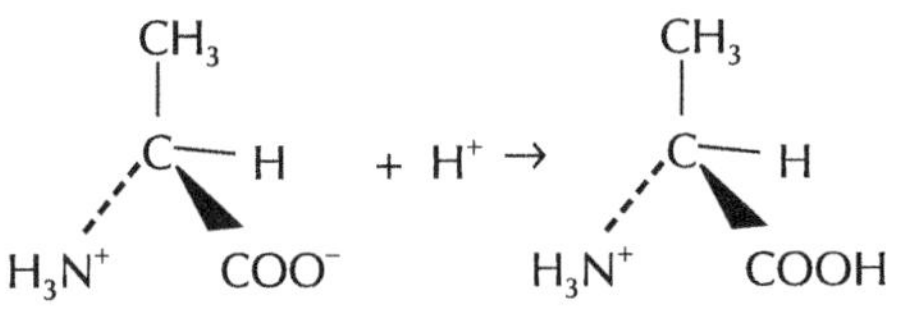

c

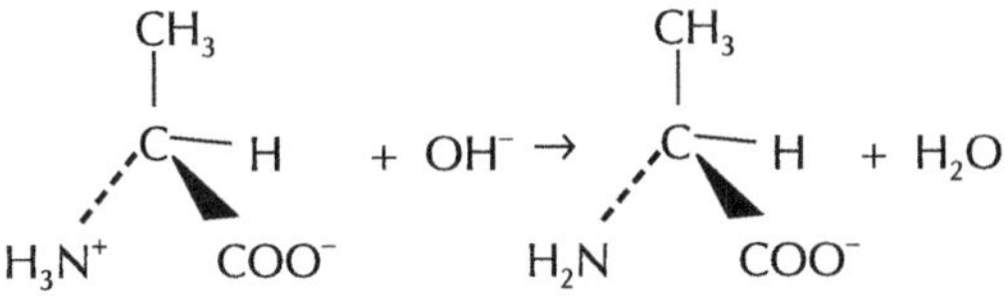

ISBN: 9780170355544   PHOTOCOPYING OF THIS PAGE IS RESTRICTED UNDER LAW.

4 a

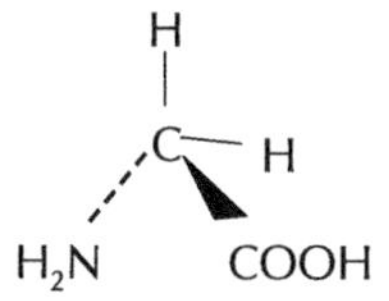

(structure: central C with H top, H right, H₂N and COOH bottom)

b

(structure: two carbon centres with H₂N, OC—NH circled, COOH)

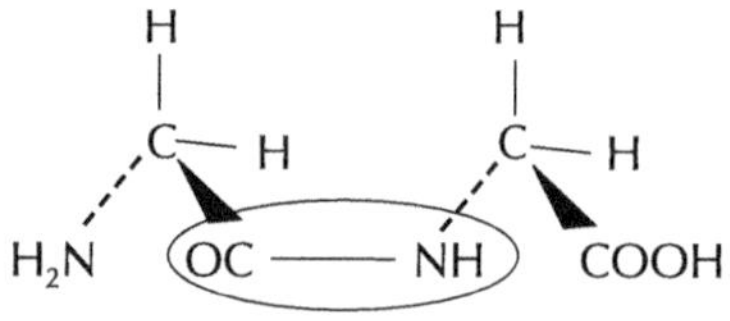

c Two large molecules (amino acids) combine with the elimination of a small molecule (water).

d

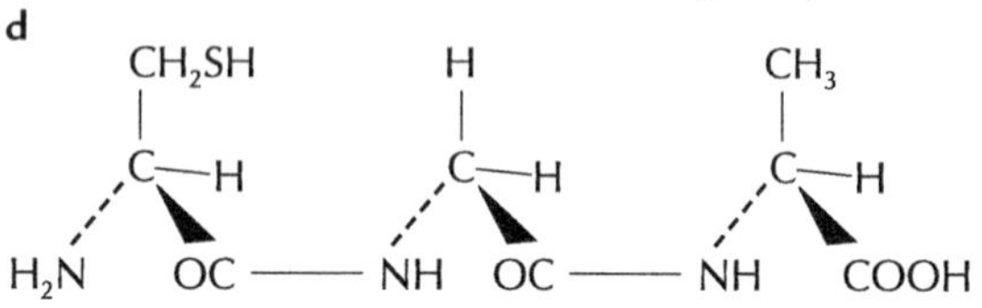

(structure: three carbon centres with CH₂SH, H, CH₃ groups; H₂N, OC——NH, OC——NH, COOH)

e 27 possible tripeptides (3 options in first position x 3 in second x 3 in third = 27)

5 a

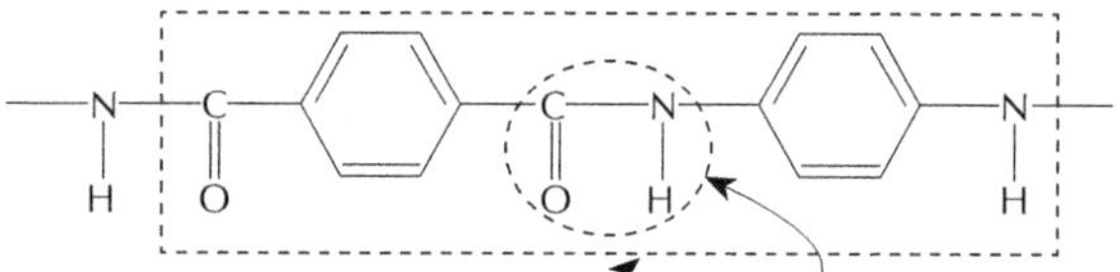

(polymer structure)

b repeating unit

c amide bond

6 a 1,6-diaminohexane

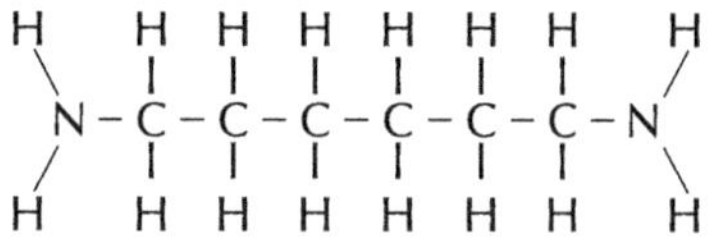

(structure: N-C-C-C-C-C-C-N chain with H's)

1,4-butanedioic acid

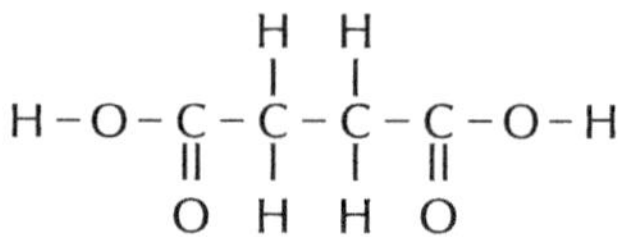

$$H-O-C-C-C-C-O-H$$
(with O, H, H, O below)

b

(polymer structure)

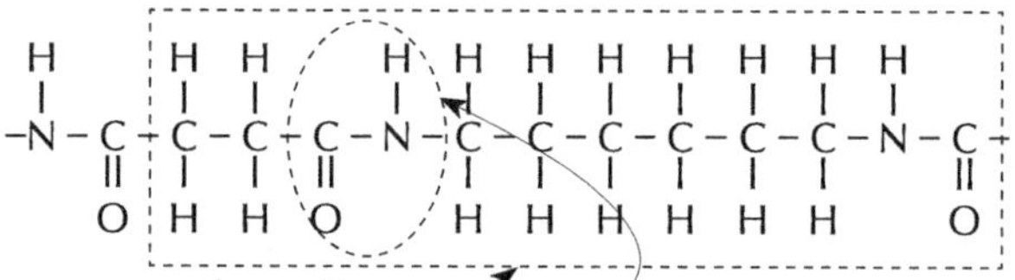

repeating unit

amide bond

7 a

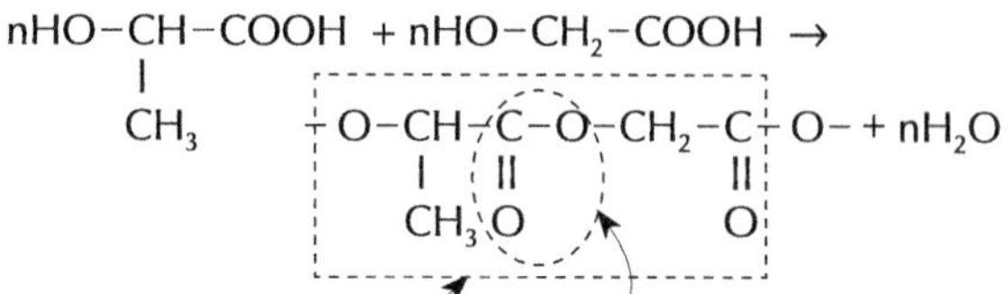

nHO−CH−COOH + nHO−CH₂−COOH →
| (CH₃)
+O−CH−C−O−CH₂−C+O− + nH₂O
(with CH₃, O, O)

b repeating unit

ester bond

8 a terephthalic acid

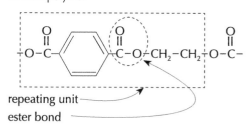

HOOC —〈ring〉— COOH

1,2-dihydroxyethane

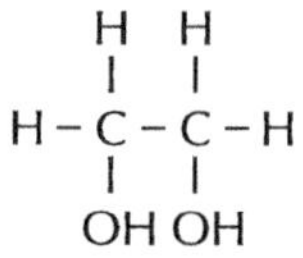

$$H-C-C-H$$
(with H H top, OH OH bottom)

b dacron polymer

(polymer structure)

+O−C−〈ring〉−C−O−CH₂−CH₂−O−C−
(with O's)

repeating unit

ester bond

Revision Five

Question I

1 A: N,N-dimethylethanamide; B: pentan-3-ol; C: propyl methanoate

2 a C; it has two carboxylic acid groups.

b i cinnamaldehyde

ii

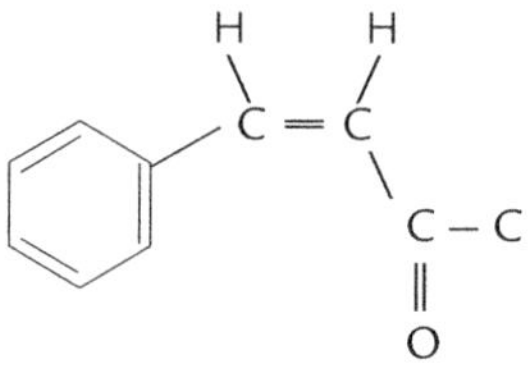

(two structures)

cis cinnamaldehyde 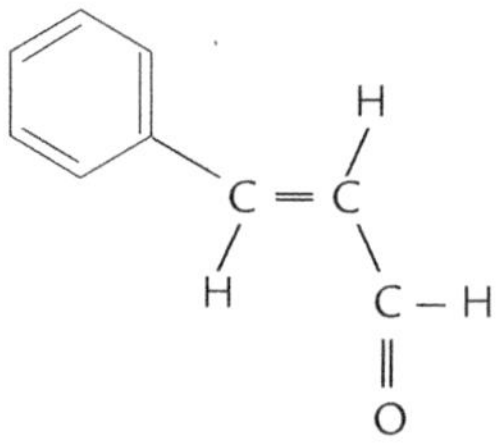   trans cinnamaldehyde

iii They have different melting points.

c 〈ring〉— CH₂CH₂CH₂OH

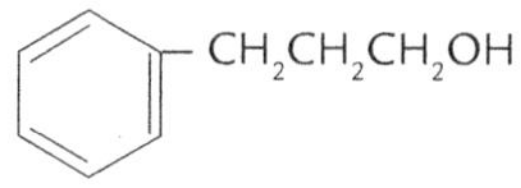

d i C

ii

CH₂COOH (C—H with HOOC, OH)   HOOCH₂C (H—C with HO, COOH)

iii They rotate plane polarised light in opposite directions.

e C. It has a secondary alcohol group which reacts with permanganate. B has a secondary alcohol group but it has a –CHO which would react with Benedict's solution. A has a double bond, but also a –CHO group.

3 A: $CH_2=CH-CH_2CH_3$
B: $CH_3CH(OH)CH_2CH_3$
C: $HOCH_2CH_2CH_2CH_3$
D: $CH_3COCH_2CH_3$
E: $CH_3CH_2CH_2COOH$
F: $CH_3CH(OH)CH_2CH_3$

ISBN: 9780170355544 PHOTOCOPYING OF THIS PAGE IS RESTRICTED UNDER LAW.

G: $CH_3CH_2CH_2COCl$
H: $CH_3CH_2CH_2COOCH_3$
X: CH_3OH
1 Addition, 2 Oxidation, 3 Reduction, 4 Substitution,
5 Condensation (or Esterification)

4 a

$$\begin{array}{cccc} CH_3 & H & CH_3 & H \\ | & | & | & | \\ -C & - \; C \; - & C & - \; C- \\ | & | & | & | \\ H & OCOCH_3 & H & OCOCH_3 \end{array}$$

repeating unit

b i 2-amino-4-methylpentanoic acid

ii

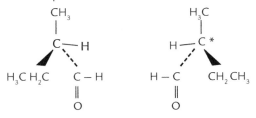

$$H_3C - CH - CH_2 - CH - C - N - CH - COOH$$
$$\quad\quad |\qquad\qquad\quad |\quad\ ||\ \ |$$
$$\quad\quad CH_3\qquad\qquad NH_2\ \ O\ \ H\ \ CH_3$$

or $\ H_3C - CH - CH_2 - CH - COOH$

iii amide (or peptide)

Question II

1 a A: 2-methylbutanal; B: dimethylaminoethane or N,
N-dimethylethanamine; C: ethyl methanoate

b Compound A

iii They rotate plane polarised light in opposite
directions.

2 a $CH_3CH_2CH_2CH_2OH$, butanol
$CH_3CH(OH)CH_2CH_3$, butan-2-ol

$$\begin{array}{c} CH_3 \\ | \\ H_3C - C - CH_3 \\ | \\ OH \end{array}$$ methylpropan-2-ol (or
2-methylpropan-2-ol)

b Butanol forms $CH_3CH_2CH_2COOH$ (butanoic acid)
or $CH_3CH_2CH_2CHO$ (butanal)
Butan-2-ol forms $CH_3COCH_2CH_3$ (butanone)
Methylpropan-2-ol gives no reaction.

3 a

$$\begin{array}{cc} \underset{\text{cis}}{\overset{CH_3 \quad CH_2CH_3}{\underset{H \quad\quad H}{C = C}}} & \underset{\text{trans}}{\overset{CH_3 \quad H}{\underset{H \quad\quad CH_2CH_3}{C = C}}} \end{array}$$

b They have different boiling points.

4 A: $CH_3CH(Cl)CH_2CH_3$
B: $CH_3CH_2CH_2CH_2Cl$
C and D: $CH_2=CHCH_2CH_3$ and $CH_3CH=CHCH_3$
E: $CH_3CH_2CH_2CH_2OH$
F: $CH_3CH_2CH_2CH_2OCOCH_3$
X: NaOH
1 Addition, 2 A ⟶ C and D Elimination, B ⟶ E
Substitution, 3 Condensation (or Esterification),
4 Hydrolysis (or Saponification)

5 a Water. $CH_3CH_2CH_2Cl$ does not mix with water.
Two colourless layers are observed. CH_3CH_2COCl
reacts vigorously with water. The reaction is
exothermic.

b Tollens' reagent. After gentle warming, colourless
CH_3CH_2CHO forms a silver mirror on the side
of the container. With CH_3COCH_3 the mixture
remains colourless.

6 a i

$$\begin{array}{cccccc} CH_3 & H & CH_3 & H & CH_3 & H \\ | & | & | & | & | & | \\ -C & - \; C & - \; C & - \; C & - \; C & - \; C- \\ | & | & | & | & | & | \\ H & H & H & H & H & H \end{array}$$

ii addition

b

c i $-N\,(CH_2)_4 - N - C - (CH_2)_4 - C - N -$
$\quad\ \ |\qquad\qquad |\ \ ||\qquad\qquad\quad ||\ \ |$
$\quad\ \ H\qquad\qquad H\ \ O\qquad\qquad\quad O\ H$
(repeating unit is underlined).

ii $SOCl_2$

iii amide

Question III

1 A: N-methyl-2-methylpropanamide; B: 3-methylbutan-
2-one; C: ethyl propanoate

2 a b

c The solution would have no effect on polarised
light.

3 a A is methylpropan-2-ol, $(CH_3)_3COH$
B is butan-2-ol
C is butanone, $CH_3COCH_2CH_3$
D is butanal, $CH_3CH_2CH_2CHO$, or methylpropanal,
$(CH_3)_2CH\ CHO$
E is butanoic acid or methylpropanoic acid

b The colourless alcohol reacts with the purple
reagent to give a colourless solution. Oxidation.

c Two colourless solutions form a silver mirror on the
sides of the container. Oxidation.

4 I is $CH_2=CH-CH_2-CH_3$
II is $CH_3CHClCH_2CH_3$
III is $CH_2ClCH_2CH_2CH_3$
IV is $CH_2=CHCH_2CH_3$
V is $CH_3-CH=CH-CH_3$
1 Elimination or dehydration, 2 Addition, 3 Elimination

5 a $CH_3-CH_2-O-CO-CH_2CH_3$. Condensation (or
esterification).

b $CH_3CH_2COO^-\ NH_4^+$. Neutralisation (or acid-base
reaction).

c $CH_3CH_2CONH_2$. Condensation.

6 a

b

$$CH_2HS \qquad CH_2SH$$

$$\begin{array}{cc} C-H & C-H \\ H_2N \quad C-N & C-OH \\ \parallel \quad \mid & \parallel \\ O \quad H & O \end{array}$$

7 a

$$\begin{array}{c} -N-(CH_2)_6-N-C-(CH_2)_4-C-N- \\ \mid \qquad\qquad \mid \; \parallel \qquad\qquad \parallel \; \mid \\ H \qquad\qquad H \; O \qquad\qquad O \; H \end{array}$$

(repeating unit is underlined)

b amide; bond is circled in **a**

c condensation

Question IV

1 A: N-ethylethanamide; B: propyl propanoate;
C: 3-methylbutanoyl chloride

2 a

$$\begin{array}{c} CH_3 \\ \mid \\ H_3C-CH-C-CH_2-CH_3 \\ \parallel \\ O \end{array}$$

b

$$\begin{array}{c} OH \quad NH_2 \\ \mid \qquad \mid \\ H_3C-CH-CH-C-OH \\ \parallel \\ O \end{array}$$

3 a i

$$\begin{array}{cc} H_3CH_2C & CH_2CH_3 \\ \mid* & \mid \\ C-H & H-C \\ H_3C \quad C-Cl & Cl-C \quad CH_3 \\ \parallel & \parallel \\ O & O \end{array}$$

ii One is the non-superimposable mirror image of the other.

b i

$$\begin{array}{cc} CH_3(CH_2)_5 \;\; (CH_2)_7COOH & CH_3(CH_2)_5 \quad H \\ \backslash \quad / & \backslash \quad / \\ C=C & C=C \\ / \quad \backslash & / \quad \backslash \\ H \qquad H & H \;\; (CH_2)_7COOH \\ cis & trans \end{array}$$

ii They have different melting points.

4 a Bromine water. The colourless $CH_3CH=CH_2$ gas decolourises the orange bromine water, but the colourless $CH_3CH_2CH_3$ gas does not.

b Litmus paper. CH_3CONH_2 has no effect on litmus paper. $CH_3CH_2CH_2NH_2$ turns red litmus paper blue.

c Acidified potassium permanganate. $CH_3CH(OH)CH_2CH_3$ reacts with the purple reagent to form a colourless solution. $CH_3COCH_2CH_3$ gives no reaction.

d Tollens' reagent. With $CH_3CH_2CH_2OH$, there is no reaction and a colourless solution remains. With CH_3CH_2CHO, a silver mirror forms on the inside of the container.

5 A is $CH_3CH(NH_2)CH_3$
B is $CH_2=CHCH_3$
C is $CH_3CH(OH)CH_3$
D is $CH_3CH_2CH_2OH$
1 Substitution, 2 Elimination, 3 Addition

6 a

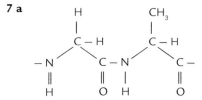

(2 repeating units underlined)

b addition

7 a

$$\begin{array}{cc} H & CH_3 \\ \mid & \mid \\ C-H & C-H \\ -N \quad C-N & C- \\ \parallel \qquad \parallel \; \mid & \parallel \\ H \qquad O \; H & O \end{array}$$

b peptide (or amide)

c condensation

Unit 27 Spectroscopic data in organic chemistry

1 a Carboxylic acid.
Strong broad peak around 300–3500 cm^{-1}, therefore –O–H
Strong peak at 1710 cm^{-1}, therefore –C=O. (–C–O stretching is in the fingerprint region.)

b Alcohol
Strong broad peak at 3350 cm^{-1}, therefore –O–H.
Strong peak at 1065 cm^{-1}, therefore C–O

c Carbonyl (ketone)
Strong peak at 1710 cm^{-1}, therefore –C=O
Not a strong peak at 1050–1150; an ester would have one.

d Aldehyde
Strong peak at about 1720 cm^{-1}, therefore –C=O.
Two medium peaks around 2700 and 2800, therefore aldehyde.

2 a Methyl ethanoate would have one less peak in the low ppm region.

b i A 2, B 3, C 3, D 5
  ii C and D

c

Compound	Structural formula	No. of peaks in C-13 NMR = no. of C environments	Peak at > 100 ppm: Y/N
Butanol	$CH_3CH_2CH_2CH_2OH$	4	N
Methyl propanoate	$CH_3O \; C \; CH_2CH_3$ with $\parallel$ O below	4	Y
Butene	$CH_2=CHCH_2CH_3$	4	N
Butanal	$CH_3CH_2CH_2C=O$ with H below	4	Y
Methyl propanol	CH_3 / CH_3CHCH_2OH	3	N
Butanone	$CH_3 \; C \; CH_2CH_3$ with $\parallel$ O below	4	Y
Methyl propan-2-ol	CH_3 / $CH_3 \; C \; CH_3$ / OH	2	N

ISBN: 9780170355544 PHOTOCOPYING OF THIS PAGE IS RESTRICTED UNDER LAW.

3 a i $CH_2=CH-CH_2CH_3$ 56 g mol^{-1}

ii

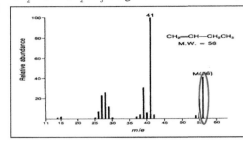

iii m/z = 27 is $[CH_2=CH]^+$
m/z = 29 is $[CH_3-CH_2]^+$
m/z = 41 is $[CH_2CH=CH_2]^+$

b i m/z = 72 gives the molecular ion
$[CH_3CH_2CH_2CH_2CH_3]^+$
m/z = 57 gives $[CH_3CH_2CH_2CH_2]^+$
m/z = 43 gives $[CH_3CH_2CH_2]^+$
m/z = 29 gives $[CH_3CH_2]^+$

ii m/z = 32 gives the molecular ion $[CH_3OH]^+$
m/z = 31 gives $[CH_3-O]^+$ and $[CH_2OH]^+$
m/z = 15 gives $[CH_3]^+$

Revision Six

1 a Alcohol
Strong broad peak around 3200–3500 cm^{-1}, therefore O-H
Strong peak around 1050 cm^{-1}, therefore C-O

b 4

c i 74 gmol^{-1}
ii $CH_3CH_2CH_2CH_2OH$ or $CH_3CH(OH)CH_2CH_3$

2 a Alcohol
Strong broad peak around 3200–3500 cm^{-1}, therefore O-H
Strong peak around 1100 cm^{-1}, therefore C-O

b Aldehyde
Strong peak around 1750 cm^{-1}, therefore C=O. Also two medium peaks around 2700 and 2800 cm^{-1}

c Carboxylic acid
Strong, very broad peak around 3200–3600 cm^{-1}, therefore O-H. Strong peak around 1720 cm^{-1}, therefore C=O. Strong peak around 1200 cm^{-1}, therefore C-O

d i 60 gmol^{-1}
ii A is propanol, B is propanal, C is propanoic acid
iii m/z = 31 $[CH_2OH]^+$, m/z = 29 $[CH_3CH_2]^+$
(Note the peak at m/z = 42. Alcohols may lose H_2O. Propanol can give $[CH_3CH=CH_2]^+$.)

3 a Compound 1 is ethanol, compound 2 is ethanoic acid
b i Mass spectrum 1 is ethanoic acid; m/z = 60 gives molecular ion so molar mass is 60 gmol^{-1}, m/z = 45 $[COOH]^+$, m/z = 43 $[CH_3CO]^+$, m/z = 15 $[CH_3]^+$
ii Mass spectrum 2 is ethanal; m/z = 44 gives molecular ion so molar mass is 44 gmol^{-1}, m/z = 29 $[CHO]^+$, m/z = 15 $[CH_3]^+$

4 a i 1670 cm^{-1}
ii 3300 cm^{-1} and 1600 cm^{-1} (overlaps with carbonyl C=O peak)
b CH_3- 10 ppm, $-CH_2$- 22 ppm, $-C(O)NH_2$ 200 ppm
c i m/z = 73 molecular ion $CH_3CH_2C=O^+$
$|$
NH_2

ii m/z = 44 $[C(O)NH_2]^+$ **iii** m/z = 29 $[CH_3CH_2]^+$
iv m/z = 57 $[CH_3CH_2CO]^+$

5 a i Compound A is butanoic acid,
$CH_3CH_2CH_2COOH$
IR spectrum has the following absorbance peaks:
— strong broad peak at 3200–3500 cm^{-1}, O-H
— strong peak at 1700 cm^{-1}, C=O
NMR spectrum shows four different carbon environments:
— peak at 188 ppm is **C** in **C**=O in the acid
Mass spectrum shows:
— molecular ion m/z 88 $[CH_3CH_2CH_2COOH]^+$
— m/z 87 $[CH_3CH_2CH_2COO]^+$
— m/z 45 $[COOH]$

ii and iii
The IR spectra show that compounds B and C are esters. Their mass spectra show a molecular ion with a mass of 88 gmol^{-1}. Possible esters are methyl propanoate, ethyl ethanoate and propyl methanoate.

ii Compound B is methyl propanoate
$CH_3 O C CH_2CH_3$
$\parallel$
O
IR spectrum has the following absorbance peaks:
— strong peak at 1300 cm^{-1}, C-O
— strong peak at 1750 cm^{-1}, C=O
— peaks around 3000 cm^{-1}, C-H
NMR spectrum shows four different carbon environments:
— peak at 178 ppm is **C** in **C**=O in the ester
Mass spectrum shows:
— molecular ion m/z 88 $[CH_3O CO CH_2CH_3]^+$
— m/z 15 $[CH_3]^+$
— m/z 31 $[CH_3O]^+$
— m/z 57 $[CO CH_2CH_3]^+$
— m/z 59 $[CH_3O CO]^+$
— m/z 29 $[CH_2CH_3]^+$

iii Compound C is ethyl ethanoate $CH_3CH_2O CO CH_3$
IR spectrum has the following absorbance peaks:
— strong peak at 1240 cm^{-1}, C-O
— strong peak at 1750 cm^{-1}, C=O
— peaks around 3000 cm^{-1}, C-H
NMR spectrum shows four different carbon environments:
— peak at 170 ppm is **C** in **C**=O in the ester
Mass spectrum shows:
— molecular ion m/z 88 $[CH_3CH_2O CO CH_3]^+$
— m/z 15 $[CH_3]^+$
— m/z 73 $[CH_3CH_2O CO]^+$
— m/z 29 $[CH_3CH_2]^+$
— m/z 59 $[O CO CH_3]^+$
— m/z 45 $[CH_3CH_2O]^+$
— m/z 43 $[CO CH_3]^+$
— m/z 29 $[CH_2CH_3]^+$

b Of the three types of spectra, the mass spectra give the most useful information. Knowing the functional group (IR) is also important. Without knowing the pattern of fragmentation of molecules in a mass spectrometer, we cannot assign some of the peaks. When we have a possible structure, we can look at possible fragments that structure would give. We can calculate their molar masses and look for corresponding lines in the spectrum.

Index

ISBN: 9780170355544
PHOTOCOPYING OF THIS PAGE IS RESTRICTED UNDER LAW.

ISBN: 9780170355544
PHOTOCOPYING OF THIS PAGE IS RESTRICTED UNDER LAW.

ISBN: 9780170355544 PHOTOCOPYING OF THIS PAGE IS RESTRICTED UNDER LAW.

The Periodic Table

Legend (each cell): Molar mass of element · Symbol · Atomic number · Name · Electron configuration

Metals	Non-metals

Main table

Period	Gp 1	Gp 2	Gp 3	Gp 4	Gp 5	Gp 6	Gp 7	Gp 8	Gp 9	Gp 10	Gp 11	Gp 12	Gp 13	Gp 14	Gp 15	Gp 16	Gp 17	Gp 18
1	1.008 H 1 Hydrogen 1																	4.00 He 2 Helium 2
2	6.94 Li 3 Lithium 2,1	9.01 Be 4 Beryllium 2,2											10.81 B 5 Boron 2,3	12.01 C 6 Carbon 2,4	14.01 N 7 Nitrogen 2,5	16.00 O 8 Oxygen 2,6	19.00 F 9 Fluorine 2,7	20.18 Ne 10 Neon 2,8
3	22.99 Na 11 Sodium 2,8,1	24.31 Mg 12 Magnesium 2,8,2											26.98 Al 13 Aluminium 2,8,3	28.09 Si 14 Silicon 2,8,4	30.97 P 15 Phosphorus 2,8,5	32.06 S 16 Sulfur 2,8,6	35.45 Cl 17 Chlorine 2,8,7	39.95 Ar 18 Argon 2,8,8
4	39.10 K 19 Potassium 2,8,8,1	40.08 Ca 20 Calcium 2,8,8,2	44.96 Sc 21 Scandium 2,8,9,2	47.90 Ti 22 Titanium 2,8,10,2	50.94 V 23 Vanadium 2,8,11,2	52.00 Cr 24 Chromium 2,8,13,1	54.94 Mn 25 Manganese 2,8,13,2	55.85 Fe 26 Iron 2,8,14,2	58.93 Co 27 Cobalt 2,8,15,2	58.71 Ni 28 Nickel 2,8,16,2	63.55 Cu 29 Copper 2,8,18,1	65.37 Zn 30 Zinc 2,8,18,2	69.72 Ga 31 Gallium 2,8,18,3	72.59 Ge 32 Germanium 2,8,18,4	74.92 As 33 Arsenic 2,8,18,5	78.96 Se 34 Selenium 2,8,18,6	79.90 Br 35 Bromine 2,8,18,7	83.80 Kr 36 Krypton 2,8,18,8
5	85.47 Rb 37 Rubidium 2,8,18,8,1	87.62 Sr 38 Strontium 2,8,18,8,2	88.91 Y 39 Yttrium 2,8,18,9,2	91.22 Zr 40 Zirconium 2,8,18,10,2	92.91 Nb 41 Niobium 2,8,18,12,1	95.94 Mo 42 Molybdenum 2,8,18,13,1	98.91 Tc 43 Technetium 2,8,18,14,1	101.07 Ru 44 Ruthenium 2,8,18,15,1	102.91 Rh 45 Rhodium 2,8,18,16,1	106.42 Pd 46 Palladium 2,8,18,18	107.87 Ag 47 Silver 2,8,18,18,1	112.41 Cd 48 Cadmium 2,8,18,18,2	114.81 In 49 Indium 2,8,18,18,3	118.71 Sn 50 Tin 2,8,18,18,4	121.76 Sb 51 Antimony 2,8,18,18,5	127.60 Te 52 Tellurium 2,8,18,18,6	126.90 I 53 Iodine 2,8,18,18,7	131.29 Xe 54 Xenon 2,8,18,18,8
6	132.91 Cs 55 Caesium 2,8,18,18,8,1	137.33 Ba 56 Barium 2,8,18,18,8,2	La 57 Lanthanum	178.49 Hf 72 Hafnium 2,8,18,32,10,2	180.95 Ta 73 Tantalum 2,8,18,32,11,2	183.84 W 74 Tungsten 2,8,18,32,12,2	186.21 Re 75 Rhenium 2,8,18,32,13,2	190.23 Os 76 Osmium 2,8,18,32,14,2	192.22 Ir 77 Iridium 2,8,18,32,15,2	195.08 Pt 78 Platinum 2,8,18,32,16,2	196.97 Au 79 Gold 2,8,18,32,17,2	200.59 Hg 80 Mercury 2,8,18,32,18,2	204.38 Tl 81 Thallium 2,8,18,32,18,3	207.20 Pb 82 Lead 2,8,18,32,18,4	208.98 Bi 83 Bismuth 2,8,18,32,18,5	208.98 Po 84 Polonium 2,8,18,32,18,6	209.99 At 85 Astatine 2,8,18,32,18,7	222.02 Rn 86 Radon 2,8,18,32,18,8
7	223.02 Fr 87 Francium 2,8,18,32,18,8,1	226.05 Ra 88 Radium 2,8,18,32,18,8,2	Ac 89 Actinium															

Lanthanides

La 57 Lanthanum	Ce 58 Cerium	Pr 59 Praseodymium	Nd 60 Neodymium	Pm 61 Promethium	Sm 62 Samarium	Eu 63 Europium	Gd 64 Gadolinium	Tb 65 Terbium	Dy 66 Dysprosium	Ho 67 Holmium	Er 68 Erbium	Tm 69 Thulium	Yb 70 Ytterbium	Lu 71 Lutetium

Actinides

Ac 89 Actinium	Th 90 Thorium	Pa 91 Protactinium	U 92 Uranium	Np 93 Neptunium	Pu 94 Plutonium	Am 95 Americium	Cm 96 Curium	Bk 97 Berkelium	Cf 98 Californium	Es 99 Einsteinium	Fm 100 Fermium	Md 101 Mendelevium	No 102 Nobelium	Lr 103 Lawrencium

PHOTOCOPYING OF THIS PAGE IS RESTRICTED UNDER LAW.

ISBN: 9780170355544